UNDERSTANDING SCIENTIFIC PROSE

RHETORIC OF THE HUMAN SCIENCES

General Editors

John Lyne
Donald N. McCloskey
John S. Nelson

Understanding Scientific Prose

EDITED BY JACK SELZER

THE UNIVERSITY OF WISCONSIN PRESS

The University of Wisconsin Press
114 North Murray Street
Madison, Wisconsin 53715

3 Henrietta Street
London, WC2E 8LU, England

Library of Congress Cataloging-in-Publication Data
Understanding scientific prose / edited by Jack Selzer.
    406 p.   cm. — (Rhetoric of the human sciences)
    Includes bibliographical references and index.
    ISBN 0-299-13900-X   ISBN 0-299-13904-2(pbk.)
    1. Gould, Stephen Jay. Spandrels of San Marco and the Panglossian
Paradigm. 2. Evolution (Biology)—Authorship. 3. Scientific
literature. 4. Rhetoric. 5. Lewontin, Richard C., 1929–
I. Selzer, Jack. II. Series.
QH371.G6843U53   1993
808'.0665—dc20   93-21870

CONTENTS

Contents

FIGURES

TABLES

ACKNOWLEDGMENTS

This book is the result of the cooperative efforts of a great many people, and I would like to take this opportunity to thank as many of them as I can.

Most of all, of course, I am indebted to the sixteen other contributors to this book. "Coauthors" would be a better term to use in describing them, for their efforts have made this much more a *book* than a collection of essays. I am grateful to them for their wisdom and eloquence and punctuality, but also for their cheerful readiness to revise in the interest of creating a coherent volume. On behalf of all the contributors, I offer special thanks to Stephen Jay Gould and Richard C. Lewontin, whose good-natured willingness to permit a scholarly anatomy of their work made the book possible. That Professor Gould was willing not only to stand for that but to contribute a chapter of his own (reflecting on his writing and on the analyses of his prose collected here) has made the book infinitely more interesting: I'll not forget that, nor some memorable correspondence and conversation about baseball and biology.

I first became interested in the writing that goes on in the area of evolutionary biology because of my friendship with my Penn State colleague Andrew G. Stephenson, whose own writing on the subject is worthy of scrutiny and admiration. Andy arranged personal conversations with Richard Lewontin and John Maynard Smith that gave me an invaluable perspective on "The Spandrels of San Marco," and he and Carl Schlichting took time to explain to me some of the background information necessary for understanding the intricacies of arguments on evolution. I am grateful to them for their patience, interest, and encouragement. My friend, former student, and collaborator Gay Gragson first called my attention to "The Spandrels of San Marco" and let me work on it with her, and she remains a long-distance source of good sense and good will. I also thank John A. Campbell and Ernst Mayr, whose critiques of the first draft of the book improved it materially, and John Lyne and Allen Fitchen, who encouraged the project on behalf of the University of Wisconsin Press.

I also want to thank the professional staff associated with the University of Wisconsin Press for their expertise, particularly Raphael Kadushin, Carol Olsen, and Sharon Ihnen. At Penn State I received excel-

Acknowledgments

lent clerical and research assistance from Kathy Leitzell, Jody Auman, and Eva Christensen; I thank my department head, Robert Secor, for making that assistance available. Thanks also to Sage Publications for permission to reprint parts of chapter ten, which originally appeared in *Written Communication* 7 (1990): 25–58; and to S. J. Gould, R. C. Lewontin, and the Royal Society of London for permission to reproduce "The Spandrels of San Marco and the Panglossian Paradigm: A Critique of the Adaptationist Programme," which appeared in the *Proceedings of the Royal Society of London, Series B: Biological Sciences* 205 (1979): 581–98.

Finally, thanks to Linda Selzer for wise counsel, countless suggestions, and unfailing support. I said it once and I'll say it again: "Nothing else is."

CONTRIBUTORS

Charles Bazerman teaches at Georgia Tech. His interests are in writing as a social activity and in the rhetoric of science. The co-editor of *Textual Dynamics of the Professions,* he has written *Shaping Written Knowledge: The Genre and Activity of the Experimental Article in Science* as well as several textbooks on writing.

Davida Charney, associate professor of English at Penn State, teaches courses in technical writing, research methods, and the rhetoric of science. The co-editor of *Constructing Rhetorical Education* (1992), she has published many articles on document design and the processes of reading.

Barbara Couture is associate professor of English at Wayne State University, where she has directed the compostion program and coordinated the technical writing program. Her publications include *Cases for Technical and Professional Writing* (with Jone Rymer) and many chapters and articles. She is also editor of *Functional Approaches to Writing: Research Perspectives* and *Professional Writing: Toward a College Curriculum.*

Jeanne Fahnestock teaches writing and rhetoric at the University of Maryland at College Park. The co-author of *The Rhetoric of Argument* and co-editor of *Readings in Argument,* she has published many essays in *College Composition and Communication, Written Communication,* and various other journals and collections. She is currently working on a book on style.

Stephen Jay Gould teaches biology, geology, and the history of science at Harvard University. A classical music enthusiast and the winner of the National Book Award in 1981 for *The Panda's Thumb: More Reflections in Natural History,* he is also the author of *Wonderful Life* (1989), *Time's Arrow, Time's Cycle* (1987), *The Mismeasure of Man* (1981), and several other books and hundreds of essays. He is a lifelong Yankees fan.

Contributors

Gay Gragson, who received a master's degree in technical writing at Penn State, currently lives in Athens, Georgia. She has assisted her husband with anthropological field work among the Yaruro Indians of southern Venezuela and has worked as a technical writer and editor.

S. Michael Halloran, professor of communication at Rensselaer Polytechnic Institute, has special expertise in classical rhetoric and the rhetoric of science and technology. He is currently studying the history of rhetoric in America.

Carl Herndl, who has published a number of articles on corporate and technical writing and the practice of ethnography, teaches rhetoric and writing courses at New Mexico State University. He is currently studying writing on environmental issues.

Debra Journet is professor of English at the University of Louisville. Her research centers on the relation of rhetorical and literary theory to scientific discourse. The author of many essays, she also coedited *Research in Technical Communication: A Bibliographic Sourcebook.*

John Lyne chairs the department of communication studies at the University of Iowa. His scholarly work has focused on the study of Charles Sanders Peirce's philosophy of rhetoric and on the rhetorical practice of evolutionary biologists.

Carolyn R. Miller is professor of English at North Carolina State University where she teaches a variety of courses in rhetoric and technical writing. Her primary research interest is the rhetoric of science and technology, on which she has published numerous scholarly essays.

Greg Myers works in the Linguistics Department at the University of Lancaster, Lancaster, England. His book *Writing Biology: Texts in the Social Construction of Scientific Knowledge* was published in 1990.

Georgia Rhoades joined the faculty at Appalachian State University after receiving her Ph.D. from the University of Louisville. She teaches a variety of courses in writing, literature, and women's studies.

Mary Rosner teaches graduate and undergraduate courses in technical and scientific writing, rhetoric, and writing pedagogy at the University of Louisville. Her research has been concerned with Victorian literature, the history of rhetoric, and the writing characteristic of technical and scientific communities.

Jack Selzer, associate professor of English at Penn State and a co-founder of the Penn State Conference on Rhetoric and Composition, has published many essays on technical and scientific discourse in various collections and in journals such as *College Composition and Communication, Rhetoric Review,* and *Written Communication.*

Susan Wells has a special interest in the interconnections among science, language, and culture. A professor of English at Temple University, she has published (among other things) *Dialectics of Interpretation* (1986).

Dorothy A. Winsor is associate professor of communication at GMI Engineering and Management Institute in Flint, Michigan. She conducts research on the influence of social context on the writing of engineers and has published many articles relating to that subject.

UNDERSTANDING SCIENTIFIC PROSE

1 INTRODUCTION

JACK SELZER

The allusion in the title of this book is so obscure that probably only I recognize it, so I had better begin by explaining what is behind *Understanding Scientific Prose.*

In 1938 Cleanth Brooks and Robert Penn Warren published their famous textbook anthology, *Understanding Poetry.* What made the book so famous was that it illustrated for an entire generation a method of practical criticism known as New Criticism: *Understanding Poetry* demonstrated and thereby domesticated the theoretical tenets of literary formalism—an effort to objectify texts and to purify criticism from "extrinsic" concerns like politics, biography, psychology, social setting, and the history of ideas—that had taken shape in the 1920s and 1930s, thanks to the efforts of William Empson, Allen Tate, I. A. Richards, F. R. Leavis, R. P. Blackmur, Rene Wellek, and W. K. Wimsatt. By the time Rene Wellek and Austin Warren formalized the tenets of New Criticism in their *Theory of Literature* in 1942, those tenets were already well on their way to becoming tacit knowledge on account of the influence of *Understanding Poetry;* the practical criticism demonstrated in *Understanding Poetry* and later in the journals associated with New Criticism succeeded so well that New Criticism dominated literary studies through the 1960s, and students and scholars were convinced that they were indeed "understanding poetry" better than they had before.

Something analogous is being attempted here. The other contributors to this book and I propose to accomplish a similar task (though of course without perpetuating the tenets of formalism): we wish to demonstrate and domesticate new methods of practical criticism in order to "open up" another kind of literature, scientific discourse. We have noticed that in the past decade or so a number of new methods of rhetorical analysis have been proposed, but those methods remain largely theoretical, untried. By introducing readers of this book to new methods of analysis and then applying them to a single text, we aim to further the effort of rhetorical criticism that has so much momentum today. By introducing methods of analysis implied by cultural studies and the concept of intertextuality, feminism and the sociology of science, structuralism and deconstruction, reader-response theory and historicism, Perelman and Burke and Habermas (among others), we mean to extend the range of analytical approaches that are available to any critic of any kind of discourse. Thus, the forebears of this book include not only *Understanding Poetry* but also more recent and more rhetorically based efforts like Edwin Black's *Rhetorical Criticism: A Study in Method* (1965), Edward P. J. Corbett's *Rhetorical Analysis of Literary Works* (1969), James R. Andrews's *The Practice of Rhetorical Criticism* (1983), Sonja Foss's *Rhetorical Criticism: Exploration and Practice* (1989), and Roderick Hart's *Modern Rhetorical Criticism* (1990)—all of which seek to expand the repertoire of analytical methods at the disposal of rhetorical critics. This book differs from them in proposing new methodologies, particularly postmodern ones (though not every approach here is postmodern and though some newer approaches are also discussed and illustrated in Hart and in Foss). More important, in its applications this book focuses not on the forms of oral public discourse that typically earn scrutiny in those other books (the exception is Corbett, who covers belletristic prose) but on written documents—and more specifically on written documents in academic science, in particular one essay on the subject of evolutionary biology.

The contributors to this book also wish to add to the developing body of scholarship on the rhetoric of science. That is why we chose to illustrate each method of analysis with a consideration of a single essay on the subject of evolutionary theory. Of course, to aim at a rhetorical analysis of scientific prose would have been impossible to imagine two decades ago, except by renegades like Kenneth Burke. Until the 1970s both science and poetry were considered to be outside the domain of rhetoric. Just as the New Critics considered poems arhetorical—as timeless, well-wrought urns insulated from writers (hence the "intentional fallacy") and from readers (hence the "affective fallacy")—so too most academicians until the late 1970s considered scientific discourse to be a special kind of discourse operating outside the realm of rhetoric. By coincidence, Brooks and Warren's 1938 introduction to *Understanding Poetry*, reprinted in successive editions into the 1970s, itself articulates attitudes toward language and science that prevailed until recently. In contrasting scientific writing and poetry, Brooks and Warren articulated the familiar "windowpane theory" of scientific language:

> The primary advantage of the scientific statement is that of absolute precision. . . . Such precision . . . can be gained only by using terms in special and previously defined senses. The scientist carefully cuts away from his scientific terms all associations, emotional colorings, and implications of attitude and judgment. (4)

Scientific discourse, they felt—certain and demonstrably true, clear and syllogistic, rational and purely denotative—"tends indeed toward the condition of mathematics" (4). It differs from "ordinary language" (i.e., rhetoric, the language of the rough-and-tumble world of practical affairs) as much as does poetry, a similarly rarified and high-minded special language that aspires not to the rational denotativeness of science but to an associative, imaginative, emotional, complex, and ambiguous language different in degree from the language of everyday rhetoric.[1] Associated with the Agrarian movement in politics and culture, the New Critics were no friends of science; but they shared with science positivist notions about language and a contempt for rhetoric that prevailed in the academy during the first two-thirds of this century.

These days, of course, we think we know better about the nature of language, rhetoric, and science. Beginning about 1970 philosophers and historians of science, among them Stephen Toulmin and Thomas Kuhn, called into question the separation of the sciences from rhetoric, contending that scientists do indeed argue (not just describe) in order to win adherence to probable claims and maintaining that the premises of science are more the product of debate than demonstration. Rhetoricians began to agree: Kenneth Burke, for example, persisted in his insistence that rhetoric was fundamentally involved in every human undertaking, including science (and poetry, for that matter); and Chaim Perelman likewise stated that "argumentation, conceived as a new rhetoric or dialectic, covers the whole range of discourse, . . . whatever the audience addressed and whatever the subject matter" (*Realm of Rhetoric*, 5).

At first, rhetoricians intrigued by the rhetoric of science concerned themselves with the rather conservative claim that scientists use stylistic tropes in their expression (e.g., Wander, Overington). But since 1980 rhetoricians have begun to study a great many aspects of scientific prose, including scientific ethos and the rules for conducting argument in various sciences, in an effort to understand more fully the rhetorical dimensions of scientific discourse. In this effort they have sometimes been joined by scientists themselves. Clifford Geertz's *Works and Lives: The Anthropologist as Author,* for example, uses the techniques of rhetorical analysis to demonstrate how the most influential anthropologists have used language to shape their discipline; and Stephen Jay Gould in *Time's Arrow, Time's Cycle* and *The Mismeasure of Man* shows that science is not so much a dispassionate description of reality as it is a set of arguments that create knowledge in a social context. As a consequence, today we have more than the beginning of a developing literature on the rhetoric of science: Michael Halloran's influential study of Watson and Crick's writing about DNA; the analyses by John Campbell, Greg Myers, John Lyne, and Henry Howe of arguments in the biological sciences; elaborations by Bruno Latour, Robert Merton, and Charles Bazerman of some of the social and rhetorical dimensions in the physical sciences; Alan Gross's account of the social construction of knowledge in the natural sciences; feminist critiques of scientific prose by Donna Haraway, Evelyn Fox Keller, and others; and the "literary" analyses of documents in the history of science collected by Peter Dear. And of course there have been studies of "the rhetoric of the human sciences" in the other volumes of the series that now includes this book and in *Rhetoric in the Human Sciences,* a collection of wide-ranging work by scholars from English, speech communication, rhetoric, and the sciences edited by Herbert W. Simons.[2]

In spite of this work, however, arguments for the rhetorical dimensions of science still remain more theoretical than practical, more general than specific. While there is now general agreement that science is indeed a rhetorical enterprise, specific demonstrations of that contention have been rare. Halloran's 1984 complaint that "very little attention has been paid to particular cases of scientific rhetoric" (70) still holds; as Charles Bazerman noted in his recent *Shaping Written Knowledge,* "Few rhetoricians have attempted serious studies of scientific language. While a few interesting propositions have been put forward, substantiated claims based on examination of actual language practices in science have been rare. We need thoroughgoing and wide-ranging research into the historical and current rhetoric within the sciences . . . to gain a grasp of the range of practices" that prevail in science (332).

The Contents of *Understanding Scientific Prose*

This book is calculated to help rectify that shortcoming and to contribute more broadly to scholarship on rhetoric in science. First, it is designed to equip analysts to understand better specific instances of "current rhetoric within the sciences" by introducing and demonstrating appropriate ways of performing rhetorical analyses—ways that are likely to be especially productive in the case of scientific discourse. Scientific writing is an especially important discourse in our culture, probably as important as the discourse of the law courts and legislative assemblies in ancient Greece that inspired the first rhetorical tracts and treatises and textbooks more than two thousand years ago.

But unlike the discourse that Aristotle and others considered, scientific discourse today is typically carried out not in public but in more private communities, communities with their own insider's language, generic conventions, and rules of argument and organization and style. For this reason it is difficult for outsiders (including professional rhetoricians) to find a way to gain access to and subdue scientific prose. This book therefore suggests a dozen or so methods of analysis that are likely to be productive in understanding scientific prose specifically. True, many of these methods will be useful in understanding other kinds of writing; indeed, several have been appropriated from methods in productive use in other disciplines. But the contributors here have sought to explain and demonstrate how their methods are appropriate specifically for making more transparent the rhetoric of science. In the decade since Michael Leff edited for the *Western Journal of Speech Communication* a "special report" on methods of rhetorical criticism, developments in hermeneutics, rhetorical theory, and critical theory have suggested a number of new analytical approaches that need to be justified, illustrated, and tested. Hence, a primary aim of *Understanding Scientific Prose* is to illustrate—and argue for—a pluralistic range of methods based on Foucault and Habermas, on gender and cultural studies, on deconstruction and reader response, on sociolinguistics and structuralism, on sociology and the stasis theory newly emerging from studies of classical rhetoric. Each essay in this collection thus includes patient introductory explanations, clarifying notes, and references that will permit others to explore and exercise the method under discussion—and thereby to contribute more themselves in the future to knowledge about the rhetorical dynamics of science.

Second, this book furthers knowledge about "the actual language practices" of scientific communities because each method of analysis is applied to a "particular case of scientific rhetoric," a single essay in evolutionary biology by S. J. Gould and R. C. Lewontin, "The Spandrels of San Marco and the Panglossian Paradigm: A Critique of the Adaptationist Programme," which was published in the *Proceedings of the Royal Society of London* in 1979. Any scientific document would have served for the analyses in this book. But "Spandrels" was chosen because it is daring and interesting and substantial enough to sustain multiple analyses without being exhausted itself and without exhausting thereby the readers of this book. By limiting itself to one example, *Understanding Scientific Prose* gains more than a measure of coherence (indeed, the chapters often speak to each other in interesting ways) and, more important, offers unprecedented depth and breadth to our understanding of one particular instance in the rhetoric of science.

(Adding to the coherence of the book, many of the contributors discuss some of the same related scientific articles, including several other items published with "Spandrels" in the same *Proceedings* volume.) "The Spandrels of San Marco," which is reprinted in the Appendix to this book for the convenience of readers and which I will introduce in more detail in a few pages, is offered here not as a "typical" piece of scientific prose—although as a publication from the *Proceedings of the Royal Society of London*, it is representative enough of the work of professional scientists. Nor is it offered as an *exemplum* of the way scientific argument ought to be conducted. "Spandrels" is simply one instance among many that could have been offered as an opportunity for us to demonstrate methods of rhetorical analysis and to offer detailed, even comprehensive insights into one rhetorical performance on a single issue in a single scientific community.

The contents of this book generally fall into two parts, though I emphasize "generally" since my primary goal has been to arrange the chapters in an order that will be most agreeable to readers who proceed through the chapters consecutively. I have tried to arrange the contributions into a coherent conversation about issues in the rhetoric of science. Chapters 2 through 7 take broadly contextual approaches to "The Spandrels of San Marco"; each defines a given context (or *setting* or *culture* or *environment* or whatever you want to call it) and then locates "Spandrels" within that context. Chapters 8 through 14 reverse the process somewhat in the sense that they begin closer to the text, offer more relentless textual analyses, and only then proceed outward to the cultural settings of "Spandrels." In that sense the final chapters are perhaps a bit more "traditional" rhetorical analyses. But none of the chapters confine themselves to the "bounded text" in the way that classic formalists (and structuralists) recommended, and none of them is truly traditional in the sense of displaying conventional analytic methods. Every contribution to this book plays off both text and contexts; each one proclaims in its own way the importance of context to contemporary efforts to understanding text, the importance (in other words) of thoroughly *rhetorical* analysis.

But the first six chapters do indeed consider "Spandrels" as part of broader frameworks of one kind or another. Charles Bazerman, for example, reads Gould and Lewontin's essay against part of what the European psychoanalyst, feminist, and semiotician Julia Kristeva would call its "intertext"—that web of other, interconnected texts against which any piece of writing is inevitably constructed and interpreted and in cooperation with which it accomplishes its social action. That is, Bazerman considers how (and how accurately) Gould and Lewontin embed within their essay references to previous scientific discourse—the works they cite in their notes and the other contributions to the scientific symposium on evolutionary theory of which "Spandrels" was a part—and references to Voltaire's *Candide* that are alluded to in the title, introduction, and other parts of "Spandrels." According to Kristeva, "Any text is constructed as a mosaic of quotations" (66); far from being fixed or self-contained, every text incorporates a dialogue, an intersection, a network among readers and writers and other texts. Bazerman shows how extravagantly Kristeva's words describe "Spandrels," and in so doing turns up evidence that "Spandrels" is not so much an argument against an explicitly named "adaptationist programme" as it is an implicit attack on the sociobiology movement of the 1970s. Susan Wells's and Carl Herndl's essays are easily read with Bazerman's since they too deal (though in very different ways) with Gould and Lewontin's attitudes toward sociobiology. Wells, with the help of Habermas's *Theory of Communicative Action* and *Philosophical Discourses of Modernity,* reads "Spandrels" as a modernist and postmodernist narrative; more precisely, she sees the essay as inhabiting the moment of modernism "when it comes to understand itself as postmodern." "Spandrels," in short, is both part of and apart from "the master plot of natural history" as it is recounted in Gould's other published work (including his critiques of sociobiology). Her analysis therefore explains the tension or balance in "The Spandrels of San Marco" between its reliance on lyricism within the master narrative and its use of more analytical and nonnarrative lines of argument as well. Herndl too studies "Spandrels" as a cultural event, as an epitome of contemporary connections among science and culture. After the example of the cultural studies theorists Raymond Williams and Louis Althusser, he considers how "Spandrels" participates in the practices of its own discipline (or subculture) and those of the larger culture—how it undermines disciplinary consensus, locates itself within the debates of its scientific subculture, and finally connects itself to larger, public culture. Like Wells and Bazerman, Herndl traces how "Spandrels" intervenes in cultural practice by posing an implicit critique of the sociobiology movement and the work of its leader, E. O. Wilson.

Chapters 5, 6, and 7 are also cultural—i.e., contextual—critiques in their own way. Mary Rosner and Georgia Rhoades mount a feminist critique of "The Spandrels of San Marco" in the tradition of Haraway and Keller that reveals a doubleness in Gould and Lewontin's rhetoric. On the one hand, Gould and Lewontin seem to sanction implicitly a more feminist version of scientific practice through their stylistic choices, in their encouragement of a questioning attitude, and by means of their arguments for wholeness, pluralism, and openness. On the other hand, contend Rosner and Rhoades, Gould and Lewontin also seem to reinforce patriarchal attitudes and values by reveling in confrontation and hierarchy (not cooperation and equality) and by inscribing power onto males while making females invisible. Next, Carolyn Miller and Michael Halloran also read "Spandrels" against the culture of science, in this case the specific scientific subculture that Gould and Lewontin belong to. Through historical and cultural research (in one sense traditional, with a foothold in classical rhetoric and its concept of ethos; in another sense quite untraditional, in searching "Spandrels" for a distinctively Foucauldian "field of discursivity"), Miller and Halloran turn up evidence for the existence of a "historical science" that has so far gone unremarked by rhetoricians. "The Spandrels of San Marco," they establish, exemplifies the reasoning and rhetoric of historical sciences such as geology, cosmogony, and paleontology (in contrast to that of experimental-predictive sciences like physics, chemistry, and molecular biology): it displays the citation patterns characteristic of essays in historical science and the stylistic markers typical of historical science as well. Miller and Halloran help to explain Gould and Lewontin's antipathy for "the adaptationist programme" in biology; its more mechanistic, less flexible ways contrast vividly with the values inherent in historical science. In chapter 7 Dorothy Winsor, like Miller and Halloran and Bazerman, studies citation patterns. But in a kind of "reception analysis" that connects her essay to a later reader-oriented chapter by Charney, Winsor looks not at citations *by* Gould and Lewontin but citations *of* Gould and Lewontin's "Spandrels" in subsequent scientific articles. In the tradition of the sociological studies of scientific texts by Bruno Latour, Winsor's essay reveals how Gould and Lewontin attempt to "make knowledge" through certain reference patterns and statement types that negotiate their own ideas higher in a hierarchy from conjecture to scientific dogma. Winsor not only provides a powerful typology for understanding statement types but also demonstrates a method for deciding whether a particular article has been successful in creating knowledge.

Winsor's chapter and the one that follows it, by John Lyne, provide a kind of fulcrum to the book; their chapters together comprise a sort of transition from more broadly contextual studies to more strictly textual ones. Just as Winsor begins with a close analysis of statement types and closes with a broader account of the reception of "Spandrels" by the scientific community, Lyne uses Burke's theory of dramatism to throw into relief specific textual aspects of "Spandrels"—e.g., Gould and Lewontin's inventional strategies—and then closes with a broad application of adaptationist metaphors to (of all things) the dynamics of American electoral politics. Lyne's essay demonstrates how a Burkean perspective, particularly a meditation on the concepts of *agency* and *purpose,* can repay anyone interested in a better perspective on just about any sort of discourse.

The next chapters stick very closely indeed to textual analysis. Jeanne Fahnestock exploits the possibilities of classical rhetoric, particularly its reincarnation in Perelman and Olbrechts-Tyteca's *The New Rhetoric,* to understand the arguments and certain stylistic patterns in "The Spandrels of San Marco." If Susan Wells in chapter 3 sees "Spandrels" as a "discourse of modernity," Fahnestock here prefers to consider it as a "discourse of antiquity," since many of the argumentative and stylistic strategies employed by Gould and Lewontin, she argues, are captured by classical theories. Fahnestock is particularly good at exposing and explaining the presence and appropriateness of tropes such as opposition, part/whole arguments, bracketing (i.e., presenting two terms as interchangeable to suggest their equivalence), register shifts (i.e., the conscious use of an unusual voice that works against the norms of a given community), and variations in lists. Then Gay Gragson and I use the method of textual analysis suggested by the first generation of reader-response theorists to examine a number of other stylistic moves in "Spandrels." By comparing their essay with another one presented at the same symposium, we note how Gould and Lewontin create an implied reader within the text—a professional scientist who is nevertheless more than a mere scientist; a worldly wise, partisan, and passionate colleague who is encouraged to take issue with arguments, not a dispassionate, reserved observer who merely takes note of ideas—that is rather different from the reader implied in most scientific arguments. Gragson and I do our best to command one brand of reader-response criticism as a method of practical criticism for understanding scientific prose.

Davida Charney does the same with another, second-generation version of reader response. She uses the technique of protocol analysis, developed in the social sciences, to turn up evidence about the response of real readers to "Spandrels." In demonstrating how protocol analysis can be employed by reader-response critics to turn up new and crucial information about what she calls textual "hot spots," Charney also indicates how a careful study of real readers can offer a commentary on the work of analysts who are more text-based. That is, Charney in her essay not only promotes a powerful new method of rhetorical analysis; by turning the "affective fallacy" into a sort of virtue, she also demonstrates how protocol analysis can serve as a corrective to other brands of criticism, particularly in that it attends to behaviors of real readers who are themselves culturally and rhetorically situated. Then, in turn, Debra Journet's deconstruction of "The Spandrels of San Marco" offers an implicit commentary on Charney's chapter—on all the previous chapters, for that matter. Journet in Derridean fashion looks to the margins of the text and especially to its metaphors to expose the places in the text where meaning "deconstructs," escaping the intentions of authors and readers and instead disseminating endlessly through the inevitable elusiveness of language. In its close attention to language, Journet's analysis of the metaphors in "Spandrels" would return us in some ways to the text-bound days of formalist criticism— except that Journet in her conclusion also deconstructs deconstruction and thereby offers insights into the limits of a deconstructive analysis and into the relation of scientific writing to other cultural artifacts.

Deconstruction derives ultimately from an extension of the work of linguist Ferdinand de Saussure. The final two rhetorical analyses in *Understanding Scientific Prose* also derive their inspiration from linguistics. In chapter 13, Greg Myers demonstrates how "politeness theory" developed by linguists in the late 1970s can lead to a deeper understanding of specific textual formulations and in the process radically undermine any vestiges that remain of the notion of scientific impersonality. Through his careful textual analysis, Myers reveals how science is both agonistic and cooperative, how its practitioners construct knowledge by making alliances and enemies with their language choices. And in the end Myers too refuses to be constrained by narrowly textual boundaries: he compares the politeness tactics in "Spandrels" with those used in other publications by Gould and by the other contributors to the same symposium; and his conclusion indicates why politeness analysis, like any other kind of analysis, must in the end depend on a contextual perspective. Barbara Couture's structuralist analysis hearkens back to an earlier period in linguistics; it more nearly lives according to structuralist assumptions first articulated by Saussure. But Couture avoids the narrowly text-based focus of Saussure and formalist critics by calling on and contributing to a more "social" brand of structuralism— "systemic functional linguistics"—derived from the work of M. A. K. Halliday. Couture's original and individual extension of Halliday's systemics—Halliday in fact has encouraged such extensions in the conviction that a theory is worth its salt only if practitioners can apply it and other theorists extend it successfully—permits her to interpret "Spandrels" both as a structure of linguistic features (grammatical and lexical) and as a text embedded in the larger contexts of the Royal Society Symposium in particular and the culture of science in general.

Together the contributors to *Understanding Scientific Prose*, owing to their various methodologies, comment on a wide variety of textual features in "The Spandrels of San Marco." Indeed, there is very little overlap among the essays, and very little within "Spandrels"—from its title to its list of references—that escapes scrutiny. Together the thirteen essays that I have just described amount to a remarkably detailed collaborative analysis of one interesting instance of the rhetoric of science.

Of course, one important reason that the collaboration is so successful is that, despite the varieties in method, the contributors all share the conviction that science is indeed fundamentally rhetorical, drenched as it is in language. Together the contributors dismiss (once and for all?) the notion that science is conducted in some sort of noiseless, transparent, predictable medium designed to facilitate the cooperative and efficient transmission of absolute representations of reality. Far from being transparent and noiseless and predictable, the contributors assert, science is thoroughly human—messy, unpredictable, and inevitably colored by its social and political circumstances. Science is a cooperative venture, yes; but it is just as much competitive, agonistic. It seeks truths, of course; but those truths are probablistic, not certain. Scientific discourse, that is, is less an impersonal and faceless demonstration than it is a set of competitive beliefs laid before a disciplinary jury. Its goal is suasion, not description; its methods include argument as well as data, narrative as well as logic. Scientific communication, writes Jean-François Lyotard—and the contributors to this book would probably agree—amounts to a "language game." It resembles the "taking of tricks" more than it does the regulated transmission of tokens from hand to hand (9–10):

> A scientist is before anything else a person who 'tells stories'. . . .
> [B]y concerning itself with such things as undecidables, the limits of precise control, conflicts characterized by incomplete information, 'fracta,' catastrophes, and pragmatic paradoxes, [postmodern science] is theorizing its own evolution as discontinuous, catastrophic, . . . and paradoxical. (60)

Some Background on "The Spandrels of San Marco"

Understanding Scientific Prose contains two additional items besides the various rhetorical analyses. One, the Appendix, is a reprint of "The Spandrels of San Marco and the Panglossian Paradigm: A Critique of the Adaptationist Programme"; it is included to permit the readers of this book to follow better the various analyses of "Spandrels" and to encourage readers to measure their own sense of "Spandrels" against the critical accounts that they will encounter here. In the other, chapter 15, Stephen Jay Gould contributes his own retrospective on "Spandrels," its genesis, and its reception; and he offers commentary on the critical analyses of his work that are published in this book. In the first two sections of his chapter, Gould offers some fascinating background on the composition of "Spandrels" that I recommend as a sort of second orientation to this volume. Here I want to elaborate on that orientation to prepare readers for what they will encounter in "Spandrels" and in *Understanding Scientific Prose.*

As Gould explains, "The Spandrels of San Marco and the Panglossian Paradigm: A Critique of the Adaptationist Programme" was presented in December 1978 at a blue-ribbon Royal Society of London Symposium that was organized with a colleague by John Maynard Smith (arguably the preeminent evolutionary biologist in Great Britain). The following year the article appeared in print in the *Proceedings of the Royal Society of London* as the final article among the other nine (one by Maynard Smith) presented at the symposium. Gould, the primary author of "Spandrels," is a paleontologist of great accomplishment who has won numerous awards for his books on the history and conduct of science. His Harvard colleague R. C. Lewontin, the coauthor of "Spandrels," is perhaps the most prominent evolutionary geneticist in America. (The details of the collaboration between Gould and Lewontin are spelled out in the first half of Gould's chapter in this book.)

The topic of the Royal Society Symposium—"The Evolution of Adaptation by Natural Selection"—was chosen to bring together for discussion some of the most influential scholars currently working with the central concepts in natural science, adaptation, and natural selection. According to evolutionary theory, *adaptation* and *natural selection* are key operations through which a species optimizes its chances of surviving and flourishing. Especially through the process of natural selection, those species whose traits make them fittest to survive in a certain environment produce more offspring and contribute to their offspring those traits most likely to be useful in the future. The many species of finches that Darwin saw on the isolated Galapagos Islands provide a classic example of adaptation: all of the species are derived from a common ancestor, all of them fill roles and niches that in more conventional and crowded circumstances would be filled by several kinds of bird families, and all of them are now distinguished by the adaptation of their bill types for different diets. Over a period of many generations, some of the finches, in order to survive on the limited but various resources of a small archipelago, exploited their ancestral short, stout beaks to crush small seeds. Other species developed bills strong enough to break larger seeds or long, pointy bills suitable for feeding on the prickly pear. Yet another has a parrotlike beak appropriate for feeding on buds and fruits. The most curious species of all uses twigs and cactus spikes to extract insects from crevices in trees. To understand the diversity of nature, scientists belonging to "the adaptationist programme" first consider the characteristics of organisms (such as the beaks of Darwin's finches) and their environments (such as the feeding opportunities and competition present on the Galapagos Islands), and then describe or predict the direction of evolution.

Though puzzling out the details of evolutionary theory occupies the research efforts of many scientists—not only biologists but geologists, anthropologists, geneticists, medical researchers, even economists and other social scientists—many of the dynamics of evolution remain mysterious or in dispute. Of particular controversy is the perennial question of just how much adaptation and natural selection explain about the diversity of nature. According to one school of thought within the field (in his chapter, Gould offers the charged term "hardline position" to characterize this school; another term would be the "optimization position"), natural selection is the driving force behind adaptation, evolution, anatomy, and behavior. Natural selection explains nearly everything about evolution and nature's diversity: species optimize their chances of surviving and flourishing by selecting those features and behaviors most likely to pay off in a given environment.

Gould and Lewontin, by contrast, occupy a different position among evolutionary theorists (a position that Gould calls "nuanced" or "pluralist"). While admitting the undeniable importance of natural selection as a guiding force on adaptations, Gould and Lewontin also argue for the recognition of other factors. In particular, along with scientists in France, Germany, and elsewhere they assign a far greater role to chance and to random processes (such as mutation), acknowledge a variety of constraints on natural selection that prohibit the optimization of a great many traits and behaviors, and give credence to "formal" and "structural" explanations for anatomical diversity. In arguing, in effect, that some or even many variations within a species can be neutral or even maladaptive, they thus support a "pluralistic approach" to understanding the agents of evolutionary change, an approach that emphasizes alternative explanations to natural selection as the single cause of evolutionary change. Rhetoricians might call it a more "sophistical" approach: in offering a narrative of multiple causality, they have less faith in chains of events that can always be rationally understood, insist

Jack Selzer

on plurality, and in general offer a less tidy, more disruptive explanation for things.

"Spandrels," then, is a critique of one school of thought within the Darwinian tradition by scholars in another school within that same tradition. Gould and Lewontin begin their critique with an elaborate and unusual architectural analogy comparing elaborately decorated spandrels—tapering triangular spaces (as in St. Mark's Cathedral) formed when four columns support a dome—to the elaborations of natural anatomy; it is their contention that adaptationists too often look at secondary epiphenomena (like spandrels and various anatomical traits in nature) as the cause of nearly every form in nature—rather than as the effect of structural systems (or other factors) in nature. Gould and Lewontin go on in their second part to make their point explicit, alluding to Voltaire's notorious Dr. Pangloss: "We wish to question a deeply engrained habit of thinking among students of evolution. We call it the adaptationist programme, or the Panglossian paradigm . . . : the near omnipotence of natural selection in forging organic design and fashioning the best among possible worlds" (150). Gould and Lewontin in their next two sections characterize the adaptationist program in detail (especially through detailed references to certain scientific articles), critique its fundamental procedures and argumentative moves (referring to still more exemplars), and summon Darwin as the advocate of a more "pluralist" explanation for natural phenomena. Finally (in part 5) Gould and Lewontin offer their own "alternatives to the adaptationist programme"—a set of alternative explanations that, if adopted, "could put organisms, with all their recalcitrant, yet intelligible, complexity, back into evolutionary theory" (163).[3]

"The Spandrels of San Marco" is an eloquent and powerful and interesting argument. But it was chosen as the object to be studied in this book not because its argument is "correct" or because the contributors wish to sanction its rhetorical choices. *Understanding Scientific Prose* does privilege Gould and Lewontin's voices as against others in the debate over adaptation (though the other voices of the symposium are included as well in many of the chapters) because theirs is the language under scrutiny here. But the other contributors and I are not attempting to canonize "The Spandrels of San Marco" or its authors. Indeed, we are quite aware that many outstanding scientists (e.g., Mayr) find Gould and Lewontin guilty of inconsistencies, of false analogies, of various violations of logic, and especially of consistently caricaturing the adaptationist position—that is, of representing extreme versions of the adaptationist position as the typical one. The aim of this book is not to judge the effectiveness of "Spandrels" but to understand it; the primary focus of *Understanding Scientific Prose* is not evolutionary theory but rhetorical theory and practice. The contributors to this book are here together not to praise "Spandrels" or to blame it but simply to demonstrate how various new methodologies can profitably be brought to bear on the rhetoric of science.

With that disclaimer, let me turn you at last to the substance of this book, the thirteen analytical methods on display here. In the study of the rhetoric of science, we need analytical techniques as sophisticated and as powerful as the texts they are designed to describe, and we need opportunities to reflect on the adequacy of the methods we use to understand science. This book is an attempt to provide both.

NOTES

1. Wellek and Warren's *Theory of Literature* elaborates even more fully the distinctions that New Critics pressed between "ordinary language" (the realm of rhetoric) and the language of science and of poetry:

> The main distinctions to be drawn are between the literary, the everyday, and the scientific uses of language. . . . The ideal scientific language is purely "demonstrative": it aims at a one-to-one correspondence between sign and referent. . . . [Its] sign is transparent; that is, without drawing attention to itself, it directs us unequivocally to its referent. Thus scientific language tends toward such a system of signs as mathematics or symbolic logic. Literary language . . . abounds in ambiguities. . . . In a word, it is highly "connotative". . . . More difficult to establish is the distinction between everyday and literary language. . . . No doubt, everyday language wants most frequently to achieve results, to influence actions and attitudes. . . . It is . . . quantitatively that the literary language is to be differentiated from the everyday. . . . The pragmatic distinction between literary language and everyday language is clear. We reject as [something that is not] poetry or label as mere rhetoric everything which persuades us to a definite outward action. Genuine poetry affects us more subtly. [It takes us] out of the world of reality. (24–25)

2. For a rather complete accounting of work on the rhetoric of science, see the bibliographies at the end of this book, at the end of Bazerman's *Shaping Written Knowledge*, and in Dear's *The Literary Structure of Scientific Argument*.

3. Page numbers for all references to "The Spandrels of San Marco" throughout this book are noted in parentheses. The text of "Spandrels," with corresponding page numbers, is reprinted in the Appendix.

Jack Selzer

WORKS CITED

Andrews, James R. *The Practice of Rhetorical Criticism.* New York: Macmillan, 1988.

Bazerman, Charles. *Shaping Written Knowledge: The Genre and Activity of the Experimental Article in Science.* Madison: University of Wisconsin Press, 1988.

Black, Edwin. *Rhetorical Criticism: A Study in Method.* New York: Macmillan, 1965.

Brooks, Cleanth, and Robert Penn Warren. *Understanding Poetry.* New York: Holt, Rinehart, and Winston, 1938.

Campbell, John A. "The Polemical Mr. Darwin." *Quarterly Journal of Speech* 61 (1975): 375–90.

Campbell, John. "Scientific Revolution and the Grammar of Culture: The Case of Darwin's *Origin.*" *Quarterly Journal of Speech* 72 (1989): 351–76.

Corbett, Edward P. J., ed. *Rhetorical Analysis of Literary Works.* New York: Oxford University Press, 1969.

Dear, Peter, ed. *The Literary Structure of Scientific Argument.* Philadelphia: University of Pennsylvania Press, 1991.

Foss, Sonja, ed. *Rhetorical Criticism: Exploration and Practice.* Prospect Heights, Ill.: Waveland, 1989.

Geertz, Clifford. *Works and Lives: The Anthropologist as Author.* Palo Alto: Stanford University Press, 1988.

Gould, Stephen Jay. *The Mismeasure of Man.* New York: Norton, 1981.

Gould, Stephen Jay. *Time's Arrow, Time's Cycle.* Cambridge: Harvard University Press, 1987.

Gross, Alan. *The Rhetoric of Science.* Cambridge: Harvard University Press, 1989.

Halloran, S. Michael. "The Birth of Molecular Biology: An Essay in the Rhetorical Criticism of Scientific Discourse." *Rhetoric Review* 3 (1984): 70–83.

Haraway, Donna. *Primate Visions: Gender, Race, and Nature in the World of Modern Science.* New York: Routledge, 1989.

Hart, Roderick. *Modern Rhetorical Criticism.* Glenview, Ill.: Scott Foresman, 1990.

Keller, Evelyn Fox. *Reflections on Gender and Science.* New Haven: Yale University Press, 1985.

Keller, Evelyn Fox. "Women Scientists and Feminist Critiques of Science." *Daedalus* 116 (1987): 77–91.

Kristeva, Julia. *Desire in Language.* Ed. Leon S. Roudiez. Trans. T. Gora, A. Jardine, and L. Roudiez. New York: Columbia University Press, 1980.

Kuhn, Thomas J. *The Structure of Scientific Revolutions.* 2nd ed. Chicago: University of Chicago Press, 1970.

Leff, Michael, ed. "Rhetorical Criticism: The State of the Art." Special report, *Western Journal of Speech Communication* 44 (1980): 264–349.

Lyne, John, and Henry F. Howe. " 'Punctuated Equilibria': Rhetorical Dynamics of a Scientific Controversy." *Quarterly Journal of Speech* 72 (1986): 132–47.

Lyotard, Jean-François. *The Postmodern Condition: A Report on Knowledge.* Trans. Geoff Bennington and Brian Massumi. Minneapolis: University of Minnesota Press, 1984.

Mayr, Ernst. "How to Carry Out the Adaptationist Programme?" *American Naturalist* 121 (1983): 324–34.

Mayr, Ernst. Personal correspondence. Dec. 31, 1991.

Myers, Greg. "The Social Construction of Two Biologists' Proposals." *Written Communication* 2 (1985): 219–45.

Myers, Greg. "Stories and Styles in Two Molecular Biology Articles." In *Textual Dynamics of the Professions,* ed. Charles Bazerman and James Paradis, 45–75. Madison: University of Wisconsin Press, 1991.

Myers, Greg. *Writing Biology: Texts in the Social Construction of Scientific Knowledge.* Madison: University of Wisconsin Press, 1990.

Overington, Michael. "The Scientific Community as Audience: Toward a Rhetorical Analysis of Science." *Philosophy and Rhetoric* 10 (1977): 143–64.

Perelman, Chaim. *The Realm of Rhetoric.* Notre Dame, Ind.: Notre Dame University Press, 1982.

Perelman, Chaim, and Lucie Olbrechts-Tyteca. *The New Rhetoric.* Trans. J. Wilkinson and P. Weaver. Notre Dame, Ind.: University of Notre Dame Press, 1969.

Simons, Herbert W., ed. *Rhetoric in the Human Sciences.* Newbury Park, Calif.: Sage, 1989.

Toulmin, Stephen. *Human Understanding: The Collective Use and Evolution of Concepts.* Princeton: Princeton University Press, 1972.

Wander, Philip. "The Rhetoric of Science." *Western Journal of Speech Communication* 40 (1976): 226–35.

Wellek, Rene, and Austin Warren. *Theory of Literature.* New York: Harcourt Brace, 1942.

2 INTERTEXTUAL SELF-FASHIONING

GOULD AND LEWONTIN'S

REPRESENTATIONS OF

THE LITERATURE

CHARLES BAZERMAN

Modern scientific articles almost universally represent themselves as part of a literature through explicit citation and through discussion of other texts identified as being closely related. In the last two centuries, reviews of the literature, both as self-contained syntheses and as embedded introductions of research articles, have told coherent stories about prior work and have thereby established frames of meaning for new work (Bazerman, "How Natural Philosophers"; Myers, "Stories and Styles"; Swales). Thus each new finding, argument, or claim locates itself upon its own reconstruction of an explicit intertextual field. The constant reformulation of the prior literature is one of the mechanisms by which consensus is reached on the value and meaning of the claims of published work (Cozzens; Messeri; Small) and on the standard sets of associations that comprise codified knowledge (Merton; Ziman). Citation studies have been used to identify not only the influence and canonical standing of various articles but also the cognitive and social networks of evolving research specialties (Mullins et al.; see also essays by Winsor, chapter 7, and by Miller and Halloran, chapter 6, in this book).

Representation of the intertext—the web of texts against which each new text is placed or places itself, explicitly or implicitly—is thus a strategic site of contention, for it is the site at which communal memory is sorted out and reproduced, at which current issues and communities are framed, and dynamics established, pushing the research front toward one future or another.[1] Readers who accept findings or local analysis without accepting the situating intertext admit only a limited part of an argument, rejecting the meanings that attach the findings to a dynamic of specialty knowledge. Especially if we are to convince readers of fundamentally new positions, at odds with existing thought, we must somehow uproot the intertext upon which current audience perceptions rest. It is not sufficient for readers to integrate a finding into their existing mental framework; we must engage them in a radically new line of discussion by discrediting the former discussion with all its implied dynamics and intellectual freight.

In "The Spandrels of San Marco and the Panglossian Paradigm," Gould and Lewontin seek just such a disciplinary reorientation. They attempt to reconstruct the intertext against which the article will be read. They have little quarrel with any specific finding or analysis presented in the articles they discuss, but only with the framework those findings are put into, a framework that keeps asserting the need for an adaptationist account of the specific survival benefit of each feature. The weight of this framework is so strong in eliciting these stories of adaptive value, Gould and Lewontin argue, that even when the evidence falsifies existing adaptationist accounts, evolutionary biologists will go to great lengths to reassert another adaptationist story, creating a never-ending trail of ad hoc alternatives that keep the adaptationist impulse alive and undamaged.

However, as we look into the intertextual struggles set up in the text, we find several unusual features that suggest that the divisions are both not as deep as they appear to be and, in another sense, deeper than that. The divisions portrayed in the article are only intelligible in light of issues that are never explicitly brought to the surface. Only by analyzing the intertext that Gould and Lewontin construct, and by comparing that to other representations of the same and closely related intertexts, can we see exactly how and why Gould and Lewontin are trying to persuade readers of a particular view of evolution, nature, and human nature. Because the intertext is such a strategic site of contention—the battlefield for control of the cognitive universe within which new claims will be read—analysis of intertextual representations lets us see not only the rhetorical game being played, but also the struggle to define the rules and limits and stakes of that game.

The Rhetoric of Intertextual Struggle in "Spandrels"

From the first sentence of the abstract that prefaces the "Spandrels of San Marco and the Panglossian Paradigm," Gould and Lewontin set themselves against a literature, a body of statements, a dominant text-producing research program: "An adaptationist programme has dominated evolutionary thought in England and the United States during the past 40 years." The next few sentences characterize the common argument of this literature. By naming it and identifying its dominance, Gould and Lewontin immediately set that literature as an edifice to be discredited, uprooted, and replaced by another discourse: "We criticize this approach and attempt to reassert a competing notion (long popular in continental Europe) that organisms must be analysed as integrated wholes. . . ." The remainder of the abstract details their alternative, their specific criticisms of adaptationism, and the alignment of their project with the Ur-intent of the founding father: "We support Darwin's own pluralistic approach to identifying the agents of evolutionary change." In their eyes, the literature has gone astray and they have come to set it aright.

The abstract is direct and blunt and contained within the professional discourse of evolutionary biology. The abstract is no summary of the rich counterpoint of the article, which strays far beyond descriptive biology and evolutionary theory to encompass voices from the French Enlightenment, medieval Venetian art and architecture, contemporary international urban intellectual life, anthropology, forensic science, mathematical modeling, and statistical genetics. The abstract does not even begin to give a clue about the allusions or arguments of the article's primary title, "The Spandrels of San Marco and the Panglossian Paradigm" (although the subtitle does fit precisely with the abstract).

The article proper begins, curiously, with an architectural travelogue, with "the great central dome of St. Mark's Cathedral in Venice." The first section then wanders far from the halls of biological meetings, which are the venue of the abstract, into realms of medieval churches and Aztec cannibalism and eighteenth-century philosophic polemic. This first section, however, has everything to do with the article's primary title. Indeed, words like *evolution* and *biology* are barely mentioned, showing up only in the cadences at the midpoint and end of the section, where the analogy between discourse about art and human action and the discourse of evolutionary biology is made.

Even as the essay moves in the second section into the domain of evolutionary biology, the ironies, alienations, and resonances set up in the first section never let the argument settle down comfortably into a purely biological discourse. The article does not even present an orderly exposition of the points outlined in the abstract. While the abstract may be said to gather together the major claims of the essay, and to present those claims in an order that plausibly forms the coherent argument of the essay, the claims do not appear in the article in the same order or with the same linkages. Rather, the abstract is an orientational metacommentary, a reconstruction of the argument that straightforwardly places it in battles over the evolutionary literature. The strategy of the article is more oblique: to re-place a discussion that it claims has defined the dominant discourse and project of evolutionary biology, overtly by displacing one programmatic argument and replacing it with another, but even more by repositioning the argument into broader social discourses—first by analogy, but ultimately by suggesting continuities of biology with other discourses and by hinting at darker ideological currents within narrowed forms of biological discourse.

The present analysis will examine Gould and Lewontin's strategies of representing the intertextual field in ways that identify oppositions, draw battle lines, and discredit the enemy. From there we will compare Gould and Lewontin's representation of the intertext against the original texts and the surrounding literatures they are claimed to represent. That is, we will look at what kind of transformations Gould and Lewontin make of their literary resources. We will then examine the literary field that forms the immediate rhetorical context of the Gould and Lewontin statement, to interrogate whether their historical reconstruction of the genesis of current issues seems consistent with the current debate on those issues. At this level it becomes clear that not all see the debate as Gould and Lewontin wish to see it, that they have not reached a focussed joining of issues, an intersubjectively agreed-upon stasis with their colleagues. Indeed, most see the issues as much smaller and less monumentally significant, more a matter of local adjustment of claims within the existing discourse.

All this will leave us with the question as to why Gould and Lewontin see their issues in such fundamental terms of total re-placement, why they feel so much is at stake. The answer lies in themes buried in this text but spelled out in other publications by the coauthors, self-cited in "Spandrels." This analysis exposes the fluidity of the intertext and shows how representation of the intertext can become a resource to reconstruct the issues before a field, and even the location of a field in relation to other discourse fields. There is a tension revealed in Gould and Lewontin's article, a tension between the need and desire to enclose a literature—thereby creating a communal project—and the need and desire to open the literature up so as to introduce new issues and arguments excluded by the enclosures.

Charles Bazerman

Defining the Enemy

In the second part of "Spandrels," after the excursion into strange realms beyond the recognized borders of biology, Gould and Lewontin get down to their overt business by announcing, "We wish to question a deeply engrained habit of thinking among students of evolution. We call it the adaptationist programme, or the Panglossian paradigm." Of course the questioning, in typical academic fashion, is a heavily ironicized rejection, indicating that the authors and anyone else who looks at it sensibly (that is, through their eyes) will clearly know better.[2] All ironies are based on someone knowing better than someone else. Here, the misguided adaptationists are clearly the ones who know less, so much less that they can't distinguish between habit and legitimate academic argument or between popularized notions and professional thought. They lack self-knowledge and critical objectivity and thus play the fools. Their foolishness is ironized by Gould and Lewontin: the teeming, fluid world is anatomized into discrete objects called traits (which are never defined), and then when that fracturing fails to produce an adequate account, the notion of wholeness is recreated under the idea of trade-off.[3]

As warrant for their critique, Gould and Lewontin characterize what they claim is the typical argumentative style of the adaptationists. What Gould and Lewontin present, however, is not the argumentative style for positively asserting adaptationist claims, but rather the tactics of adaptationists when reacting to the apparent failure of an adaptationist argument. Gould and Lewontin construct a strategy of ad hoc repairs that they claim adaptationists pursue: replacing the failed adaptationist account with a new one; assuming that a new adaptationist account must exist; claiming that the lack of an account is the fault of imperfect knowledge; and excluding data that might imply anything other than an adaptationist account. The generic expectation of adaptationist accounts is so strong, Gould and Lewontin in essence claim, that a secondary genre of explanations arises to uphold respect for the genre even when it is not easily fulfilled.

Gould and Lewontin have thus described a literature based on a set of simplifications and reductions that distort the phenomena being studied, but that resists challenges to its underlying simplifying account by a combination of obduracy and willed blindness. These characterizations of the literature are only partially substantiated by detailed critique of specific articles.[4]

Gould and Lewontin attribute the Anglo-American narrowness and reductionism to a false turn by two of the early disciples of the evolutionary doctrine, Wallace and Weismann. The false notions of false disciples were not shared by the founding father Darwin, as Gould and Lewontin assert by citing two passages (and a third by implication) where Darwin explicitly extended the bounds of legitimate discourse beyond natural selectionist accounts. Following this pluralistic lead, Gould and Lewontin suggest five alternative nonadaptationist evolutionary accounts for the literature to pursue, which they substantiate through openings already in the literature, including much self-cited work.

They also identify a more satisfactory tradition of evolutionary thought pursued among continental biologists that accommodates their concerns with the overall interaction between all the features of the organism and the context within which the organism thrives. They cite a largely German and Austrian literature as the locus of this argument for considering an organism's *Bauplan*, which conceives of the organism as a whole rather than as an assemblage of atomized parts. By citing this Central European literature, Gould and Lewontin foster a self-conscious split between misguided Anglo-Americans and wiser Europeans. They reject one literature and establish a new program, enlisting both Darwin and a contemporary tradition of truer Darwinian disciples from beyond the borders of the reviled adaptationist program.

A Larger Cultural Frame for the Local Silliness

To help the readers step outside the blinders of the narrowed Anglo-American discourse, Gould and Lewontin go beyond alternate legitimate biological voices—Darwin and the Europeans—to enlist other highly legitimated cultural voices outside of biological conversation altogether. Indeed the article opens outside biology, within a great cathedral arching above it. The spandrels of San Marco, represented by picture and description, remind us of the wisdom of the original artisans, who properly (from Gould and Lewontin's perspective) understood the relationship of local features to architectonic structure and constraints.

The aesthetic rightness of this understanding is echoed in the voice of the art critic that speaks in the first paragraphs of the introduction. The voice, by invoking a familiar cultural genre, implies an intertext of the great body of aesthetic analysis. In the implied aesthetic humanistic conversation, form, structure, and appearances are all given weight in

assessing meaning and value. Meaning is treated as complex, and reductionism is always transparently out of place. That transparent inap-
propriateness of reductionism and the tendency of art to see beyond
simple surface meanings are directly encapsulated in the comments
that move the discussion from appearances to context in typical critical
fashion:

> The design is so elaborate, harmonious, and purposeful that we
> are tempted to view it as the starting point of any analysis, as the
> cause in some sense of the surrounding architecture. But that
> would invert the proper path of analysis. The system begins with
> an architectural constraint: the necessary four spandrels and their
> tapering triangular form. (148)

From analysis of architecture the discussion proceeds to an analysis
of human behavior (after a brief passage through the literary philosophic world of *Candide,* to which we will return in a moment). Here we
again start out with a reductionist account, this time of cannibalism.
This account is so out of keeping with the usual gist of anthropological
discussion that it immediately seems discordant and shocking: the cultural ritual and practice of cannibalism treated as simply a matter of
protein. The reductionism is from culture to nutrition, from anthropology to biology. Whatever the truth might be about the claims, the
emotional valences of the arguments are clear: anthropology is a re-
specter of human complexity and difference, a chronicler of the varieties of human spirit, creativity, and life; nutrition is a matter of chemistry that cuts across all complexities to reduce us all to the same beast.
Although nutrition and chemistry are forever with us all, it strikes us as
discordant and insufficiently respectful of humankind to reduce cultural practice (especially a remarkably repellent and therefore fascinating
cultural practice) to a least common denominator. Gould and Lewontin's
review of the ensuing anthropological literature in fact follows the path
back from reduction to complexity, from nutrition back through culture,
class, and conspicuous consumption. Reductionism is presented as foolishness (to be treated ironically with Gould and Lewontin's comments
about poor butcher shop management) because it is blind to the full
range of facts and a reasonable assessment of the whole picture.

Pangloss the Fool and Voltaire the Wise

Pangloss, present as a puzzle from the beginning in
the article's enigmatic title, does not fully enter the text until a discus-

sion of the wisdom of various approaches to interpretation of features. There he is presented as the paradigmatic fool who forgets the full range of facts and the whole picture: "Anyone who tried to to argue that the structure exists because of the alternation of rose and portcullis makes so much sense in a Tudor chapel would be inviting the same kind of ridicule that Voltaire heaped on Dr. Pangloss" (149). Pangloss becomes the leitmotif emblem of foolish reductionism and atomism throughout the article. He returns in the critique of ad hoc adaptationist explanations, as an epigram to a discussion of optimistic teleology, and as a shadow beneath all the tales of foolishness. His presence lies especially underneath Galton's anecdote about Spencer, who is overtaken by a fit of teleological a priori reasoning; Spencer is as rapidly deflated by Galton as Pangloss is by Voltaire.

The effect of establishing the battle lines of the article as congruent with Voltaire's ridicule of Pangloss is twofold. Not only does it help establish the adaptationists as narrow, blind, optimistic, reductionist, teleological fools, but it also puts Gould and Lewontin in the position of Voltaire—the wise, truthful philosopher who is the scourge of fools. Biological correctness is left behind as the issue is transformed into philosophic wisdom. The humane understanding of the human world is now at stake, and the evil to be scourged is beyond silliness or error: it is the evil of bad doctrine.

The emotional weight attached to Pangloss and Voltaire comes from many other features of the novel *Candide* beyond Voltaire's critique of the Panglossian doctrine that all is for the best in this best of all possible worlds, and that each detail of this world has a simple optimistic teleological explanation. In the novel the problem with Pangloss is not his silly, simple-minded doctrine, rather his excessive cleverness: he is a manipulative hypocrite who uses his doctrine for self-aggrandizement, deception, oppression, moral disengagement, and a host of other wickednesses. He is ridiculed and reviled for his hypocrisy, not his simple-mindedness. The resolution in the book is not victory over teleological optimism, rather the act of transcending it, as Candide invites both Pangloss and Steven (Pangloss's philosophically pessimistic opposite) into the modest garden. They can keep up their silly debate as long as they get on with the planting. Candide's journey in the book leads, indeed, not to philosophic truth, but away from it, to a philosophic modesty that creates the glow of humane wisdom attached to Voltaire despite his cold satiric eye.

This is not to suggest that Gould and Lewontin want to make detailed, explicit connections between their critique of the adaptationists and the complex position of the novel. Quite the contrary; they simply want to label the adaptationists with a damning label of philosophic foolishness. Properly so: all analogies are incomplete and limited, for narrow purposes. On the other hand, the persuasive effect of this analogy as the leitmotif of the article, as the enriching chord beneath the overt melody of the argument, depends precisely on the richness, ironic complexity, and humanity we associate with the novel—and by extension attribute to Gould and Lewontin.[5]

Similarly, the affective weight of the preliminary framing of the argument—in terms of aesthetics and cultural examples—suggests that cultural wisdom warns against the reductionist strategies of the adaptationists, who are portrayed as mistaking local epiphenomena for the fundamental patterns that produce them. The wiser, more deeply human approach, the approach that incorporates humanistic and social scientific thinking, undertakes to understand the complexity of creativity that produces cultural rituals, universities, cathedrals, and satire.

The Text of the Intertext: Unacknowledged Complexities

Such are the argument and the enemies and allies in the theater of the intertext as Gould and Lewontin represent them. But not everyone may read the literature in the same fashion. Other readers with different interests and perspectives might not select the same set of texts as the most relevant nodes of discussion, nor might they find the same stances and divisions in those texts. To identify the perspective through which Gould and Lewontin reconstruct the literature, I examined as many as I could obtain of the texts that they cite. I was able to examine all but three[6] of the forty-one publications on Gould and Lewontin's reference list. I then compared Gould and Lewontin's characterization or use of each publication with my own readings of the relevant parts of the cited material. Since I am not a biologist, nor do I have a detailed familiarity with the evolving evolutionary debates, my readings are obviously limited in their validity. Nonetheless, in the case of issues that Gould and Lewontin claim were raised in the published material, I was usually able to determine whether they were indeed specifically raised and whether the cited text's position approximated Gould and Lewontin's representation. When I was uncertain, I erred on the side of accepting Gould and Lewontin's interpretation. Although I found that the articles generally conformed to the claims Gould and Lewontin suggest, I found a number of interesting variances worth noting: in implications drawn by Gould and Lewontin, in hardening ambiguous oppositions, or in revealing other underlying issues.

Of the thirty-eight source texts I examined, twenty-eight unproblematically conform to Gould and Lewontin's use or characterization,[7] but ten raise issues of interpretation,[8] almost always in determining the article's position on the adaptationist–antiadaptationist issue, or even whether they recognize that issue as relevant.

Several researchers identified as fully and simply adaptationist (e.g., Coon, Shea, Jerison) actually present more complex positions that take into account some of the alternative or wider-ranging accounts that Gould and Lewontin argue for. On the other side, Gould and Lewontin enlist to their cause several publications that have adaptationist leanings (Lande 1978),[9] are more cautious in their commitments (Waddington),[10] or seem to be outside the debate altogether (Galton).[11] Understandably, middle cases make categorization hard. But the very point is Gould and Lewontin's creation of sharp divisions—intellectual, geographic, and temporal—to define the terms of their argument.

The most revealing and significant neatening of battle lines concerns Gould's four earliest self-cited publications, which discuss size and timing of maturation of individuals of a species (allometry and heterochrony). In the "Spandrels" article he sees allometry as supporting alternative evolutionary mechanisms that go beyond adaptationism. In his cited 1966 article, however, Gould treats allometry simply as a subtopic of adaptationism. The article in fact explicitly announces its task as provision of an adaptationist account of size variation. The second self-cited article (1971) also argues that allometry is within adaptationism, but a more complex version of it. The 1974 article is a comprehensive review of allometry, including errors and misapplications; Gould points out evolutionary implications, but without taking an explicit stand on adaptationism. The cited 1977 book, *Ontogeny and Phylogeny,* complicates allometry by adding the heterochronic issue of retardation of development, a condition that allows for education and family bonding in human development. Although the "Spandrels" article sees these very cultural issues as going beyond adaptationism, the 1977 book still sees this opening to culture and cumulative achievement as part of adaptation. In 1977 Gould approvingly quotes Krogman ("Man is programmed to learn to behave, rather than to react via an imprinted determinative instinctual code") and Alexander Pope ("A longer care man's helpless kind demands, / That longer care contracts more lasting bands"). Although two years later in "Spandrels" Gould would label learning and bonding as being outside adaptationism, in the 1977 book he still repeatedly refers to the "adaptive significance of retarded development."

Charles Bazerman

Apparently Gould, as he looked more deeply into issues of allometry and the related timing of development, came to an increasingly open-ended and creative version of adaptationism. Eventually he changed his mind altogether about whether his evolutionary accounts were fully adaptationist at all. That is fine. He is allowed to change his mind. What is curious is that in the "Spandrels" article, even while self-citing his earlier discussions, he does not reveal that change, nor renounce his earlier positions. He simply enlists his earlier articles in his current position to keep sides neatly divided. He in effect rewrites those articles after the fact by quietly assimilating them into an opposite position.

The Intertext of the Current Debate: The Symposium

Divisions may be kept neat not only by the manner of representing individual texts, but also by selectivity of the literature cited. A more comprehensive examination of the literature may not support the representation of direct opposition drawn from a carefully selected subset of texts. Not being widely read in the evolutionary literature, however, I cannot begin to pass judgment as to whether Gould and Lewontin's characterization of the hard-core adaptationist program plausibly reflects Anglo-American publications on evolution or whether researchers define themselves explicitly along the divisions that Gould and Lewontin point toward.

The Royal Society Symposium in which Gould and Lewontin's argument was first presented does, nonetheless, provide a limited sample of the evolutionary literature into which Gould and Lewontin are injecting their argument. Of course, the symposium is not a random sample of the literature; it was a planned, shaped event.

Given the range of the ten primary papers and one commentary presented in the symposium entitled "The Evolution of Adaptation by Natural Selection" and given that the organizers, J. Maynard Smith and R. Holliday, are responsible for two of the most self-avowedly adaptationist papers in the meeting, the panel seems intended to show the wide range and power of work developed under the agenda of adaptationism. Moreover, the placement of the Gould and Lewontin paper at the end of the session suggests that their paper was intended by the organizers as the critical or cautionary voice that frequently caps such occasions, while not undermining the generally positive implications of the papers overall.

Yet although all the papers are geared to an adaptationist symposium and although all except Gould and Lewontin's propose adaptationist arguments, they do not uniformly reflect the narrow adaptationism excoriated by Gould and Lewontin. In fact, examination indicates that only two of the ten primary papers fall into Gould and Lewontin's characterization of adaptationism. The remainder move across the spectrum, some ultimately adopting positions rather close to Gould and Lewontin's, although still remaining under the banner of adaptationism.

The two papers that most closely fit Gould and Lewontin's characterization of adaptationism are those of the symposium organizers. Maynard Smith's article, "Game Theory and the Evolution of Behaviour," argues that game theory gives a functional account of fighting behavior consistent with an evolutionary selection of that behavior. This argument closely parallels the cannibalism–nutrition argument. A behavior that depends on the coordination of many physical features and is usually thought of as a complex causal entity is identified as a single evolutionary feature and through analysis (nutritional or game theoretical) is characterized as adaptive. R. Holliday and T. B. L. Kirkwood's "The Evolution of Ageing and Longevity" argues that even an apparently nonadaptive quality such as aging can be seen as adaptive. An account of accumulating cell deterioration through genetic variation over time suggests that aging provides a mechanism for balancing between the flexibility needed for genetic diversity and the consequences of genetic deterioration. Both these articles provide what Gould might categorize as "just-so" stories to bring even apparently nonadaptive nonfeatures under adaptationist accounts.

Two further articles, although overtly adaptationist, do not fully fit Gould and Lewontin's characterization of a reductionist program. "Selection in Vitro" by L. E. Orgel brings adaptationism down to the molecular level by identifying adaptive chemical features at the binding site in prebiotic RNA proteins. This paper, while clearly making an adaptationist argument, does so at so fundamentally biochemical a level that it is difficult to distinguish between isolatable features and the *Bauplan*. The adaptive feature here examined is not an evolved, mature feature so much as a structural element that has consequences for later, complex organisms. Even though the argument is overtly adaptationist, because the adaptation concerns the very character of the genetic material, the consequences can hardly be maintained at the level of isolatable surface feature. "The Evolutionary Genetics of Sexual Systems in Flowering Plants" by D. and B. Charlesworth, while attempting to explain complex sexual systems in adaptationist terms, admits difficulties without proposing easy solutions or ad hoc accounts to preserve a neat adaptationist story.

Charles Bazerman

Two other articles, while following general adaptationist lines, move even closer to the kinds of considerations that Gould and Lewontin wish were better attended to. "The Evolution of Enzyme Structure" by B. S. Hartley finds adaptation occurring as the result of multiple forces and complex contexts, not as a result of simple, isolatable selection of single independent features. "The Evolution of Genetic Diversity" by B. C. Clarke, even while wishing to provide adaptationist accounts of diversity, nonetheless moves away from a single feature–single creature explanation into one of a species-wide *Bauplan*. The existence of genetic variety, the paper argues, reduces predation by decreasing density of individual species.

The last three primary articles of the symposium (other than Gould and Lewontin's) accept much of Gould and Lewontin's argument, although not in all cases seeing the consequence as a rejection of adaptationism. R. Dawkins and J. Krebs in "Arms Races Between and Within Species" present adaptive evolution occurring in a complex interactive environment of intra- and interspecies pressures, driving complexity expressed at the *Bauplan* level. Dawkins and Krebs do not rule out rapid punctuated evolution, nor do they claim to argue for a naive perfectionism: evolution works when it is good enough. T. H. Clutton-Brock and P. H. Harvey in "Comparison and Adaptation" argue that the definition of adaptationism should shift from causal explanations to explanations of consequences, opening the way as well for nonadaptive changes and interactive constraints of varying features. Although their final account of variation closely mirrors Gould and Lewontin's, they claim their account only modifies, but does not reject, the adaptationist program. G. C. Williams in "The Question of Adaptive Sex Ratio in Outcrossed Vertebrates" even after much effort cannot find adaptative causes for sex ratios, which he identifies finally as random. Nonetheless, he remains puzzled by the lack of a viable adaptive explanatory alternative.

Thus if the symposium is any indication of the field of Anglo-American evolutionary research and thought (indeed eight articles are British-authored and the last, other than Gould and Lewontin's, is American), then the field is much more varied, with fewer hard boundaries, than Gould and Lewontin represent. The adaptationist–antiadaptationist boundary seems to exist only when someone wants to make an issue of it. In the primary symposium papers, only Clarke does so, by trying to take the issue of genetic diversity from the antiadaptationists by arguing for diversity's adaptive value.

A. J. Cain's discussion paper at the end of the symposium, however, does mount a direct adaptationist attack on Lewontin. The first half of the discussion paper praises, with minor caveats, the growing knowledge gained by adaptationism. Midway, however, the discussion curiously turns to an adaptationist account of the survival value of emotionalism and an irrational confidence in one's own power to control one's destiny. Such overestimates of one's ability to overcome obstacles gives the adolescents of a species the necessary courage to face the rigors of a threatening environment. Cain finds the same attributes of emotionalism and unwarranted confidence in one of Lewontin's earlier publications. Lamarckism, he quotes G. B. Shaw as noting, while not necessarily correct, is more human. Objectivity is only adaptive for the mature of the species, who embody the true spirit of science. Beyond the clever ironies of this ad hominem attack, it is worth noting how Cain counteracts Gould and Lewontin's privileging of humanistic creativity by raising the humanistic stakes onto the moral planes of dispassionate understanding and maturity. Science, characterized as the realm of mature contemplation, displaces art, characterized as adolescent emotionality, as the pinnacle of human wisdom.[12]

The great variety of positions actually expressed in the symposium suggests that Gould and Lewontin's characterization of the battle is overdrawn, but the last response by Cain, who redraws the same battle lines, suggests that there is something more at stake in observing those lines. That issue is one's belief about human nature and the proper moral stance on the human condition. Cain's argument is sociobiological. Rationality and irrationality are only sociobiological consequences of the need to survive. Cain implies that the nominally adult Lewontin would be acting more adaptively if he left his adolescence behind him and accepted the sociobiological reality that we have little control over our fate. "Spandrels" rejects implicitly, however, precisely that quietist position, which denies creativity, will, and action.

Sociobiology—the Real Stake

That rejection becomes clearer through an examination of Gould and Lewontin's most recent publications self-cited in "Spandrels." These three articles (Gould 1978; Lewontin 1978, 1979) are unproblematically consistent with the positions taken in the "Spandrels" article, but they place the issues directly within the sociobiology controversy. None is directly within the evolutionary biology literature or closely related disciplinary literature, suggesting that Gould and Lewontin were not prepared to wage a full-scale attack on sociobiology within the evolutionary literature. Rather the attack on adaptationism aims at the technical underpinnings, the explanatory mechanism by which sociobiological accounts gain the appearance of plausibility. Only as they removed themselves from the overt, disciplinary biological literature were Gould and Lewontin ready to tip their hand in a more direct way. The two 1978 articles are from semipopular journals, the *New Scientist* and *Scientific American*. Both adopt genres (the book review and the expository overview) that allow a more direct discussion of general issues, rather than a focussed technical debate. The 1979 article is from a social science journal, allowing Lewontin to take on the role of general expositor, interpreter, and critic of an outside disciplinary perspective.

Each of these three articles is self-cited only as the source of specific pieces of evidence, and no hint is given of their broader arguments. Lewontin 1978 is cited only as the source of the chin example, but in fact it presents the full set of complaints about adaptationism, winding up with the same list of five alternatives. Gould 1978 is cited as the primary source for a critique of Barash's studies of aggression in mountain bluebirds. Although Gould and Lewontin do not mention it, Gould 1978 rehearses much of the critique of adaptationism, accusing it of telling just-so stories open to frequent ad hoc revisions. Gould 1978 then goes one step further by linking this adaptationist explanatory style with sociobiology—that is, the account of behavior as biologically and genetically determined. In fact, the critique of Barash is used exactly as a transition between the critique of adaptationist accounts and the critique of sociobiology. Gould goes on to argue that genetic adaptationism breaks down with human beings, who adapt through cultural evolution, and that sociobiological accounts are therefore inappropriate. Gould asserts that the "grandest goal . . . of human sociobiology, . . . the reduction of the behavioral sciences to Darwinian theory," must fail in recognition of the impact of cultural evolution, which is faster, more modifiable, and more easily transmitted. He argues that cultural history and Darwinian theory must stand beside each other as two independent disciplines.

Lewontin 1979 is also cited in "Spandrels" only as the source of a specific example, but it too in fact gives an extensive critique of adaptationism as a preliminary to opposing sociobiology as an inappropriate extension. After pointing out that adaptationism is an a priori explanatory strategy, and thus not a testable theory, Lewontin presents sociobiology as "one manifestation of the adaptationist program, concentrating on the behavioral aspect of the phenotype." Sociobiology is undermined by the argument that no specific genetic grounds or mechanisms have been identified for the claimed behavioral adaptations.

After listing the now familiar five alternatives to adaptationism, Lewontin further argues that sociobiology makes four errors: arbitrary ag-glomeration, reification, conflation, and confusion of levels. Each of these errors, according to Lewontin, is based on an inadequate under-standing of human creativity and culture. Arbitrary agglomeration re-fers to the problem of identifying the natural suture lines between fea-tures, especially troublesome when dealing with human mind, memory, and culture. The problem of reifying mental or conceptual categories is the second error. Mental constructs, Lewontin argues, are created in human minds and not directly heritable traits. The third, conflation, again refers to the mistake of taking culturally created categories and attributing them to animal behavior when the animals themselves are not operating under those explicit categories. Lewontin's last critique, of the confusion of levels, points out that sociobiologists have no real concept of the nature or influence of human culture, but in fact describe only individual behavior. Aggression is treated, for example, only as a programmed individual behavior rather than the product of complex social forces. "Educational and political systems and the creation of ideologies become nothing but the collective manifestation of individ-ual drives for conformity and indoctrinability." As Gould argued in his 1978 review, creativity and culture are the very human things that so-ciobiologists neither honor nor give place to.

Culture and creativity are precisely what are given honor in "Span-drels," and the fools like Pangloss, who are blind to humans' abilities to intervene creatively within the constraints of history, architecture, and other structures, are precisely those who are rejected in the article. Indeed, the first quotation from Pangloss exhibits confusion between human cultural creativity and natural determinism: "Our noses were made to carry spectacles, so we have spectacles. Legs were clearly in-tended for breeches, and we wear them." Throughout the article the favored mode of explanation is creative response to circumstances, add-ing complexity that confounds reductionist explanations. The arguments presented throughout "Spandrels" gain more specificity and force when read as rejections of sociobiology and not merely as complaints about some methodological difficulties in evolutionary research and debate.

Biological constraints are shown as delimiting pathways and structur-ing development rather than as determining individual choice, whether of the mosaicist of San Marco or the mollusc elaborating its shell. Crea-tions of powerful art and spirit are shown to be responses to constraints rather than the simple determined result of constraint. Not only archi-tecture but also cannibalism is best understood by refraining from "im-posing biological biases." Even more, biology is shown as best under-stood by not "imposing biological biases," given that plants and animals exhibit complex creative responses to environment, experience, and genetic constraint. The gist of the *Bauplan* argument is the total response of the organism to the complex environment. Even the tyrannosaur's use of its small forelimbs is a creative response to needs and opportunities, whether for sexual foreplay or the desire to get up from a prone position—neither one an easy task for a tyrannosaur.

Panglossian determinism becomes an especially delusive evil in light of this pervasive creative response that Gould and Lewontin find among all creatures and humans, for Pangloss's philosophy leads to a quietism, an acceptance of the way things are in this best of all possible worlds. Thus, as Gould and Lewontin recall, Pangloss accepts his own venereal disease because Columbus's adventure into the New World brought chocolate as well as syphilis. The later quotation concerning the Lisbon earthquake appears in the novel as a rationalization for Pangloss's failure to save a drowning sailor. Gould and Lewontin might say that the objectivity that Pangloss displays is precisely that displayed by Cain when he calls Lewontin's sense of control over life an unavoidable adolescent adaptation to the overwhelming difficulties of the world.

The implied moral argument against sociobiology comes closest to the surface in the cannibalism example, marked by the reference to E. O. Wilson as one who has taken up the nutritional explanation of cannibalism. Although sociobiology is not overtly discussed, Wilson is so associated with the sociobiology movement that it would be foolish for Gould and Lewontin to mention him if they did not want to raise the spectre. Wilson's attempted reduction of cannibalism from complex cultural behavior to protein source removes the moral stigma from a socially repellent custom. By reasserting the cultural explanation, Gould and Lewontin recuperate the moral responsibility for individual and social choice. Participants of a culture are held responsible for their choices. They could have done otherwise. Moreover, the social waste and lack of nutritional necessity that Gould and Lewontin cite also categorize the activity as needless killing and savagery. Evil is not biologically determined but is a real social ill for which individuals are responsible. Even more, we should note—as Carl Herndl does later in this book—that Gould and Lewontin counter the sociobiological account of the behavior with a Marxist account of classes, power, and conspicuous consumption. Evil here is done precisely as form of dominance, oppression, and selfishness.

The critique of adaptationism in the first half of "Spandrels" cites many examples of overt behavior, for which adaptationists are presented as foolishly missing the point of choice, creativity, and responsibility within structured circumstances; the latter half of the article, however, offers few behavioral examples as attention turns to the structural frameworks within which evolutionary opportunities are elaborated. The research program suggested by Gould and Lewontin thus turns away from individual characteristics to a concern for *Bauplan* and structure as providing opportunities for response. We are reminded of the now familiar statement of Marx that men make their own history but not in conditions of their own making.

The Marxian theme of constrained responsibility and creativity sounds once more in the form of the final anecdote reprising the Panglossian fool in the person of the social Darwinist Herbert Spencer (even though, as with Wilson, the target ideology is not mentioned). Social Darwinism as the ideological justification for late nineteenth-century laissez-faire competitive capitalism, red in tooth and claw, is presented as the radically individualist and quietist teleological optimism that jumps to reductionist conclusions with little close attention to the complexity of details.

Conclusion: Levels of Intertext and the Strategies of Knowledge Containment

In "The Spandrels of San Marco and the Panglossian Paradigm," Gould and Lewontin have created an intertextual drama that makes most sense in terms of another submerged intertextual struggle over sociobiology. In that struggle the control over representation of the intertext is a crucial strategic weapon, for whoever controls the intersubjective intertext (that is, the widely accepted representation of the intertext) controls the communal memory and thereby the framework of knowledge. If Gould and Lewontin can construct the history of evolutionary discourse as a struggle between foolish adaptationism and a wiser pluralism, they can knock the communal underpinnings out from sociobiology, which they consider morally, politically, and intellectually repellent. Beneath the overtly scientific discussion of appropriate investigatory method and explanation lies a struggle of philosophic ideas and human commitments. It is on those most fundamental human conversations that representations of the scientific intertext are built.

It is a standard strategy of disciplinary debate to reach back into the history of a field to open up questions that were thought to be settled years ago, just as Gould and Lewontin reach back to identify what they consider a false turn in the movement of the evolutionary word from

Charles Bazerman

Darwin to the disciple Wallace. But Gould and Lewontin implicitly reach back even further to question the enclosure of biology itself by setting biological explanation side by side with aesthetic and cultural explanation. They do this because their cause ultimately reaches beyond biology. They must remove the protection of biology from the sociobiology they believe is wrapping itself in untenable biological reductions. To wean us, the readers, from sociobiology, they must first wean us from "our biological prejudices." These prejudices have led us down a path that denies our very nature and responsibilities. They have led us to unwisdom and foolishness.

Every scientific and disciplinary literature is built upon a series of containments, constructing a literature set apart from other discussions, a literature that follows its own questions and listens to its own special evidence and arguments. The history of disciplinary writing is filled with instances of literatures setting themselves apart from each other, often with stunning results for the development of new and powerful knowledges. Yet every discipline can be traced, without too complex a genealogy, back to one branch or another of philosophy, and philosophy is at root a wisdom literature.

Every disciplinary writer relies on those containments in order to construct a local intertext that will be persuasive and useful for carrying forth the writer's own work. But issues can drive one deep into the commonly accepted intertext and its sedimentation of taken-for-granted knowledge. Old texts can be revived to create new representations of the intertext and old containments can be cast aside to propose new linkages. Disciplines, when interrogated by deep enough issues, can be disintegrated into wisdom literatures. But writers can only propose those realignments. It is up to the readers, who constantly reformulate the discipline by the representations of the intertext of knowledge they accept, to see the wisdom in the new philosophy.

NOTES

1. At least two studies indicate that even though representations of the intertext are not persuasive or acceptable to certain kinds of readers, they may find the ensuing work credible or of interest (Bazerman, *Shaping Written Knowledge*, chap. 8; Schwegler and Shamoon), but that is only because the readers themselves have powerful intertexts in their heads, against which they place the immediacies of the current findings and argument. If the new material does not place within the reader's intertext in terms of its implied aims, methods and warrants, and body of findings or knowledge, then the new

work will not be creditable or even intelligible. For an introduction to the concept of intertextuality and for applications of the concept to the analysis of various kinds of texts, see Orr; Devitt; and Selzer, "Intertextuality."

2. See Myers, *Writing Biology* and "Rhetoric of Irony."

3. The one concession Gould and Lewontin make to the adaptationists is their admission of nonadaptive evolutionary mechanisms such as genetic drift and allometry. However, Gould and Lewontin immediately minimize the import of this concession by claiming that adaptationists, while admitting nonadaptive evolutionary mechanisms in general, rarely consider them significant in any real case. Again they assert that the generic expectation of adaptationist accounts is so strong that there is little room to assert contrary accounts, even when they are in principle possible. In this manner the literature is kept homogeneous.

4. The first characterization of the process of adaptationist research (atomization and trade-off repairs) is substantiated by no specific articles as evidence. The first claim of atomization refers only to prior critiques making the same point. The second point concerning the trade-off ploy immediately shifts into the *Candide* analogy.

Concerning the characterizations of the repair strategies, substantive evidence from the literature is put forth. Three cases in which older, failed adaptationist arguments have been replaced by new ones are cited, as is one case of an appeal to an assumed (but as yet undiscovered) adaptationist account and one case of appeal to unknown details of environment. Concerning this last appeal to ignorance, it is worth noting that the appeal comes from Wallace, near the beginning of a putative research program, where indeed an appeal to ignorance might be warranted. Finally, concerning the exclusion of evidence, the supporting example is drawn from the popularizations of the Boston Museum of Science rather than from professional journals, although a disclaimer from Gould and Lewontin asserts that similar arguments appear throughout the professional literature.

The most extended evidence of adaptationist obduracy and willed blindness comes in the form of a two-page account of Gould's critique of Barash for narrowness in considering alternative nonadaptationist accounts and a summary of a follow-up study that further calls into question Barash's results and story. The summarized critique, however, maintains an adaptationist stance.

5. For more on the intertextual relation between "Spandrels" and *Candide,* see Debra Journet's paper, chapter 12.

6. Baer, Davitashvili, and Wallace.

7. Barash; Costa; Darwin 1872, 1880; Falconer; Gould 1978; Grasse; Gregory; Harner; Lande 1976; Lewontin 1978, 1979; Morton; Ortiz de Montellano; Remane; Rensch; Reidl 1975, 1977; Rudwick; Romanes; Sahlins; Schindewolf; Seilacher 1970, 1972; Stanley; Sweeney; Thompson; Wilson.

8. Coon; Galton; Gould 1966, 1971, 1974, 1977; Jerison; Lande 1978; Shea; Waddington.

9. Lande's argument concerning limb loss and reexpression appears on the

Charles Bazerman

surface to be adaptationist, even though he is aware of structural constraints and mechanisms. Nonetheless, by interpretations that emphasize the weighing of constraints against adaptive gains, the article can be seen as providing a way of testing nonadaptive hypotheses and thus fitting Gould and Lewontin's characterization.

10. Although Waddington indeed opens up the possibility of the interpretation given by Gould and Lewontin, he himself is agnostic as to causal factors that might account for the patterns he notes.

11. Galton's anecdote is taken from a context totally outside the adaptation, or even evolutionary, debates. It is part of a discussion of the value of anthropomentrics; the point is simply to show the importance of gathering facts. This is not to say that the anecdote isn't appropriate to Gould and Lewontin's argument. It is. However, the context may make it seem that Galton fought the same battle. He didn't.

12. For more on Cain's discussion paper and its relation to "Spandrels," see Gould's essay, chapter 15.

WORKS CITED

Bazerman, Charles. "How Natural Philosophers Can Cooperate." In *Textual Dynamics of the Professions,* ed. Charles Bazerman and James Paradis, 13–44. Madison: University of Wisconsin Press, 1991.

Bazerman, Charles. *Shaping Written Knowledge: The Genre and Activity of the Experimental Article in Science.* Madison: University of Wisconsin Press, 1988.

Cozzens, Susan. "Comparing the Sciences: Citation Context Analysis of Papers from Neuropharmacology and the Sociology of Science." *Social Studies of Science* 15 (1985): 127–53.

Devitt, Amy. "Intertextuality in Tax Accounting." In *Textual Dynamics of the Professions,* ed. Charles Bazerman, 336–57. Madison: University of Wisconsin Press, 1991.

Maynard Smith, John, and R. Holliday, organizers. "The Evolution of Adaptation by Natural Selection: A Discussion." *Proceedings of the Royal Society of London, B: Biological Sciences* 205 (1979):435–604.

Merton, Robert K. *The Sociology of Science.* Ed. Norman Storer. Chicago: University of Chicago Press, 1973.

Messeri, Peter. "Obliteration by Incorporation." Paper delivered at the meeting of the American Sociological Association, San Francisco, September 1978.

Mullins, Nicholas, L. Hargens, P. Hecht, and E. Kick. "The Group Structure of Cocitation Clusters: A Comparative Study. *American Sociological Review* 42 (1977): 552–62.

Myers, Greg. "The Rhetoric of Irony in Academic Writing." *Written Communication* 7 (1990): 419–55.

Myers, Greg. "Stories and Styles in Two Molecular Biology Articles." In *Textual*

Dynamics of the Professions, ed. Charles Bazerman and James Paradis, 45–75. Madison: University of Wisconsin Press, 1991.

Myers, Greg. *Writing Biology: Texts in the Social Construction of Scientific Knowledge*. Madison: University of Wisconsin Press, 1990.

Schwegler, Robert, and Linda Shamoon. "Meaning Attribution in Ambiguous Texts in Sociology." In *Textual Dynamics of the Profession*, ed. Charles Bazerman and James Paradis, 216–33. Madison: University of Wisconsin Press, 1991.

Selzer, Jack. "Intertextuality and the Writing Process: An Overview." In *Writing in the Workplace: New Research Perspectives*, ed. Rachel Spilka. Carbondale: Southern Illinois University Press. In press.

Small, Henry G. "Cited Documents as Concept Symbols." *Social Studies of Science* 8 (1978): 327–40.

Swales, John. *Genre Analysis: English in Academic and Research Settings*. Cambridge: Cambridge University Press, 1990.

Ziman, John. *Public Knowledge*. Cambridge: Cambridge University Press, 1968.

3

"SPANDRELS," NARRATION,

AND MODERNITY

SUSAN WELLS

In S. J. Gould and R. C. Lewontin's "The Spandrels of San Marco and the Panglossian Paradigm: A Critique of the Adaptationist Programme," we find a critique of adaptationist narratives that is tightly focused and broadly significant, written for a relatively small audience, for very high stakes. The authors are calling into question the central explanatory paradigm of their own discipline, and this intervention appears in the *Proceedings of the Royal Society of London,* a very prominent and prestigious publication. Everything about the essay, from its illustrations to the joke about titillated tyrannosaurs in the abstract, indicates that here we are not about biological business as usual. It is rare in any discipline, and almost unknown in the natural sciences, for researchers to question their central research program while it is still productive; that is exactly the project of "Spandrels," which also argues for an alternate approach of analyzing organisms as "integrated wholes."

Within this intellectual project, narrative and the critique of narrative play a central role; the authors' critique is shaped by the notion of falsifiability. Falsifiability, the criterion for scientific explanations proposed by Karl Popper (and subsequently questioned by others), implies that a theory that cannot be disproved by any conceivable experiment, that cannot be "falsified" by any experimental test, also cannot be verified. If all possible features of all possible organisms can be explained by plausible adaptationist stories, then no possible feature of any organism is exempt from adaptationist explanation. Since adaptationism can explain anything, it cannot in principle be refuted, and a theory that cannot be refuted cannot be proved. Gould and Lewontin play this argumentative card exuberantly, parodying adaptationist stories, replicating them, proposing extreme analogies from disciplines cognate and remote. Adaptation is attacked, first of all, as a theoretical error that requires the rethinking of an entire body of research; it is also, by implication, attacked as a bad story, a narrative that is too simple, lacking sufficient complication and resistance to interpretation.

The narrative of adaptation is associated, through Pangloss, with the smug invocation of the best of all possible worlds. Gragson and Selzer's essay in chapter 10 shows how determined "Spandrels" is in its attack on the norms and expectations of scientific discourse, how inventive the authors are in soliciting the audience to undertake a new reading of their discipline. The authors' initial literary reference to Pangloss opens that project of rereading. Gould and Lewontin place their text under the sponsorship of Voltaire, one of the central writers of the Enlightenment, associating it with the initiating moments of modernity, when the sciences appeared as effective opponents of all received ideas. Adaptationism, therefore, is associated with religious and political pieties: what had been insurgent has become moribund, so that a second Enlightenment is required, and will emerge from the satiric energy of the authors. The text establishes, outside its own boundaries, a new set of narrative vectors; it locates itself within the project of modernity. It emerges, therefore, on the terrain that Jürgen Habermas, the German critical theorist, has cultivated, terrain that this essay will inhabit.

Such a narrative repositioning is a complex move, but it is quite characteristic of at least one of the authors. Stephen Jay Gould, the writer of compelling scientific popularizations, has taught S. J. Gould, the coauthor of "Spandrels," a great deal about narratives and their structure. Let us review those lessons, to see how they have been adapted to differing demands of a scientific audience.

Stephen Jay Gould's writing is remarkable for its narrative energy, for the skill with which he translates analytic concepts into agents whose projects arc through time gracefully, but with minimal anthropomorphism. Gould can write about embryo mites that devour their mothers before birth, exploit all the creepiness inherent in the story, and keep on telling a story that is still, blessedly, only about mites (*Panda's Thumb*, 91–96).

Many of Gould's narratives follow a central scheme, a way of ordering events, a kind of master plot of natural history. In this scheme, the action builds very slowly through a range of collateral examples, often with complications and subtle variations. Stasis is represented here as tension and potential rather than homeostatic calm. Suddenly there is a serious crisis or reversal, a stochastic leap, an explosion of forces held in suspension. Then follows a very long, complicated, and ambiguous resolution, in which many of the narrative's initial tensions reemerge in translated form. Thus, Darwin studies finches and barnacles, works in obscurity for years, learns of a rival, publishes *Origin*, and spends the rest of his life in a long rearguard action against misunderstanding

Susan Wells

(*Ever Since Darwin*, 21–38). Thus, the long accumulation of blue-green algae (*Panda's Thumb*, 217–26), the tentative and obscure movement into more complex forms (*Panda's Thumb*, 245–56), the stochastic leap of the Cambrian explosion (*Ever Since Darwin*, 126–35), and the long after-elaboration into species (*Flamingo's Smile*, 230–44). Thus, in fact, speciation itself (*Panda's Thumb*, 204–13).

This plot has several implications. Although it is not a specifically modernist narrative, it articulates an experience of time that is distinctly modern. It is not triumphalist: neither the tentative opening action nor the very long and complicated resolution lends itself to gee-whiz teleologies. Neither, however, is it a gradualist scheme: the story is not a matter of the slow accumulation of increments but includes the possibility—indeed, the necessity—of radical breaks, of qualitative changes, of ruptures for good or ill. This narrative form is not, of course, Gould's invention. We find it throughout nineteenth-century fiction, and cognate forms have been identified in Darwin's work (Landau). One of its most striking formulations is to be found in Marx[1]:

> Proletarian revolutions, like those of the nineteenth century, criticize themselves constantly, interrupt themselves continually in their own course, come back to the apparently accomplished in order to begin it afresh, deride with unmerciful thoroughness the inadequacies, weaknesses and paltriness of their first attempts, seem to throw down their adversary only that he may draw new strength from the earth and rise again, more gigantic, before them, recoil ever and anon from the indefinite prodigiousness of their own aims, until a situation has been created which makes all turning back impossible, and the conditions themselves cry out:
> *Hic Rhodus, hic salta!*
> *Here is the rose, here dance!* (Marx, 19)

And I cannot help noticing, although I would not want to make too much of it, that Stephen Jay Gould is giving an account of change that is homologous to one central historical plot: the Leninist story of revolution after long preparation, a revolution that generates its own series of problems, its own possibilities of derailment. This is no anarchic insurrection, no liberal gradualism, but an address to the central problems that Marxism posed to our understanding of history; a constitution of history as a series of projects undertaken by human agents, without the aid of any transcendent teleology. It is deeply uneven in its tempo, so that time itself varies in the unfolding of the narrative. It is the time of modernity, and as we shall see, it is the time in which modernity comes to understand itself as postmodern.

But understanding this narrative form is not enough; it is not enough even to understand it as more complicated and satisfying than the simple adaptationist story of traits and their reproductive benefits. We need also, I think, question what is becoming a truism in the study of scientific discourse, a truism that I have had my own part in establishing: the contention that since narrative is unavoidable in scientific writing, all scientific texts can be read as competing narratives. Greg Myers's *Writing Biology: Texts in the Social Construction of Scientific Knowledge* may stand as the magisterial, even heroic, exemplar of the very great lengths to which this method can take us. But I hold that the reliance on narrative analysis needs to be balanced with an understanding of the elements of scientific texts that are held in tension with narrative: elements that may be called, as figures of thought, analytic, or as figures of language, lyric. I want to argue for the inadequacy of narrative as an analytic framework for three reasons: 1. Under conditions of modernity, narrative comes under very great pressure from alternate methods of explanation, differentiated disciplinary forms in which narrative schemes are relatively etiolated. To see scientific explanations as essentially narrative is to lose sight of this historical tension, described most broadly in Jürgen Habermas's *Theory of Communicative Action.* 2. Under conditions of modernity, differentiated disciplines become self-reflexive. Thus, the rhetorical analysis of narrative is not distinct from a critique internal to a discipline: Myers's informants are not at all surprised by his remarks on their writing; Gould's own narrative practice demonstrates all the sophistication we could wish for. To indicate that a scientific text is narrative does not add to our knowledge of the text. Rather, it is in many ways simply a codification of knowledge available to practitioners of a discipline; to begin to add to that knowledge, we must also understand what is not narrative in the text, and the ways that the two elements are related. 3. Finally, under conditions of modernity, the energy of narrative desire is very often brought to bear upon a project of escaping narrative, or of frustrating normal narrative satisfactions. Disciplinary constraints that lead to analysis rather than narrative, therefore, intersect with the self-reflexive project of modernity, rendering suspect any such overt narrative moves as anthropomorphized agents or happy endings. The pleasure of the text is invested in contingency, retarding action, complication—all those representations of events that derail the narrative train. There is an extensive body of literary criticism that relates such tropes to the postmodern: my own reading follows Habermas in placing them as extensions of the modernist project of self-reflection (*The Philosophical Discourse of Modernity*).

In sum, if we read scientific texts as generally narrative in their force, we neglect the ways in which those texts evade narrative, or reflect on it and transform it in serious ways. We miss something of both the intellectual rigor and the textual pleasure of an essay like "Spandrels." I will take up all three of these themes, drawing on Habermas and beginning with an explanation of his central ideas.

Habermas understands sociality as controlled by two complementary spheres: that of the system, which includes nonlinguistic means of control, such as money and power, and that of the lifeworld, in which customary and nonreflective modes of organization prevail.[2] At the intersections between these two means of control, conflicts emerge. If all goes well—and, of course, it need not go well at all—those conflicts are resolved in discourse. Those who are affected by a policy, or who are interested in describing how the world is structured and how social life is organized, undertake to resolve disputes and understand the world. They use whatever specialized forms of knowledge they can develop, and dispute among themselves according to the norms that their projects have suggested to them, always controlled by a set of abstract and counterfactual rules that Habermas characterizes as those of the "ideal speech situation": disputes are settled by argument, rather than force; all participants have a right to enter the discourse in any role; any proposition is open to criticism on any level. These rules are not necessarily followed at every moment of the discourse, but the perception that they have been violated leads the participants in discourse to rectify their communicative relations, to reconstruct the terms of the conversation. Habermas's theory is a relentless description of how social arguments might be undertaken and of why we might be unsatisfied by their current forms.

Central to this theory is the notion of narrative as the paradigmatic explanatory structure of the lifeworld. Customary and received arrangements are explained, in a way that generally satisfies the participants, by stories. Religion, one form within which the lifeworld is codified, can be seen as an exemplary compilation of such stories. And many of us move toward narrative whenever we wish to enshrine some norm in the collective life of our communities ("We do it like so because, long ago, we had a problem . . ."). In all those instances where the norms of the lifeworld are undisturbed and uncontested, where no competing interests call them into question, narrative works quite well as an explanatory structure.

Such instances, however, are not the rule, and they are becoming rarer as the incursions of system upon lifeworld intensify. In such a society, there are many distinct terrains of conflict and distinct discursive forms that address them. It is, for Habermas, a distinguishing mark of a fully developed and differentiated discursive formation that it breaks from overt narrative as a device for explanation and argument. Natural history becomes evolutionary biology; historical linguistics becomes semiotics. While a discipline may continue to use narrative persuasions (man arising from the apes), it also will elaborate an analytic frame that is not narrative in its force (alleles combine and recombine). "Spandrels" repeats this break from narrative by presenting two alternatives to the adaptationist narrative. First, the text presents an analytic matrix, combining and recombining adaptation and selection, so that the two processes vary independently. The combinations generated by this matrix will remind literary theorists of the Greimas rectangle, and classical rhetoricians of the square of oppositions; they represent an analytic, nonteleological alternative to the narrative program of adaptation. More surprising is the text's second alternative: the lyric invocation of the whole, the organic, the complex, the individuated. Such recourse to the lyric, while uncommon, is not unknown in scientific texts. A similar combination of lyric evocation and structural analysis can be found in Darwin:

> He who will go thus far, if he find on finishing this treatise that large bodies of facts, otherwise inexplicable, can be explained by the theory of descent, ought not to hesitate to go further, and to admit that a structure even as perfect as the eye of an eagle might be formed by natural selection, although in this case he does not know any of the transitional grades. (160)

Darwin's reader is invited to forgo the narrative satisfactions that a survey of the transitional grades might provide in favor of the explanatory satisfactions of the theory of natural selection, such as its ability to place and explicate "large bodies of facts"; that rather dry pleasure is rendered palatable by the lyric contemplation of the perfection of the eagle's eye. All three elements—the promised but elided narrative, the lyric contemplation, and the analytic explanation—coexist; they depend upon one another, in fact, for both their expository force and their persuasive power.

Such a loosely Habermasian account of the relations between the disciplines and narrative is borne out by Nigel Gilbert and Michael Mulkay's *Opening Pandora's Box*, which argues persuasively that the discourse of scientists is informed by narrative, especially in their conversational, "contingent" register, and that there is a constant interplay between the contingent register and the "empiricist" register of written scientific argument. But although scientists labor to excise narrative from the formal expressions of the empiricist repertoire, the two regis-

Susan Wells

ters interpenetrate. Narrative structures remain in empiricist texts, and the contingent repertoire is also vulnerable to interruption, subversion, and falsification by the empiricist register. Scientists who thought they were enmeshed in a story of "scientific politics" are seen to have been convinced by evidence to change their minds. Disciplinary discourse interrupts narrative, even the narrative of the discipline's lifeworld.

The practitioners of disciplines know that narrative forms can organize their discourse and that those narratives are subject to disruption. Such practitioners, in fact, are the writers most likely to disrupt powerfully a prevailing story, and "Spandrels" is an instance of precisely such an intervention. Indeed, the project of modernity has been marked since its inception by self-reflection: the critique of Enlightenment arises simultaneously with Enlightenment; no sooner does reason address domination, than the domination of reason is itself called into question. This structure of interruption and dislocation is not a happy one, but then modernity is not an especially happy period.

The self-critical and self-reflective force of disciplines may be derived from their status as the products of discursive communities. If a discursive community poses for itself the project of understanding the world or social life, and constructs a problem that can organize its investigations, then a field has been established in which questions and reflections are not easily restricted. Disciplines do, in fact, rule ideas out of order for all sorts of bad reasons: ideas are suspect if they are new, or their proponents are of low status, or if they would be disruptive. But such ideas cannot be refuted *on the grounds that* they are new, or lack prestigious proponents, or are inconvenient. To put it crudely, you cannot happily debate with a loan officer about what is fair, operating within the nonlinguistic sphere of the system, or with your mother, operating within the narrative forms of the lifeworld. But you can engage in such a debate in a philosophy classroom, and a classroom is likely to operate according to the counterfactual, fantastically utopian rules of academic discursive communities: it pretends to itself that all its participants are equal, that all their ideas are worthy of respect and hearing.

Such a norm against restriction of debate implies that no claim within a disciplinary framework is permanently secure. Since lifeworld and discourse interpenetrate continually, and since discourse communities are unevenly differentiated from the lifeworld, participants in disciplines are free to recognize any element of theory, any piece of received wisdom or of disciplinary lore, as a "story," and to bring the analytic norms of the discipline to bear against it; such a move is potentially very productive and both intellectually and professionally rewarding.

This moment of the disrupted story is Stephen Jay Gould's special domain: he brings to the public the good news of the contingency, temporality, and replaceability of all scientific stories, the possibility of substituting for them less plausible, less narratively inevitable, more lyrical forms of explanation. Consider these quotations, at varying levels of generality, from *Ever Since Darwin: Reflections in Natural History:*

> When it [*Australopithecines africanus*] was discovered in the 1920s, many evolutionists believed that all traits should change in concert within evolving lineages—the doctrine of "harmonious transformation of the type". . . . But, as modern evolutionary theory developed during the 1930s, this objection . . . disappeared. Natural selection can work independently upon adaptive traits in evolutionary sequences, changing them at different times and rates. . . . Paleontologists refer to this potential independence of traits as "mosaic evolution." (58)

> I will try to illustrate this thesis with two examples drawn from the "classical" data for continental drift. Both are old tales that had to be undermined while drift remained unpopular. (162)

> In the conventional model of scientific "progress," we begin in superstitious ignorance and move toward final truth by the successive accumulation of facts. In this smug perspective, the history of science contains little more than anecdotal interest. . . . It is as transparent as an old-fashioned melodrama: truth (as we perceive it today) is the only arbiter and the world of past scientists is divided into good guys who were right and bad guys who were wrong.
>   Historians of science have utterly discredited this model during the last decade. (201)

Given the regularity and persuasiveness with which disciplinary discourses produce analyses of their own narratives, our work as rhetoricians of scientific inquiry requires some analysis of both the stories disciplines tell and the way they discard them, some movement beyond the identification of narrative in scientific texts to an engagement of the moment when narratives give way to the solicitations of the lyric, to nonnarrative forms of critique, or to the urgings of a more complicated narrative.

Finally, it is in the identification of those discursive forms that frustrate, retard, or complicate narration that much of the pleasure of the scientific text is to be found. An analysis that focuses on narrative as the central figure in scientific argument is immune to that pleasure. Let us consider Gould's critique of Barash's monogamous bluebirds as a case in point. Barash is seen, first of all, as the narrator of a stupidly simple story—although Gould does not, at any point, argue that his story is not plausible. To discern the biological roots of male anticuckoldry, Barash composes a story that counts episodes of male aggression toward a model bluebird, and relates those episodes to the received narrative of reproduction: the further along the process of reproduction, the less likely an aggressive response to the model. Such a crudely constructed story is quite characteristic of Barash. In a talk Barash gave at the American Academy for the Advancement of Science symposium on sociobiology in February 1978, which Gould also attended, we find two stories with identical structures, stories that acknowledge and discount exactly the issues of falsifiability that Gould raises so forcibly in "Spandrels." Barash acknowledges, "Anything can be made to seem adaptive, if we are ingenious enough" (212), and then proposes experiments that will test whether adaptive behaviors follow predicted paths. Both experiments plot the incidence of protective behaviors against the narrative of reproduction: male damselfish repel predators more energetically before they have attracted a female; sparrows attack a dummy predator more vigorously if their species broods only once. Both experiments, moreover, derive a great deal of their rhetorical power from unblushing anthropomorphism. Barash imagines the female damselfish evaluating males for their efficiency at repelling invaders: "Of course, a possible female counter-strategy is to watch males at a greater distance—that is, when they don't think they are being watched. Females might also simply devalue a certain amount of the behavior of all males, as mere bravado put on for the benefit of the ladies" (219). Surely, in this example, biological explanation is being driven by the culturally available narrative of female skepticism—we are inches from following the damselfish into a ladies' room and listening to them gossip about their dates.

Gould responds to Barash's story within the context of the sociobiology debate, a controversy related to "Spandrels" (as Bazerman and Herndl also demonstrate in different ways in chapters 2 and 4, respectively, of this book). Gould anticipated his critique of Barash's bluebirds in an essay, roughly contemporaneous with "Spandrels," that also takes up the analysis of scientific explanation as narrative. This essay, "Sociobiology: The Art of Storytelling," is a version of Gould's paper for the American Academy for the Advancement of Science symposium on sociobiology. Gould's "Sociobiology," in both its versions, anticipates some of the material used in "Spandrels"; we find there not only the critique of Barash's monogamous bluebirds, but also the ironic invocation of Darwin as a divine authority, and, more generally, the critique of adaptationism as story-telling. In "Sociobiology," however, this material is situated within a broad project of intervention in the sociobiology controversy. Gould's focus in "Sociobiology" is less methodological and theoretical than in "Spandrels": sociobiology is criticized not only on the grounds of its propensity to tell stories, but also for drawing inferences about human behavior from animal behavior, and for paying insufficient attention to the use of its research to justify retrograde social policies. The critique of sociobiological narratives is, again, a critique of nonfalsifiability: if a plausible story can be told about any feature of an organism, then no possible feature of any organism is exempt from adaptationist explanation. But adaptation, while crucial to the sociobiological case, is not the exclusive property of sociobiologists; nor is it the most provocative or, in Gould's eyes, the most politically damaging element of their theory.

In both "Sociobiology" and "Spandrels," the authors are participating in an extended controversy, and both controversies, luckily for us, have been studied by rhetoricians. In the "Sociobiology" article, Gould is taking up the central argument of E. O. Wilson's *Sociobiology: the New Synthesis*, although Wilson's work is not cited in Gould's essay. Wilson and his colleague Lewontin, the coauthor of "Spandrels," are sufficiently prominent in the sociobiology controversy to have been named the two "Colleagues in Conflict" in Ullica Segestrale's sociological account of the dispute. The terrain of "Sociobiology," then, indeed lies very close to that of "Spandrels." And both Gould and Lewontin were working very hard at understanding the implications of adaptationist stories during the composition of both essays. Greg Myers's careful account of the sociobiology controversy portrays both parties' attempts to inscribe their opponents in a narrative: sociobiology was shown to be an instance of ideology, rather than true science, while critics of sociobiology were held to be deluded by systems of belief that were probably rooted in their limbic systems (*Writing Biology*, 239). Within this argumentative context, to characterize the narrative of sociobiology as a "twice told tale" is to assign to Wilson responsibility for all the tellings of the adaptationist story and to imply that the stories of sociobiology, like those Hawthorne told, are stories of repression, of the unfinished labor of modernity.

To return, after digression, to the bluebirds in "Spandrels": for Gould, Barash's matching up of episodes and the anthropomorphism it supports provoke a parodic reversal. In the "Sociobiology" critique, Gould shows us the bluebird muttering "the avian equivalent of 'it's that damned stuffed bird again' " (531). In "Spandrels," Gould simply imagines the male approaching, testing, and discounting the model. In both essays, Barash's failure to repeat the experiment, to vary the fit between the story of aggression and the story of reproduction, provokes an iron-ically simplified, mocking parody: a story of increasingly savvy blue-birds. And in "Spandrels," the repetition of Barash's experiment by another researcher simply raises the parodic ante: "Since Barash's males can replace a potentially 'unfaithful' female, they can afford to be choosy and possessive. Eastern bluebird males are stuck with uncommon mates and had best be respectful" (155). Myers's point that direct quotation within a scientific controversy has strong potentials for parody and irony is quite telling here.

What is at issue for Gould and Lewontin is the unchecked prolifera-tion of narratives, the ability of the adaptationist program to prompt an exfoliating series of stories, so that no anomaly could force reconsidera-tion of any adaptationist explanation (let alone the theoretical hege-mony of adaptationism). Narrative energy is seen as a threat to the parsimony of scientific explanation. Indeed, Barash ends his article by discounting the problem of falsifiability and invoking the alternate cri-terion of experimental fruitfulness: "I am not interested in proving or disproving sociobiology—whatever proof means in this context—but rather, in using its insights as they allow me to ask and answer interest-ing questions about behavior" (223). In "Spandrels," Gould, rather than simply exorcising narrative from the text, stages a critique that contains narrative through parody, complication, and repetition. Over against the initial narrative frame (i.e., progress of reproduction→de-cline of aggression toward model), Gould constructs two alternate con-nections (bluebirds learn to recognize models→rise of boredom with model; decrease in availability of female bluebirds→increase in polite-ness of male bluebirds).

Gould's interest here is not at all empirical: it is irrelevant to the argument whether any of these connections describes bluebird behav-ior. What is at issue is the move from a single, uncomplicated relation between reproduction and aggression to a broad conclusion about the genetic roots of anticuckoldry. By increasing the complexity of the rela-tions between bluebirds and models, Gould disrupts Barash's simple narrative tie: we have many episodes of connected behavior, but we no longer have a story. Instead, we have a stronger sense of the need to "devise criteria to identify proper explanations among the substantial set of plausible pathways to any modern result" (154), an argument, to borrow from Stephen Jay Gould, for seeing evolution as a bush rather than as a ladder (*Ever Since Darwin*, 56–63). We have a sense, then, of the need for some form of explanation that does not present us with direct narrative satisfaction, and we have been taught to take some pleasure— satirical pleasure, to be sure—in the disruption of narrative ties.

"Spandrels" attempts to induct us into differing forms of explanation, and to present those forms as more reliable, more internally coherent, and more productive of pleasure for writers and readers than the adaptationist story. In section 5 of the essay, the authors present a typology of alternatives to the adaptationist program, a "substantial set of plausible pathways" that presents a deeply interesting transposition of the relationship between narration and analysis. We read an anthology of striking facts from the history of evolutionary biology, many of them drawn from the authors' own research, facts real (the whorls of *Cerion*) and imagined (Lewontin's hypothetical mutation). And they are presented as little jewels of natural history—which, of course, they are. *Cerion*, for example, is quite consciously introduced as an aesthetically pleasing example: "Every naturalist has his favourite illustration. In the West Indian land snail *Cerion*. . . ." (159).

But these stories are displayed in a resolutely nonnarrative frame, a matrix of possible relations between selection and adaptation. Again, the truth value of the stories is not in question: Lewontin's hypothetical example of a possible mutation is just as useful as the widely accepted allometry of vertebrate brain size. What is in question is, again, the proliferation of stories: exactly that narrative fecundity that rendered the adaptationist paradigm nonfalsifiable, and therefore useless, is here put in the service of an alternate paradigm, one that brings us "face to face with organisms as integrated wholes" (157).

That process has two movements: the narrative is first analytically decomposed into an inventory of possible episodes, and the episodes are then placed in a paradigmatic, rather than syntagmatic, relation to one another. Adaptations, that is to say, are not related to each other by being formed in series—Darwin's "transitional steps"—but by their placement in one of the alternate positions within a paradigm of choices. All the positions within the paradigm are logically equivalent to each other; we could not determine a priori, but only after a rather tricky empirical investigation, whether a particular structure was a random phenotypic variation, or an adaptation that would not be successfully selected, or an adaptation that had been selected. Since the authors can provide us with examples that fill all the positions in this paradigm, and since they can explain the plausibility of all these events within the basic assumptions of evolutionary biology, their inventory of possible episodes functions as a disruption of the adaptationist story. That story has been derailed, not by an alternate story (as we rhetoricians might say) nor by an analysis that supplants it (as Gould and Lewontin might say), but by a fragmentation of the story into episodes, a multiplication of those episodes, and a rearrangement of them in a nonnarrative, analytic form. Narrative untells itself by multiplying itself into discontinuous "turns" that cannot be resolved into a continuous story.

Within this analytic matrix, we find one of the most distinct moves of "Spandrels," the authors' evocation of "organisms as integrated wholes," a moment that will lead, later in the essay, to a discussion of the alternate paradigm, that of the *Baupläne.* This evocation is quite unusual for Gould, who generally presents organisms as instances of general theories and principles rather than simply savoring their singularity. Stephen Jay Gould would rather prompt his reader to reimagine the chin as the intersection of two embryological zones of growth than move him to marvel at its design. I am puzzled, therefore, by the evocation of the whole organism here.

I can advance several possible reasons for this evocation (others are proposed elsewhere in this book, notably in Carl Herndl's essay): it emerges by homology with a similar evocation of the individual in "Sociobiology"; it functions as a critique of the discipline of evolutionary biology; and it is an instance of the lyric disruption of narrative, in which science is seen as a direct contemplation of the organism itself, as when Darwin presented us directly with the perfected eye of the eagle. That contemplation permits certain textual pleasures that we do not yet fully understand.

Turning first, however, to the relationship between "Spandrels" and "Sociobiology," we might conjecture that, having written "Spandrels" within the context of the sociobiology controversy, the authors would be drawn to make use of figures of thought developed in that context, to gravitate to those established loci of value. In "Sociobiology," Gould's final move is the evocation of visceral, direct experience. In such experience, physiological reaction, group cohesion, and emotional and aesthetic responses fuse. Gould's example here is his participation in a performance of Berlioz's *Requiem:* the adaptationist topics of neurology and group survival cannot explain such a moment. "And I say this not to espouse mysticism or incomprehensibility, but merely to assert that the world of human behavior is too complex and multifarious to be unlocked by any simple key"—such richness, Gould concludes, is "our hope and our essence" (533). Gould has moved outside the domain of the discipline to find the object of sociobiological investigation, complex social behavior. In the Berlioz *Requiem,* he presents us with such behavior unmediated—or, rather, differently mediated, as it inheres in the lifeworld. Gould has identified an episode marked by the very physiological traits that sociobiology identified as its objects of study—a tingling spine, tears—and those privileged responses Gould reinscribes within the forms of the lifeworld, placing them outside the scope of any disciplinary investigation. The singing of Berlioz's *Requiem* cannot be "unlocked by any simple key" because it is not a lock, not an object created within a discipline to be investigated. In a move that is lyric in its force, Gould places the object beyond the domain of comparison and analysis.

In "Spandrels," similarly, the authors end by vaulting over the traditional objects of disciplinary investigation, placing against the adaptationist focus on parts a European interest in *Bauplan*, or overall form, an object of investigation that must be defended against the charge of mysticism. Mysticism, no; but lyricism, I think, yes. Consider Gould's treatment, in this final section, of the divaricate patterns of molluscs, ending in the gorgeous evocation of *C. cardissa*, a mollusc with translucent markings that transmit sunlight to the symbiotic algae beneath them. Among the many elegant examples in "Spandrels," this is the most improbable and graceful; it invokes the organism as an object of wonder. Its logical force is the more remarkable: *cardissa* in some ways contradicts the overall argument of the passage, and in others the general argument of the essay. The translucent windows of *C. cardissa* are adaptive, since they support the symbiotic algae; they are therefore complications of an argument that divaricate patterns are generally nonadaptive. But such a complication complies with the criterion of falsifiability: the argument against adaptation, based on falsifiability, would itself be equally subject to that critique if no contrary case exists or can be imagined to exist.

On another level of generalization, however, the relationship of the pattern of *C. cardissa* to adaptation is problematic. However useful those windows are to the algae, they are not adaptations of *cardissa* alone, but elements of a symbiotic system. *C. cardissa*, then, like Siamese twins, multicellular colonial cnidarians, or a cloned stand of bamboo, is a living thing that poses for Gould the question of the limits of the organism (*Flamingo's Smile*, 64–78, 79–80, 94). Precisely at the moment when the *Bauplan* of the organism is advanced as a new research paradigm, the boundaries of the organism are themselves revealed as a disciplinary convention, rather than as a fact of nature. The organism, like *C. cardissa*, is simultaneously intelligible and opaque; *cardissa* embodies the essay's most sustained and intense lyric moment, a moment that opens out into an evocation of the "recalcitrant, yet intelligible, complexity" that will shape the authors' revised, less triumphant, more complicated, story of evolution.

We can also read the authors' invocation of the organism as a whole as an interruption of the differentiated discourse of their discipline, an attempt to disrupt the discursive forms specific to evolutionary biologists, to find in a more poeticized language some way of renegotiating the terms of disciplinary differentiation. Indeed, the opening movements of both "Spandrels" and "Sociobiology" place the reader in conversation with other disciplines: "Spandrels," by way of examples from architecture and anthropology; "Sociobiology," by invoking, to the audience at the AAAS symposium, a critical scientific community: "It [adaptationist narrative] is also a procedure that has given evolutionary biology a bad name among many experimental scientists in other disciplines. We should heed this disquiet" (530). If "Spandrels" traces out a critique of the central research program of evolutionary biology, it also casts doubt on the power of that profession to effectively monitor its own production of knowledge. The traits, parts, and features of organisms that the adaptationist program created as its disciplinary objects of knowledge are, within the text of "Spandrels," eroded; disciplinary structures give way, not to those of ordinary language, nor to the narrated objects of the lifeworld, but to a heightened language, a new object of study, an organism seen as an instance of architecture: "Under these windows dwell endosymbiotic algae!" (530).

These movements are lyric in their metaphoric organization, although the metaphors in "Spandrels" are neither decorative nor, in the form known to the rhetoric of inquiry, epistemic: they are not ornaments of the text nor ways of organizing its knowledge, but openings within the text, ways of moving it to a different discursive terrain. And they are lyric in their evocation of particular speakers and particular moments—Spencer's fingerprints, Seilacher's study, the turn of evolutionary biology back to its "great historic themes." I would go further, connecting such moments of disciplinary self-reflection with postmodernism, since in "Spandrels" the discipline's practitioner, conscious of and dissatisfied with its differentiation, turns a discursive formation of modernity—in this case, the scientific article, but in other settings a novel or film or building—against itself. And this brings me to my final point, the pleasure of such lyric disruptions of narrative, and its relation to postmodernism.

It is of course open to debate whether the postmodern represents a new organization of culture, or simply an intensification of the self-reflection that has always been characteristic of modernity. I follow Habermas, connecting the two moments, rather than Jameson or Lyotard, who separate postmodernism as a distinct period. But the question of periodization is less relevant to this essay than an understanding of how the moments I have been calling lyric interrupt the narrative of modernity, disrupt the closure of differentiation.

And an analysis of such interruption and derailment can profitably bring us to the postmodern novel, the object of the last of my digressions. In texts of postmodernism such as (conventionally) those of Don Delillo or (less conventionally) those of Joseph McElroy, we find lyric uses of the vocabulary of science and technology, attempts to incorporate scientific language, over against those stories of subjects that are the traditional concern of the novel. Scientific language becomes a means, lyric in its force, of dislocating narrative: "Who knows what anyone wants to do? How can you be sure about something like that? Isn't it all a question of brain chemistry, signals going back and forth, electrical energy in the cortex?" (Delillo, *White Noise*, 45). Technical language—the language of computers, of genetics—can also function, as in McElroy's *Women and Men*, as an alternate field for working out traditional familial and relational plots. Such a translation of the traditional plot to a discourse that combines scientific, colloquial, and literary languages is of course also a transformation of the plot, its expansion into something more complex, inclusive, and self-conscious.

Or, as in McElroy's *Plus*, the activity of science is the object of narration. *Plus* recounts a hypothetical experiment in which a disembodied brain, orbiting in space, grows a body and invents a language. Here we find a rather precise complement to Gould's invocation of the organism as a whole, that lyric moment which disrupts the adaptationist narrative and moves outside the discursive forms of the discipline. *Plus* can be read as a novel that, embracing the structures of "plot" and "character," disrupts them, calling into question our assignment of languages and bodies to subjects, moving outside the discursive forms of the novel. And, like "Spandrels," it is a work of great beauty, written with great invention and attentiveness to language:

> The radius of color is not everywhere the same, he saw. It drew these certain parts of his sight together into a point as brief as the space was large that he had once found he could make by division and division when he tried to see between the white gel of a glue (or glial) cell and the twig cells that fired their bud ends from time to time across this divisible space and sometimes split into other twig cells that did not fire but only divided.
> This brief point was bright. (McElroy, *Plus*, 115)

McElroy evokes the instant, the perception of color—that central topic of Enlightenment science. Gould and Lewontin evoke the organism. In both cases, Enlightenment reflects upon itself, on its own forms of

Susan Wells

closure and opening, its own story of reason and desire. We have, in this culture, very few and very crude tools for explaining the connections among these texts. Stephen Jay Gould and S. J. Gould have given us some of the best we have.

NOTES

1. See also the analysis of this passage in Hayden White, *The Content of the Form.*

2. The works of Jürgen Habermas constitute a substantial body of essays ranging from theoretical works on philosophical issues, works in social criticism, and extended controversies with such theorists as Gadamer to interviews and essays on political issues. The most important titles include *Toward a Rational Society: Student Protest, Science, and Politics* (1970), *Knowledge and Human Interests* (1971), *Theory and Practice* (1973), *Legitimation Crisis* (1975), and two works that are central to my interests in this book: *The Theory of Communicative Action,* 2 vols. (1983), and *The Philosophical Discourse of Modernity* (1987). The standard summary of Habermas's work is still Thomas McCarthy's *The Critical Theory of Jürgen Habermas,* although important additional treatments include Seyla Benhabib, *Critique, Norm, and Utopia: A Study of the Foundations of Critical Theory;* Peter Dews, *Logics of Disintegration: Post-Structuralist Thought and the Claims of Critical Theory;* Martin Jay, *Marxism and Totality: The Adventures of a Concept from Lukacs to Habermas;* David Held, *Introduction to Critical Theory;* and John Thompson and David Held, eds., *Habermas: Critical Debates.* The collection by Thompson and Held includes a bibliography of translated work available to them; the complete bibliography is Rene Grotzen's *Jürgen Habermas: Eine Bibliographie seiner Schriften und der Sekundarliteratur, 1952–81.*

WORKS CITED

Barash, David. "Predictive Sociobiology: Damselfishes and Sparrows." In *Sociobiology: Beyond Nature/Nurture?* ed. G. W. Barlow and J. Silverberg, 209–26. Boulder: Westview, 1980.

Benhabib, Seyla. *Critique, Norm, and Utopia: A Study of the Foundations of Critical Theory.* New York: Columbia University Press, 1986.

Darwin, Charles. *On the Origin of Species and the Descent of Man.* Facsimile of the first edition, with an introduction by Ernst Mayr. 1859. Cambridge: Harvard University Press, 1964.

Delillo, Don. *White Noise.* New York: Viking, 1986.

Dews, Peter. *Logics of Disintegration: Post-Structuralist Thought and the Claims of Critical Theory.* London: Verso, 1987.

Gilbert, Nigel, and Michael Mulkay. *Opening Pandora's Box: A Sociological Analysis of Scientists' Discourse.* Cambridge: Cambridge University Press, 1984.

Gortzen, Rene. *Jürgen Habermas: Eine Bibliographie seiner Schriften und der Sekundarliterature, 1952–81.* Frankfurt: Suhrkamp, 1982.

Gould, Stephen Jay. *Ever Since Darwin: Reflections in Natural History.* New York: Norton, 1977.

Gould, Stephen Jay. *The Flamingo's Smile.* New York: Norton, 1985.

Gould, Stephen Jay. *The Panda's Thumb.* New York: Norton, 1980.

Gould, Stephen Jay. "Sociobiology and the Theory of Natural Selection." In *Sociobiology: Beyond Nature/Nurture?* ed. G. W. Barlow and J. Silverberg, 257–69. Boulder: Westview, 1980.

Gould, Stephen Jay. "Sociobiology: The Art of Storytelling." *New Scientist* 80 (1978): 530–33.

Habermas, Jürgen. *Knowledge and Human Interests.* Boston: Beacon, 1971.

Habermas, Jürgen. *Legitimation Crisis.* Boston: Beacon, 1975.

Habermas, Jürgen. "Modernity: An Incomplete Project." In *The Anti-Aesthetic,* ed. Hal Foster, 3–15. Seattle: Bay Press, 1983.

Habermas, Jürgen. The *Philosophical Discourse of Modernity.* Cambridge: MIT Press, 1987.

Habermas, Jürgen. *Theory and Practice.* Boston: Beacon, 1973.

Habermas, Jürgen. *The Theory of Communicative Action.* 2 vols. Boston: Beacon, 1983.

Habermas, Jürgen. *Toward a Rational Society: Student Protest, Science, and Politics.* Boston: Beacon, 1970.

Held, David. *Introduction to Critical Theory.* Berkeley: University of California Press, 1980.

Jameson, Fredric. *Postmodernism: Or, the Cultural Logic of Late Capitalism.* Durham: Duke University Press, 1991.

Jay, Martin. *Marxism and Totality: The Adventures of a Concept from Lukacs to Habermas.* Berkeley, University of California Press, 1984.

Landau, Misia. "Human Evolution as Narrative." *American Scientist* 72 (1984): 262–67.

Lyotard, Jean-François. *The Postmodern Condition: A Report on Knowledge.* Minneapolis: University of Minnesota Press, 1984.

Marx, Karl. *The Eighteenth Brumaire of Louis Bonaparte.* Hamburg, 1869. Reprint. New York: International, 1972.

McCarthy, Thomas. *The Critical Theory of Jürgen Habermas.* Cambridge, MIT Press, 1978.

McElroy, Joseph. *Plus.* New York: Knopf, 1976.

McElroy, Joseph. *Women and Men.* New York: Knopf, 1987.

Myers, Greg. *Writing Biology: Texts in the Social Construction of Scientific Knowledge.* Madison: University of Wisconsin Press, 1990.

Popper, Karl. *The Logic of Scientific Discovery.* London: Routledge and Kegan Paul, 1959.

Susan Wells

Segestrale, Ullica. "Colleagues in Conflict." *Biology and Philosophy* 1 (1986): 53–87.

Thompson, John, and David Held, eds. *Habermas: Critical Debates.* Cambridge: MIT Press, 1982.

White, Hayden. *The Content of the Form: Narrative Discourse and Historical Representation.* Baltimore: Johns Hopkins University Press, 1987.

Wilson, Edward O. *Sociobiology: The New Synthesis.* Cambridge: Harvard University Press, 1975.

4 CULTURAL STUDIES

AND CRITICAL SCIENCE

CARL G. HERNDL

Steven Jay Gould and Richard Lewontin spend the first half of "The Spandrels of San Marco" in a caustic and literate critique of the adaptationist program in evolutionary biology. In the second half of the essay, they offer their own, much more elaborately technical alternative. Balanced between these two moments of high culture and high science, in precisely the middle of their article, they turn to the "master's voice" and ironically "invoke God's allegiance" in their own critical project. In a punning reference to the Book of Genesis activated by their religious language, Gould and Lewontin quote Darwin from "the last edition of the *Origin*":

> As my conclusions have lately been much misrepresented, and it has been stated that I attribute the modification of species exclusively to natural selection, I may be permitted to remark that in the first edition of this work and subsequently, I placed in a most conspicuous position—namely at the close of the Introduction—the following words: 'I am convinced that natural selection has been the main, but not the exclusive means of modification.' This has been to no avail. Great is the power of steady misinterpretation. (155)

After referring to G. J. Romanes's earlier defense of Darwin, a work that "deserves a *resurrection*" (emphasis added), they quote Darwin (writing in 1880) in the act of castigating the naturalist Wyville Thomson for his misrepresentation of Darwin's theory: "Can Sir Wyville Thomson name any one who has said that the evolution of species depends only on natural selection?" (156). They do not report whether or how the unlucky Thomson replied.

The point of this ironic moment for Gould and Lewontin is to identify their alternative theory with Darwin and against the reductionism of the adaptationist program, the contemporary misrepresentation of

Carl G. Herndl

Darwin's theory. "We should cherish," they tell us, Darwin's "consistent attitude of pluralism in attempting to explain Nature's complexity" (156). Eight pages later they close their article with what, in their own ironic spirit, we might call an invocation of pluralism: "We welcome the richness that a pluralistic approach, so akin to Darwin's spirit, can provide. . . . A pluralistic view could put organisms, with all their recalcitrant, yet intelligible, complexity, back into evolutionary theory" (163). This appeal to pluralism has the capability of accommodating the complexity Gould and Lewontin value in evolutionary biology. As the unusual irony of their earlier religious language suggests, however, the invocation of pluralism is the most powerful and problematic moment in their essay. In closing this way, they return to pluralism as a scientific but also a cultural value, which they have repeatedly invoked to distinguish their theory from the adaptationist program. They have identified the adaptationists not only with the figure of Wyville Thomson, but also with Dr. Pangloss, the apologist for the status quo from Voltaire's comedy *Candide*. Their argument exposes the disciplinary and cultural hegemony of the adaptationist program, yet the pluralism they espouse is itself grounded in the transcendent authority of the *Origin*. Thus, their final gesture repeats what they have earlier described as the dominant form of argument in biology:

> You generally do not support your favored phenomenon by declaring the rivals impossible in theory. Rather, you acknowledge the rival, but circumscribe its domain of action so narrowly that it cannot have any importance in the affairs of nature. Then, you often congratulate yourself for being such an undogmatic and ecumenical chap. (151)

To anyone interested in the relations among science, culture, and history, the pluralism Gould and Lewontin advocate and the irony with which they use religious metaphors to introduce it pose a number of interesting questions. If Darwin's *Origin* has replaced *Genesis*, Darwin's voice now speaking in the place of God, what kinds of authority does science claim? Since science has largely replaced religion as a source of our culture's ideas of order and truth, how is scientific authority related to our ideas of cultural and political order? What kind of pluralism can science accommodate? And how would scientific pluralism support or challenge the increasingly bureaucratic and technological institutions (e.g., medicine, environmental policy, education) that organize our public life?

Since Darwin published the *Origin*, both biology and evolutionary theory have helped shape our definitions of culture, the nature of man, and the proper form of social organization. Throughout the late nine-teenth and early twentieth century, Social Darwinism provided a "sci-entific" rationalization for the exploitation of labor and natural resources. The relationship between science and culture has not, of course, been a one-way influence; a number of powerful metaphors have moved back and forth between biology and other fields such as economics, politics, sociology, and religion. Darwin's own formulation of evolutionary the-ory owes much to Malthus's *Essay on the Principle of Population* and to Adam Smith's laissez-faire economic models (see Kaye and Schweber). As Karl Marx commented in a letter to Friedrich Engels:

> It is remarkable how Darwin recognizes among beasts and plants his English society with its division of labor [read, diversifica-tion], competition, opening up of new markets [niches], "inven-tions" [variations], and the Malthusian "struggle for existence." (quoted in Sahlins, 101)

The cultural construction of Darwinian theory is then reversed by the movement back from a capitalist or capitalized nature to culture's rep-resentation of itself. As Engels later remarked:

> When this conjurer's trick has been performed [the transference of social order onto nature] . . . the same theories are transferred back again from organic nature into history and now it is claimed that their validity as eternal laws of human society has been proved. (quoted in Sahlins, 102–3)

This dialectical relationship between scientific and social theories of order suggests that the pluralism Gould and Lewontin advocate and their revisionary project in evolutionary biology may well be part of a cultural as well as a scientific debate. At issue is not only the social construction of biology, but also the scientific structuring of society. How Gould and Lewontin's essay participates in contemporary culture is the province of cultural studies.

Cultural Studies and Cultural Work

It would be misleading to suggest that there is a single, uncontested position or method that can be called "cultural studies." Cultural studies has always been a subversive and interdisciplinary inquiry that adapts its investigation to its subject and that operates in the space between established disciplines and methodologies. Indeed, many cultural theorists oppose the academic division of study into dis-

Carl G. Herndl

tinct disciplines of expert knowledge because it too often precludes alternate understandings; "culture," as Patrick Brantlinger points out, "is a category that transcends or transgresses various disciplinary and theoretical boundaries" (37). It is possible, however, to identify a general set of concerns about "culture, consciousness and experience, and its accent on agency" (Hall, "Cultural Studies," 58) that open the space from which cultural studies has emerged.[1] It is also possible to identify a provisional set of themes that run through many, though by no means all, cultural analyses. Typically, these studies explore and expose how particular social practices and modes of representation organize cultural power. These studies tease out the relationships between the knowledge a society produces and the material conditions and ideological structures through which it produces this knowledge. Cultural critics explicate the way cultural discourses realize and make available forms of subjectivity and possibilities for action. And, finally, they specify the way "culture" defines itself by a process of exclusion and domination. Throughout its exploration of these and similar issues, cultural studies seeks alternate social possibilities and is committed to concrete social action. For example, Paul Willis has studied the way a group of British working-class "lads" experienced their schooling and how they developed their identity in a "counter-school" culture. He understands this culture not as deviant, but as the way these working-class youths understand their social position and construct their oppositional social identity. Unfortunately, this culture encourages these youths into working-class jobs, reproducing the social conditions and labor relations from which the lads came. Willis's goal is to explain "how working class kids get working class jobs [and] why they let themselves," and to change the educational system that contributes to this cycle (Willis, 1).

Despite the wide variety of the work that goes on in cultural studies, most analyses fall within one of two general and contradictory positions: a "humanist" and a "structuralist" conception of culture and human agency. (These are, of course, provisional and evolving categories held in a strategic tension that articulates different positions on the question of the subject in contemporary theory.) The humanist position is associated with Raymond Williams's notion that "culture is ordinary" and with those moments in Marx's writings in which Marx argues that men are "engaged in the creation of the conditions of their social life" (162), and are thus social agents who can affect change. Critics working from this position tend to study the way people understand and organize their day-to-day lived experience. Such studies describe social phenomena—from education to hairstyles to novels—as human experiences through which people construct their identities, knowledge, and values, and through which they position themselves within their everyday living conditions. In *Reading the Romance,* for example, Janice Radway explores the social activity in which women engage as they read romance fiction. Because she begins by acknowledging the "constructions they [women readers] place on their behavior, the interpretations they make of their actions" (8), Radway can describe the women's reading of romances as a social process (as opposed to the interpretation of romances as isolated texts) through which women readers both express opposition to patriarchal culture and unintentionally maintain that very culture. Similarly, Dick Hebdige explores the spectacular style created by the various subcultures of British working-class youth as a "symbolic form of resistance" (80) through which the punks struggle "to embellish, decorate, parody and wherever possible to recognize and rise above a subordinate position which was never of their choosing" (139). For Hebdige and others, the outrage of style expresses the youths' lived experience of the outrage of class subordination.

In its second, "structural" moment, cultural studies analyzes experience as a secondary phenomenon, the result rather than the source of the structural relations that provide men and women with their possible modes of existence, expression, and self-representation. From this perspective, human experience or any social activity is conditioned by the ideological structures that provide the categories of thought through which social activity has significance or meaning. In Louis Althusser's terms, ideological structures produce the "representations of the imaginary relationships of individuals to their real conditions of existence" through which people come into social being (162). This kind of structural analysis tends to describe ideology as operating behind the backs of people who do not recognize the way these structures construct their lives. For example, Susan Jeffords has studied the way the representation of the Vietnam War in American film and television has contributed to the ideological construction of the masculine in opposition to the feminine. Jeffords argues that the "representational features of the Vietnam War are structurally written through relations of gender" (xi) so that the "discourse of war [through which we understand Vietnam] is linked to the process of remasculinization current in America" (185). This process, she argues, reconfigures the structural relations of patriarchy that legitimize both the war and the domination of the feminine in contemporary American culture.

Whether it proceeds from the humanist or the structuralist position (or from another kind of cultural analysis), work in cultural studies understands all social and intellectual activities as part of the production of culture in specific historical situations. Because it understands

Carl G. Herndl

culture as something produced, cultural studies maintains a critical relationship to its own concepts and methods and to the concrete studies it produces. Put more simply, critical work in cultural studies recognizes that it does not merely reflect or describe culture, but itself contributes to the making of culture. This relationship is similar to the dialectic between science and social theory that Marx and Engels identified. Cultural studies comprises a circuit in which the analyses of culture it produces become part of the lived experience through which culture changes or maintains its forms and representations.² This circular relationship requires that cultural analysts maintain a degree of self-reflexivity that acknowledges the political and cultural consequences of their own work. As we will see, the necessity of self-reflection is an issue for the critical work of "Spandrels"—and it is an issue for this essay as well.

Because the contemporary movement called cultural studies was initiated by interest in working-class life and by the need to redefine *culture* to include not only "high" literary culture but also working-class or "popular" culture, a cultural studies analysis of science may seem something of a stretch. Cultural analysis of science is relatively new, a raid on the philosophy and history of science. Such an analysis can take a variety of forms, but in the case of a scientific field like evolutionary biology, the cultural analyst might explore interconnections between biology and other cultural activities that have been obscured or denied by the discipline; or the analyst might detail the connection between the knowledge developed by evolutionary biology and its material and ideological conditions and consequences. More simply, since cultural studies assumes that no undertaking, including science, is autonomous and that any discourse is inherently ideological, cultural critics might ask how evolutionary biology participates in the whole political and social process of organizing life and legitimizing knowledge and power.

In taking up Gould and Lewontin's "The Spandrels of San Marco," my own analysis will consider the way this specific text participates in the practices of its discipline. How does the language of the text maintain or undermine the disciplinary consensus? How does the biology it constructs constitute human agency and the possibilities for individual and social change? How does the knowledge developed by this text facilitate or change other cultural practices (e.g., medicine or social policy)? How does the text locate itself within its disciplinary debates? How does it represent its connections with its discipline and with the larger culture?

The case of "The Spandrels of San Marco" is particularly interesting to cultural critics of scientific texts. Both Gould and Lewontin are themselves accomplished critics of the positivist project in biology; both have written quite extensively about connections among the sciences and other cultural institutions. "Spandrels" presents a theoretical argument in evolutionary biology, but in two specific ways the article is itself part of the cultural studies project. Some of the crucial ideas and analytic strategies that shape Gould and Lewontin's critique of the adaptationist program are developed from prior studies of the relationship between science and culture.[3] Moreover—and less overtly—their essay is an attempt to intervene in biological practice by posing an implicit critique of the sociobiology movement that came to prominence in the 1970s with the work of E. O. Wilson. In opposing the scientific and cultural program of sociobiology, "Spandrels" is doing what I call "cultural work." Because scientific ideas of order are dialectically related to other cultural images of order, Gould and Lewontin's opposition to the determinism of sociobiology becomes part of the cultural process through which people represent the relationship between man and nature and through which they construct the metaphors that legitimize our social structures. As Lewontin has argued elsewhere, "Ideas of order are profoundly ideological, so the description of evolution as producing order is necessarily an ideological one. In this sense evolution is neither a fact nor a theory, but a mode of organizing knowledge about the world" (Levins and Lewontin, 14).

Gould and Lewontin and the Project of Critique

Gould and Lewontin begin their description of the adaptationist program by identifying its two fundamental operations. They claim that in the first adaptationist move, "an organism is atomized into 'traits' and these traits are explained as structures optimally designed by natural selection for their function" (151). They omit a discussion of the problem What is a trait? but they offer an example of the adaptationist procedure:

> Our favorite example involves the human chin (Gould 1977, pp. 381–382; Lewontin 1978). If we regard the chin as a *'thing'*, rather than as the *product of interaction* between two growth fields (alveolar and mandibular), then we are led to an interpretation of its origin (recapitulatory) exactly opposite to the one now generally favored (neotenic). (151, emphasis added)

To understand the significance of this brief example and the critique of biological explanation it entails, readers need to understand something

Carl G. Herndl

of the missing argument about "traits" and the connections between this analysis and the analysis of culture from which Gould and Lewontin draw their criticism. Their analysis of the chin as a product of "interaction" rather than a "thing" recapitulates their discussion of Aztec human sacrifice and the adaptationist reduction of culture. In the example of the Aztec sacrifice, they argue that the ritual was part of the interaction of "a complex set of explicit justifications involving myth, symbol, and tradition" (150) rather than a simple pursuit of protein. The point of their argument about both the Aztec ritual and the human chin is: "Organisms are *integrated entities,* not collections of discrete *objects*" (151, emphasis added). This crucial insight rests on two concepts that operate in both cultural and biological explanation: the concept of "reification" developed by Georg Lukacs in the analysis of culture, and the subsequent idea that reducing relationships to "things" is an ideological procedure that transforms active "subjects" into "objects." Once the "thing" or object created by reification has been naturalized and accepted as part of our commonsense world, it can be controlled and dominated.

In *The Dialectical Biologist,* a book that he coauthored with Richard Levins, Lewontin takes up this biological argument and makes explicit much of what remains foreshortened in "Spandrels." Lewontin and Levins articulate in *The Dialectical Biologist* a form of biological analysis based on the work of Friedrich Engels. Their theory opposes the assumptions of the Cartesian reductionism that dominates biology and lies behind the adaptationist discussion of traits. They reject the Cartesian assumptions that an organism can be divided into discrete parts which are ontologically prior to the organism and that causes can be neatly separated from effects in biological or social explanation. They argue that the organism and the environment codetermine each other, and they reject the idea that traits exist independent of the constant "interpenetration" of the organism and the environment. Because of this interaction, traits cannot be isolated as the effects of a mechanical selection operating externally to the organism, as the adaptationist view holds. Furthermore, no trait or part exists separably from the whole organism through which it becomes meaningful. Thus, traits are not "discrete objects" caused by natural selection; rather they are aspects of the interaction of organism and environment and of the relations among parts within the organism. To treat them as real "things" caused by selection and adaptation misrepresents the two-directional nature of codetermination, and erroneously constructs organisms as mere objects lacking any active agency or innate value.

In *The Mismeasure of Man,* Gould too questions the idea of traits and argues that they are the result of "reification," of "our tendency to convert abstract concepts into entities" (24). The example he develops in *Mismeasure* is that of "intelligence" as a discrete and measurable thing. Gould argues that when people "atomize a behavioral repertory into a set of 'things,'" they perform a fallacy that has "plagued studies of intelligence throughout our century" (333). This reification has led to attempts to quantify intelligence, to measure it, and to set up social policies for dealing with the people whose intelligence is unacceptably low—low, that is, as measured by scientifically approved and socially efficacious techniques. The example of bluebird aggression that Gould and Lewontin take up in "Spandrels" is a similar example of reification (154–55). Gould and Lewontin charge that Barash and his defenders have reduced the interaction of the bird, the environment, and the broad range of possibilities in the bluebird's genetic code to a discrete and real thing, "aggression," which they claim was caused by natural selection and about which they can construct endless explanatory stories. The examples of "intelligence" in man and "aggression" in bluebirds suggest the ease with which biological reduction of relationships and concepts moves between scientific and cultural explanation.

The second step of the adaptationist program, according to Gould and Lewontin, represents the organism as essentially a product developed through economic relations within an open market. Traits develop through "trade-offs" between "competing demands" (151); they are explained as "optimizing" each part while minimizing the "expense" to other parts (151). The explicitly economic metaphors imply the politics of the adaptationist program. Moreover, those economic metaphors indicate the origins of the concept of "reification" through which Gould and Lewontin criticize traits.

In Marx and again in Georg Lukacs, *reification* refers to the ideological process by which the relations among the people who work to produce something become understood as the relations among the things produced (Lukacs, 83–103). The result of the workers' relations becomes a commodity, and workers then sell their labor for a wage. When productive relations among people are turned into "things" or commodities, the social relations between people are reduced to the "exchange value," the price of the commodity. So too with the reification of traits and the "optimizing" of parts. The trait is explained as a thing existing in reality, separate from the relations among organism, environment, and the organism's genetic information. In both biology and society, this process allows us to dominate the situation and deny the active part of organisms, or workers, in determining their biological or social existence. It allows us to "optimize" the structure and introduces efficiency as the sole criterion of value and analysis. This reification also creates a model of subjectivity, like that of the trait, as uniform, isolated, and independent of social relations. Gould and Lewontin underscore the cultural origins of the adaptationist language when they again invoke the image of Dr. Pangloss, the representative of conservative ideology. "Our world may not be good in the abstract sense, but it is the very best we could have. Each trait plays its part and must be as it is" (151).

By using concepts developed in studies of culture, then, Gould and Lewontin challenge the normal operation of evolutionary explanation and identify the ideological basis of a biological thinking that has come to seem common sense. In doing so, they expose the cultural work that the adaptationist program performs when it announces an ideological order in which each of an organism's parts "must be as it is."

Reduction and the Sociobiology Debate

> The Biologist, who is concerned with questions of physiology and evolutionary history, realizes that self-knowledge is constrained and shaped by the emotional control centers in the hypothalamus and limbic system of the brain. These centers flood our consciousness with all the emotions—hate, love, guilt, fear, and others— that are consulted by ethical philosophers who wish to intuit the standards of good and evil. What, we are then compelled to ask, made the hypothalamus and limbic system? They evolved by natural selection. That simple biological statement must be pursued to explain ethics and ethical philosophers, if not epistemology and epistemologists, at all depths. . . . In a Darwinist sense the organism does not live for itself. Its primary function is not even to reproduce other organisms; it reproduces genes, and it serves as their temporary carrier. (Wilson, 3)

In a chapter titled "The Morality of the Gene," E. O. Wilson opens *Sociobiology: The New Synthesis* by reducing organisms to mere vehicles whose function is to reproduce their genes. In his concluding chapter, Wilson asks his readers to consider man "in the free spirit of natural history as though we were zoologists from another planet" (547). "In this macroscopic view," he goes on to tell us, "the humanities and social sciences shrink to specialized branches of Biology" (547). Arguing that natural selection and adaptation have developed the genetic basis for human physical and behavioral traits, sociobiology collapses the distinction between biology and culture. Sociobiology seeks a new unified science that will envelop the study of organisms, including man, and their social arrangements. Like the unified field theory long sought after in modern physics, Wilson's master theory explains biological and cultural phenomena by identifying their base element, the gene, and the master mechanism that explains the evolution of these phenomena, adaptation through natural selection. This project assumes that it can in fact operate in the "free spirit" of natural history and with the objectivity of extraterrestrial scientists.

It is as an intervention in this scientific and cultural program that Gould and Lewontin perform the second kind of cultural work to which I referred earlier. It is specifically against the reduction of the cultural to the genetic that Gould and Lewontin speak when they say that "much confused thinking in human sociobiology arises from a failure to distinguish this mode [cultural adaptation or change] from Darwinian adaptation based on genetic variation" (159). The movement known as sociobiology had been growing throughout the 1960s and early 1970s, but when Wilson published his comprehensive and massively documented volume, sociobiology exploded as a prominent issue in both the scientific and popular press. There were headline articles in the *New York Times* and the *Chicago Tribune* and critical exchanges in journals like the *New York Review of Books, Bioscience, Science,* and the *New Scientist.* As Marshall Sahlins points out, the issue became a "media event" (ix). For cultural studies, something that permeates cultural awareness to this extent is not merely a media event; the media—excuse the uncritical generalization—are a powerful site for the production of culture and dominant ideological positions. Contrary to the "free spirit" and objectivity Wilson claims, sociobiology became so prominent and attracted so much criticism from the Left precisely because it was so thoroughly embedded in the dialectic between science and social theory that Marx and Engels described in the case of Darwin. In much the same way that Darwin found (and thus accidentally legitimized) the social relations of his English society among the beasts and plants, sociobiology provided a genetic justification for our current social relations when it collapsed cultural evolution into Darwinian evolution. And it is because "Spandrels" participates in this same dialectic, although in a much more restricted forum, that Gould and Lewontin's essay should be seen as doing cultural as well as scientific work.

"Spandrels" intervenes in this scientific and cultural dialectic in three ways. First, the panselectionism that Gould and Lewontin attack is not only the central theoretical tenet of sociobiology; the examples they use in their discussion identify sociobiology as the target of their critique.

Carl G. Herndl

At the beginning of their essay, Gould and Lewontin cite E. O. Wilson's treatment of Aztec cannibalism as "a primary illustration of an adaptive, genetic predisposition for carnivory in humans" (149). After they have described the elusive style of adaptationist argument in the second section of their essay, Gould and Lewontin offer the work of D. P. Barash as an example of the adaptationist "story." Gould and Lewontin reject this and all similar adaptationist stories because they are untestable, and "what good," they ask, "is a theory that cannot fail in careful study?" (155). Barash's article on bluebird aggression, which they use to represent adaptationist arguments, is part of the larger sociobiological vision that Barash details in *Sociobiology and Behavior.* Finally, Gould and Lewontin close their discussion of Seilacher's theory of constraints on evolutionary change by repeating an anecdote about Sir Herbert Spencer. Although Spencer worked in the latter half of the nineteenth century and was not a sociobiologist in the literal sense, the biological determinism that drove Spencer's Social Darwinism was very similar to that of contemporary sociobiology. For instance, while Spencer thought that biology provided a grounds for a "scientific morality," Wilson calls for "the genetic evolution of ethics" (563). By connecting Wilson and Barash with Spencer, Gould and Lewontin locate sociobiology in the field of social and scientific determinisms characterized by an ideological position that we recognize more clearly in the Social Darwinism of the nineteenth and early twentieth centuries.

The second way in which "Spandrels" participates in the cultural and scientific debate over sociobiology is by acknowledging the connection between cultural and scientific theory and then using this connection in the essay's own argument. By identifying the adaptationist program with the cultural conservatism of Pangloss, for whom the world "is the very best we could have," Gould and Lewontin suggest a political analysis of this program that they have made explicit elsewhere. In a critique of Wilson's sociobiology written by the Sociobiology Study Group of Science for the People, of which Gould and Lewontin were both members, the collective authors write:

> Determinist theories all describe a particular model of society
> which corresponds to the socioeconomic prejudices of the
> writer. . . . Moreover, such determinisms provide a direct justi-
> fication for the status quo as "natural". . . . [T]hese institutions
> operate as powerful forms of legitimation of past and present so-
> cial institutions such as aggression, competition, domination of
> women by men, defense of national territory, individualism and
> the appearance of a status and wealth hierarchy. (182)

Gould and Lewontin compare the adaptationists with Pangloss at five different points in the first half of their essay. But it is not only the absurd Panglossian logic that Gould and Lewontin ridicule in the adaptationist program. They also attack the cultural conservatism of Pangloss's "things cannot be other than they are" (149).

Having ridiculed the Panglossian paradigm in the first half of their essay, Gould and Lewontin balance their attack with an equally frequent appeal to pluralism in the second half. Certainly they intend this pluralism to refer to the multiplicity of possible explanations for evolutionary development. But their use of pluralism also has an ideological meaning and sets up a cultural opposition between the two theories. By using examples from architecture and literature, they affiliate their argument with the cultural power and tradition of St. Marks and *Candide*. They defend a view of "an elaborate cultural fabric" against Wilson's genetic explanation of human sacrifice. And they invoke the cultural capital of pluralism in Western democracy. This is a powerful, elegant, and, I think, successful rhetorical strategy; their position appears broad-minded, aesthetically sophisticated, and worldly, set against a narrow, doctrinaire conservatism.

The final way in which "Spandrels" participates in the science–culture dialectic is through the inherent politics of Gould and Lewontin's position; this is the sense in which people speak, with varying degrees of irony, of a theory as being "politically correct." If the adaptationist program is Panglossian, what are the politics of Gould and Lewontin's alternative? If the adaptationist program works toward culturally conservative and determinist positions, what cultural work does "Spandrels" do? What kinds of social order and subjectivities are implied in Gould and Lewontin's theory? Evaluating the politics of "Spandrels" involves asking about the consequences of their theory for biology and related scientific disciplines. And, given the dialectical relations between science and social theory we have discussed, this evaluation also includes exploring how Gould and Lewontin's theory affects our cultural common sense and the social practices it supports. Answering these questions is an uncertain and exploratory undertaking at best, first because anyone but a professional biologist is very limited in relevant scientific expertise and, second, because predicting such consequences or attributing causal relations to cultural practices usually entails one or another form of reduction. This is, of course, the objection to orthodox Marxist analyses that reduce phenomena to economic terms, even those analyses that resort to this reduction only "in the last instance." Rather than try to predict the positive consequences of Gould and Lewontin's theory, then, let me adopt a common cultural studies strategy. If we understand *ideology* (as many now do) as a cultural position that emerges only through struggle with opposing interests, we might ask the question about the politics of "Spandrels" in a negative form: What kinds of practices and positions does the theory undermine or disallow?

The theory that Gould and Lewontin propose might best be described as a theory of genetic *potential,* as opposed to genetic *determinism.* When Gould and Lewontin write that "constraints restrict possible paths and modes of change so strongly that the constraints themselves become much the most interesting aspect of evolution," they are referring specifically to structural constraints imposed by "the basic body plans of organisms" (160). I think, however, that we may take this to refer also to the constraints placed on the development of genetic traits, both by an organism's internal relations and by the organism's mutually determining interaction with its environment. Thus, their theory seems to be that an organism's genes determine a broad range of potential phenotypic characteristics and developmental paths, but that the realization of this potential depends on both the organism's activity and its codetermination with the environment. A theory such as this, which emphasizes the importance of "constraints" over genetic determination, has a more sophisticated model of causality and would undermine any practice that ignores dialectical relations or the ubiquity of multiple determination in both science and culture.

I return to sociobiology one final time, mostly because this particular case seems so clear. E. O. Wilson offers a hypothetical argument that "socioeconomic groups will be defined increasingly by genetically based differences in intelligence" (555). The critique in "Spandrels" undermines any such notion that genes determine the social hierarchy, first by denying that "intelligence" is a "part" "optimized" by adaptation and, second, by directing attention to race, class, and gender relations as a system of constraints. In the recession economy of the early 1990s, a theory such as Wilson's will be as useful in rationalizing cuts in social spending as it was in legitimizing the ever-increasing accumulation of wealth during the 1980s. If the socioeconomic hierarchy is part of the natural order written into our genes, why try to alter it? In opposing this kind of theory, "Spandrels" works to promote a more egalitarian social consciousness.

Cultural Work in Related Practices

In order to begin evaluating the politics of "Spandrels," I have described how the theory of constraints on genetic potential undermines the determinist agenda of sociobiology. To indicate adequately the cultural work that "Spandrels" performs, however, this analysis should also consider the way Gould and Lewontin's critique will influence work in other fields. Because it is part of the same dialectic between scientific and social order that Marx and Engels found in Darwin's work, the evolutionary theory that "Spandrels" proposes will have consequences that ripple through a variety of scientific and social practices. To understand the scope and importance of these consequences, and thus the cultural work of Gould and Lewontin's theory, I turn now to consider the implications of their position for medicine and the human genome project.

With their emphasis on "organisms as integrated entities, not collections of discrete objects" (151) and their closing celebration of "organisms, with all their recalcitrant, yet intelligible, complexity" (163), Gould and Lewontin take part in what Stanley Aronowitz identifies as the "new organicism" (Aronowitz, 301–16). I take "recalcitrant" here to indicate that organisms, properly recognized, cannot be explained by the unicausal determinism of Cartesian, reductionist explanation. According to the scientific and ideological position that Gould and Lewontin occupy, the qualitative and dialectical relations that constitute the life of an organism cannot be reduced to the quantitative relations between discrete parts as described by the model of economic management. This ideological resistance to quantitative analysis is, I think, the essential meaning behind Gould's title for *The Mismeasure of Man*. His general point in this case is not that incorrect or "unscientific" measurements have been used, but that the drive to "measure" man is itself ineluctably reductionist and ideological and tied to a reification of dominant cultural interests.

In opposing this general mode of thinking about organisms, Gould and Lewontin implicitly undermine much of contemporary medical practice, for which, as Aronowitz has recently argued, the "object of knowledge is defined as the human body, not principally the quality of social life" (334). In Aronowitz's analysis, "the technical division of labor within medicine *segments* parts of the body so that research is often directed to *discrete characteristics* of each part that may be susceptible to disease, without considering the *relationship* between these and the *body taken as a whole*" (335, emphasis added). In such a model, the side effects of chemical intervention on specific organs or symptoms are the "expense" and "trade-offs" Gould and Lewontin describe in the adaptationist model of optimizing traits. This model of segmentation leads medical practitioners to pursue treatment that regards the disease as a reified "thing" rather than as the result of relationships among

Carl G. Herndl

the organism, its genetic potential, and its parts, and between the organism and its environment. Thus, dominant medical practice seeks a cure or treatment for cancer rather than an understanding of "the entire matrix of social existence within our society" (Aronowitz, 334), which would redefine the disease as a social, economic, and political evil as well as a physiological problem. By explaining the development of traits through a complex set of constraints in biology, the organicism Gould and Lewontin propose suggests that we explain "disease" through a similar set of human and environmental constraints. Such a model would question our reliance on high-technology medicine and support more "ecological" models of medicine as social practice.

Gould and Lewontin's model of genetic potential and their organicist focus on constraints and interaction also challenges the reductionist ideology of genetic determinism in the human genome project, one of the most important and controversial developments in contemporary biology. The human genome project is a long-term research effort to "map (locate) the genes in the DNA of any organism and then to sequence (order) each of the DNA units, known as nucleotides, that constitute the genes" (National Research Council, 1). By mapping the 3 billion individual nucleotides in all of the approximately 100,000 genes in the human body, researchers hope to construct "a complete biological book on humankind" in order to address the "reasons we are what we are" (National Research Council, 100). While the Research Council admits that "even the complete sequence of DNA in the human genome will not by itself explain human biology," the possibility that this powerful technology will be constructed through the reductionist ideology is obvious (13). According to the Research Council, "encoded in the DNA sequence are the fundamental determinants of those mental capacities—learning, language, memory—essential to human culture" (12). The connection of this position to the panselectionism that Gould and Lewontin criticize in "Spandrels" is clear, particularly when the Research Council describes human genes as the result of adaptation through natural selection, which erases nonadaptive genetic changes: "Evolutionary biologists believe that changes in most of these sequences have occurred at one time or another during evolution, but because the changes were deleterious, the mutant individuals who carried such changes were eliminated from the population by natural selection" (30). This faith in adaptation through natural selection is denied both by the five alternatives to the adaptationist program that Gould and Lewontin offer (156–59) and by their objection to an "evolutionary biology of parts and genes, but not of organisms" (163).

The point to be made here is not that Gould and Lewontin's position ideologically opposes the project of mapping the human genome, but that their organicist theory would understand this project and its results as describing only genetic potential. In the organicist model, genetic potential becomes a general boundary condition, one constraint among others on the organism's development. As Stanley Aronowitz and Bruno Latour have argued, the laboratory has become perhaps the most powerful site for the production of culture in our technological society (Aronowitz, 272–300; Latour, 160). It is in this laboratory practice that the organicism of Gould and Lewontin's essay will do its cultural and ideological work.

"Spandrels" and Scientific Discourse

The final way to understand the cultural work of Gould and Lewontin's essay is to recognize the ways in which "Spandrels" participates in the construction of scientific subjectivity and the dominant model of science as nonideological. With illustrations of St. Mark's and King's College Chapel and a controlling metaphor drawn from Voltaire, "Spandrels" is an unusual essay for a scientific journal like the *Proceedings of the Royal Society.* Nevertheless, its critique of the ideology of reduction and determinism is largely implicit. "Spandrels" does not contain the sophisticated cultural analysis characteristic of Gould's more widely distributed essays or Lewontin's argument in *The Dialectical Biologist.* Despite its allusive style and the attack on the Panglossian paradigm, "Spandrels" maintains the cultural position of scientific discourse in a number of important ways. In a word, or two, the authors uphold the cultural and disciplinary position of the scientist as an "ecumenical chap" (151).

As we have seen, the repetition of "pluralism" throughout the second half of the essay sets up an opposition between the adaptationist's conservative ideology and the "undogmatic" position that pluralism evokes. Gould and Lewontin repeatedly refer to adaptationist arguments as "stories" that must only be "plausible" to be acceptable within the program; such stories are untestable since they can accommodate any evidence. This is the point of Gould and Lewontin's discussion of Barash's work and the steadfastly adaptationist explanation offered by Morton et al. even in the face of seemingly contradictory evidence.

By contrast, the latter half of the essay, in which they detail their alternative, bears no reference to "stories." Gould and Lewontin offer careful analyses of natural selection and adaptation and a sophisticated description of the European theory represented by Riedl and Seilacher.

Carl G. Herndl

Within the structure of an essay that balances the two theories paradox-ically on the pluralistic "spirit" of the "master's voice," the absence of any "story" from the second half constructs the scientist's subjectivity in opposition to the cultural and ideological positioning of the first half. This is, I think, a conflict within the essay's understanding of science as a social activity, an uneasy position required by the power of the disci-plinary discourse.[4] Indeed, in the introduction to *The Mismeasure of Man*, Gould himself has eloquently acknowledged the ineluctably ideo-logical nature of scientific activity:

> I do not intend to contrast evil determinists who stray from the path of scientific objectivity with enlightened antideterminists who approach the data with an open mind and therefore see truth. Rather, I criticize the myth that science is an objective enterprise, done properly only when scientists can shuck the constraints of their culture and view the world as it really is. (1)

The cultural work of maintaining the scientist's subject position in "Spandrels" is accomplished by stylistic as well as structural opposition. The adaptationist theory is described through metaphor and literary and architectural allusion. The description of Gould and Lewontin's alter-native uses careful disciplinary language without explicit cultural con-text and metaphor. The subjectivity they construct here is appropriate to their discipline's self-representation; objective and pluralist, it is a rationality freed of the unscientific power of "program." Published in the forum of the Royal Society, their essay cannot escape the discourse of professional science.

As Gould and Lewontin's essay circulates in the institutional forum of the Royal Society, the pluralism they propose operates on at least two levels: on one level it opposes the unicausal explanations offered by reductionist theories, and on the more abstract level of disciplinary discourse, it masks the ideological construction of science and scientific subjectivity. On this second level, their pluralism holds out the hope of an apolitical position that can overcome the "power of steady misrepre-sentation" that Darwin had lamented. In doing so, the essay overlooks its own insight that the institutional practice of biology is a fundamen-tally social activity caught up in the dialectic between culture and sci-ence. In such a dialectic, pluralism often masks the power of dominant ideologies to continually reify their positions as science or method.

The same criticism, it must be said, applies to my own essay if I ignore the possibility that the plurality of methods represented in this collection might obscure their differences in political commitment and the cultural work they themselves perform. The self-reflection involved in cultural critique forces me to recognize that my own essay might misrepresent the question of method and the disciplinary identity of cultural studies, for the understanding of cultural production from which cultural studies proceeds regards method and disciplinarity as historical and institutional positions. They are necessary to the work of cultural critique, but should be understood not as "things" but as relations between varying political and ideological commitments. Thus the definitions I have offered are strategic; like the notion of traits in biology, any definition of "cultural studies" reifies the relations between a variety of theories and the institutional practice of criticism.

NOTES

For his generous and copious suggestions on this essay, I thank Tom Streeter. Any errors are, of course, my own.

1. For careful definitions of cultural studies and its history, see Stuart Hall, "Cultural Studies: Two Paradigms" and "The Emergence of Cultural Studies and the Crisis of the Humanities"; and Richard Johnson, "What is Cultural Studies Anyway?" A careful exploration of the issues of ideology and culture in science appears in Stanley Aronowitz, *Science As Power: Discourse and Ideology in Modern Society.* Longer histories of the cultural studies tradition can be found in Patrick Brantlinger's *Crusoe's Footprints* and in David Punter's *Introduction to Contemporary Cultural Studies.* Finally, a very useful analysis of the alternate theories of the subject that run through cultural studies appears in Paul Smith's *Discerning the Subject.*

2. A useful visual representation of this self-reflexivity appears in Johnson's diagram of the circuit of cultural production and consumption in "What is Cultural Studies Anyway?" In this schematic, "lived experience" and "texts" and the cultural forms they generate appear at opposite places around the circuit. Since this circuit is a closed one, the diagram suggests that the texts and forms of cultural studies are conditioned by lived experience and the social relations upon which they depend, but that once these texts are produced, published, and read, they become part of the culture's lived experience. Anyone working in cultural studies must thus consider the political effectivity of his or her work once it enters this circuit.

3. Gould provides a powerful analysis of biology as cultural process in *The Mismeasure of Man* and *Ever Since Darwin.* Lewontin develops an explicitly Marxist theory of biology and biological explanation in *The Dialectical Biologist.*

4. For more on the tension between narrative and nonnarrative in "Spandrels," see Susan Wells's essay (chap. 3).

Carl G. Herndl

WORKS CITED

Althusser, Louis. *Lenin and Philosophy and Other Essays*. Trans. Ben Brewster. New York: Monthly Review Press, 1971.

Aronowitz, Stanley. *Science as Power: Discourse and Ideology in Modern Society*. Minneapolis: University of Minnesota Press, 1988.

Brantlinger, Patrick. *Crusoe's Footprints: Cultural Studies in Britain and America*. London: Routledge, 1990.

Gould, Steven Jay. *Ever Since Darwin: Reflections in Natural History*. New York: W. W. Norton, 1977.

Gould, Steven Jay. *The Mismeasure of Man*. New York: W. W. Norton, 1981.

Hall, Stuart. "Cultural Studies: Two Paradigms." *Media, Culture, and Society* 2 (1980): 57–72.

Hall, Stuart. "The Emergence of Cultural Studies and the Crisis of the Humanities." *October* 53 (1990): 11–23.

Hebridge, Dick. *Subcultures: The Meaning of Style*. London: Methuen, 1979.

Jeffords, Susan. *The Remasculinization of America: Gender and the Vietnam War*. Bloomington: Indiana University Press, 1989.

Johnson, Richard. "What is Cultural Studies Anyway?" *Social Text* 16 (1986–87): 38–80.

Kaye, Howard L. *The Meaning of Modern Biology: From Social Darwinism to Sociobiology*. New Haven: Yale University Press, 1986.

Latour, Bruno. "Give Me a Laboratory." In *Science Observed*, ed. Karin D. Knorr-Cetina and Michael Mulkay, 141–70. Beverly Hills: Sage Publications, 1983.

Levins, Richard, and Richard Lewontin. *The Dialectical Biologist*. Cambridge: Harvard University Press, 1985.

Lukacs, Georg. *History and Class Consciousness*. Trans. Rodney Livingstone. Cambridge: MIT Press, 1968.

Marx, Karl. *Grundrisse: Foundations of the Critique of Political Economy*. Trans. Martin Nicolaus. New York: Random House, 1973.

National Research Council. *Mapping and Sequencing the Human Genome*. Washington, D.C.: National Academy Press, 1988.

Punter, David, ed. *Introduction to Contemporary Cultural Studies*. London: Longman, 1986.

Radway, Janice. *Reading the Romance: Women, Patriarchy, and Popular Literature*. Chapel Hill: University of North Carolina Press, 1984.

Sahlins, Marshall. *The Use and Abuse of Biology: An Anthropological Critique of Sociobiology*. Ann Arbor: University of Michigan Press, 1976.

Schweber, Silvan S. "Darwin and the Political Economists: Divergence of Character." *Journal of the History of Biology* 13 (1980): 195–290.

Schweber, Silvan S. "The Origin of the *Origin* Revisited." *Journal of the History of Biology* 10 (1977): 229–316.

Smith, Paul. *Discerning the Subject*. Minneapolis: University of Minnesota Press, 1988.

Sociobiology Study Group of Sciences for the People. "Sociobiology–Another Biological Determinism." *Bioscience* 26 (1976): 182, 184–86.

Willis, Paul. *Learning to Labor: How Working Class Kids Get Working Class Jobs.* Farnsborough, England: Saxon House, 1977.

Wilson, Edward O. *Sociobiology: The New Synthesis.* Cambridge: Harvard University Press, 1975.

5

SCIENCE, GENDER, AND

"THE SPANDRELS OF

SAN MARCO AND THE

PANGLOSSIAN PARADIGM"

MARY ROSNER AND GEORGIA RHOADES

> *Science it would seem is not sexless; she is a man,*
> *a father and infected too.*
> Woolf, *Three Guineas* (quoted in Rose, 59)

While it speaks in many voices, feminism does seem to have several common strains: a concern for the whole and for the human part in the whole, which leads to a recognition of connections rather than conflict as an ordering principle; a challenge to the habit of dividing the world into only two categories (male/female, subject/object, mind/nature, reason/emotion, etc.); and a shifting of values that requires acknowledgement of personal experience and of multiple perspectives. Implicit in all these assumptions is the metaphor of the web. As Gilligan explains, the web represents complexity, with interconnected parts not privileged over the whole. In such a model, coexistence is more important than competition. Thus, according to Kolodny, feminists "give up the arrogance of claiming that our work is either exhaustive or definitive" (186), a position characteristic of patriarchal thinking (often represented by the metaphor of the ladder). Patriarchal thinking unfortunately infects almost all disciplines; and science, often considered the most objective of knowledge, has been especially patriarchal. Our analysis of Gould and Lewontin's "The Spandrels of San Marco and the Panglossian Paradigm" explains what feminists question about patriarchal science, how some feminists might read the essay, and how the essay might have been different if it had been written from a perspective that reflected feminism in both its science and its argument. Like Woolf, we too have found that science is not sexless: for Gould and Lewontin, it is explicitly feminist but implicitly patriarchal.

1. *Every discipline is made up of a set of restrictions on thought and imagination.*

(White, 126)

The discipline of science is still haunted by a number of inaccurate and damaging commonplaces. Science is said to be objective, "value-free, free of norms because it [is] solely determined by facts and logic in one way or another" (Hooker, 217); these facts result from "close-up" examinations of a subject, minute dissections of some part of some part of the world, based on some laws that some scientists have conceived as governing it. These laws are said to be free of bias, free of perspective, free of limitations; they accurately mirror what is actually there. As Bleier explains, "the lab coat, literally and symbolically, . . . wraps the scientist in the robe of innocence—of a pristine and aseptic neutrality—and gives him . . . a faceless authority that his audience can't challenge. From that sheeted figure comes a powerful, mysterious, impenetrable, coercive, anonymous male voice" ("Lab Coat," 62). These mirroring scientists, then, do not affect the reflection at all; the world continues "fixed, immutable, always out there, unaltered by the scientific enterprise itself" (Namenwirth, 37).

The texts these scientists write present the truth of that world through passives, nominalizations, abstractions, and mathematical and visual symbols that elevate the results of an interpretation while simultaneously subordinating both the interpreter and the interpretive process:

> [I]deally, the scientific text has no signature. The author of the modern scientific text, or the authority of modern science, is simultaneously everywhere and nowhere; on the one hand, it is manifest, self-evident, the arch-enemy of secrets and secrecy, and on the other, anonymous, uninterpretable, and unidentifiable. (Keller, "Making Gender Visible," 74)

The facts speak, not the scientist, and they speak with an impersonal authority: "The style of scientific communication . . . as reproduced in scientific journals is aimed at eliminating any traces of emotional or personal involvement: the style is cold, passive, impersonal, a jargon to be learned, a respectable mask of objective detachment, an elimina-

Mary Rosner and Georgia Rhoades

tion of the human subject" (Fee, "Is Feminism a Threat," 385). When they write, then, scientists play a role that seems to make them as passive as the world whose secrets they scrutinize in their investigations:

> [T]he belief that the job [of science] is to discover fixed relations of some sort, and that the application of observation, experiment and reason leads ineluctably to unifiable, if not unified, knowledge of an independent reality, is still with us. It is evidenced most clearly in two features of scientific rhetoric: the use of the passive voice as in 'it is concluded that . . .' or 'it has been discovered that . . .' and the attribution of agency to the data, as in 'the data suggest. . . .' Such language has been criticized for the abdication of responsibility it indicates. Even more, the scientific inquirer, and we with her, become passive observers, victims of the truth. (Longino, "Can There Be a Feminist Science," 51)

While as *writers* and as *mirrors of nature* scientists may seem to have passive roles, the authority given to science places it in a tradition that the Western culture has long associated with the powerful, with men.[1] As Keller explains, "The founding fathers of modern science . . . embraced the patriarchal imagery of Baconian science. . . . The goal of the new science [was] not metaphysical intercourse but domination, not the union of mind and matter but the establishment of the 'Empire of Man over nature'" (*Reflections*, 54). In the patriarchal society of seventeenth-century England, science was male, epitomizing "objectivity, rationality, culture, and control"; nature was female, "a mystery to be unveiled and penetrated" and dominated (Bleier, "Introduction," 9). The pretension to objectivity continues in the scientific ethos today: "In the patterns of words they choose for use in public lectures and research articles, scientists almost invariably project an image of impersonal authority and absolute confidence in the accuracy, objectivity, and importance of their observations. By all appearances, they will brook neither doubt nor vacillation" (Bleier, "Introduction," 23). In addition to the traits conventionally associated with the scientist, the topics of scientific research also privilege the male, as Hubbard explains:

> [A]mong billions of animal species, why have certain ones been studied repeatedly and in great detail, while others have been ignored? Until very recently, greater attention has been focused on the social structure among Savannah baboons than on chimpanzees or gibbons. Could this be because it has been easy to stereotype baboon social behaviour as hierarchical, with relatively rigid sex roles? Could it be because chimpanzees have very

fluid relationships with one another that are difficult to stereotype by sex except for the fact that females nourish the unweaned
young?

Is it an accident that among billions of insect species, those
whose social behaviour easily conforms to rigid roles are the ones
that have caught the imaginations of naturalists from the nineteenth century onward? The "scientific" language of the last century is still in use—ant and bee societies still contain slaves and
queens, as well as workers and soldiers. Yet we hear almost nothing about the behaviour of insects whose social arrangements
do not lend themselves to analogies reinforcing human social
arrangements that many people think of as "natural."

Turning to studies of our own species, is it an accident that
scientists have been primarily interested in exploring contracep-
tive techniques that tamper with the female reproductive system,
following the curious logic that because "fertility in women
depends upon so many finely balanced factors . . . it should be
easy to interfere with the process at many different stages . . ."?
Would it not be more sensible to conclude that it is more difficult
and riskier to tamper with a woman's reproductive system than a
man's because the woman's system is made up of "so many finely
balanced factors?" ("The Emperor," 214)[2]

While Western society may not be so completely patriarchal today, and
while "scientific" and "masculine" may no longer be entirely synony-
mous, "contemporary science nonetheless carries the history of that
conjunction [scientific and masculine] in the operative norms that guide
its choice of questions, institutional structures, and methodological and
explanatory preferences" (Keller, "Women Scientists," 79).[3]

This stereotype of the distanced and privileged (male) scientist who
works on subjects of limited interest from a limited perspective has
been challenged by many, including feminists who have traditionally
looked at sexual politics, the relationship between "aggressive males"
and "submissive females," and the locations of power in societies and
in texts. Many of these feminists have pointed out "the dissonance
between the personality characteristics of successful creative scientists
and the roles of nurturer, communicator and passive behind-the-scenes
manipulator for which women have been socialized" (Fausto-Sterling,
46). They have argued that science is not value-free, that it is "con-
stituted in part by social needs and interests that become encoded in
the assumptions of research programs" (Longino, *Science As Social*, 191)—
needs and interests that have traditionally been defined by males. And

Mary Rosner and Georgia Rhoades

they have insisted that "abstraction, reductionism, the determination to repress one's feelings to promote 'objectivity' " (Namenwirth, 32) is only one way of doing science, a way that discourages women and diminishes science:

> This authoritative demeanor is maintained even though it is anti-thetical to the nature of science, for the data and control experiments that underlie scientific 'truth' are always limited . . ., the instrumentation and analytical methodology always approximate, and alternative interpretations abundant. The hypothetical, incompletely verified, continually evolving character of scientific 'truth' is disguised by a veil of masculine authority. The weaknesses and inaccuracies, the holes in the data, are purposefully hidden as scientists interpose a shield of confident authority between themselves and the public. (Bleier "Introduction," 23)

As a result, feminists especially have urged that science be redefined:

> In a science constructed around the naming of object (nature) as female and the parallel naming of subject (mind) as male, any scientist who happens to be a woman is confronted with an a priori contradiction in terms. . . . Only if she undergoes a radical disidentification from self can she share masculine pleasure in mastering a nature cast in the image of woman as passive, inert, and blind. Her alternative is to attempt a radical redefinition of terms. Nature must be renamed as not female, or, at least, as not an alienated object. By the same token, the mind, if the female scientist is to have one, must be renamed as not necessarily male. (Keller, *Reflections,* 174–75)

Similarly, both the content and method of science have to change. Feminists have insisted that science broaden its area of studies to include both topics of specific concern to women (for instance, "menstruation, vaginal penetration, lesbian sexual practices, birthing, nursing, and menopause," the "bodily experiences men cannot have" [Harding, "Instability," 662]) and topics that have been distorted through patriarchal treatment (evolutionary theory, sex differences, brain lateralization), always noting discrepancies between the experiences of women and the results of narrow scientific studies that have seemed to ignore those experiences. Compare, for example, Perry's study of intellectual development and Belenky's; or Gilligan's study of the development of ethics and Kohlberg's; or McClintock's explanation of genes and Watson and Crick's; or fertilization stories that see the sperm as "heroic victor" with the egg as "passive prize awarded to the victor" and those that describe the egg as an active, indeed powerful, force that shapes the process (The Biology and Gender Study Group, 174–75); or the competing theories concerning "woman the gatherer" and "man the hunter":

> On their own the data are dumb, requiring [certain] assumptions in order to function as evidence. The frameworks belong to ways of seeing and being in the world that assign different degrees of reality and value to male and female activities. If female gathering behavior is taken to be the crucial behavioral adaptation, then stones are evidence that women began to develop stone tools in addition to the organic tools already in use for gathering and preparing edible vegetation. If male hunting behavior is taken to be the crucial adaptation, then the stones are evidence of male invention of tools for use in the hunting and preparation of animals. (Longino, *Science As Social*, 111)

Since it is the framework that defines what is seen, feminists have urged scientists to acknowledge that both the organisms studied and the scientists themselves are active (not passive) and complex; that very possibly more than one interpretation of complex organisms and phenomena is possible; and that a conventionally "objective" interpretation based on a positivist notion of science may not be appropriate. In place of a patriarchal framework that distorts the world by imposing unexamined assumptions on it, feminists suggest a more complex perspective, one that pays "careful attention to the dynamics of living systems [both subjects of investigation and investigators] as pieces of a larger and more awesome natural world which is constantly responding to, and responsive to, itself" (Ginzberg, 71). Consider Haraway's description of Altmann's *Baboon Mothers and Infants*, which treats the animals neither as " 'pre-discursive bodies' just waiting to validate or invalidate some discursive practice, nor [as] blank screens waiting for people's cultural projections." Instead, they "are active participants in the constitution of what may count as scientific knowledge." They "resist, enable, disrupt, engage, constrain, and display. . . . [They] yield no unique, univocal, unconstructed 'facts' waiting to be collected" (*Primate Visions*, 310–11). Similarly, McClintock views nature not "as a passive, mechanical object ruled by externally imposed law, but as alive, growing, internally ordered, and resourceful" (Fee, "Critiques," 48). As McClintock herself says, " 'Misrepresented, not appreciated, . . . [organisms] are beyond our wildest expectations. . . . They do everything we [can think of], they do it better, more efficiently, more marvelously' " (quoted in Keller, *Reflections*, 162).

To describe fairly, to predict justifiably, then, science should pay "greater

Mary Rosner and Georgia Rhoades

attention to the linkages between different levels of organization and between different aspects of the same subject. It is the WHOLE, complete with all its details and idiosyncracies and individualities that is important to a female world view" (Namenwirth, 32) and that is certainly essential to understanding what is studied. In McClintock's work with maize, for example, "the smallest details provided the keys to the larger whole. It was her conviction that the closer her focus, the greater her attention to individual detail, to the unique characteristics of a single plant, of a single kernel, of a single chromosome, the more she could learn about the general principles by which the maize plant as a whole was organized, the better her 'feeling for the organism' " (Keller, *Feeling for the Organism*, 101). Recognizing how an individual kernel acquires significance in a context that defines it as an anomaly means recognizing the complexity of both the organism and the ways it is perceived.

Just as atomizing nature distorts it, feminists say that calling science innocent and objective imposes an inaccurate frame on it; "what we can know depends upon our social practices and contexts" (Flax, 633). The ability to discover an absolute truth "would require the existence of an 'Archimedes point' outside of the whole and beyond our embeddedness in it from which we could see (and represent) the whole. What we see and report would also have to be untransformed by the activities of perception and of reporting our vision in language. The object seen . . . would have to be apprehended by an empty (ahistoric) mind and perfectly transcribed by/into a transparent language" (Flax, 633). Rejecting that "ahistoric mind," feminists ask that science examine and question its assumptions, its beliefs, its practices:

> Only in this way can we hope to produce understandings and explanations which are free (or, at least, more free) of distortion from the unexamined beliefs and behaviors of . . . scientists themselves. . . . The beliefs and behaviors of the researcher are part of the empirical evidence for (or against) the claims advanced in the results of research. This evidence too must be open to critical scrutiny no less than what is traditionally defined as relevant evidence. (Harding, "Introduction," 9)

Because "impartial" science has been shown to possess "observational and evaluative bias" (Gilligan, 6), many feminists believe that scientists should make explicit the assumptions that direct their research, and that scientists should be, in fact, part of their research: "the inquirer her/himself [should] be placed in the same critical plane as the overt subject matter, thereby recovering the entire research process for scrutiny in the results of research" (Harding, "Introduction," 9).

And because who they are affects what and how they study and what they discover as they study, scientists cannot offer absolute truth but only versions, interpretations, that together may move toward a truth:

> [N]o single individual scientist, scholar, or theorizer can produce the "whole truth" about a given phenomenon. Each of us brings to the inquiry, to the investigation of a particular phenomenon, our own life history of experiences, knowledges, and attitudes as well as our particular skills and training, and, consequently, each illuminates one or another facet of the complex phenomenon we are trying to explain. Together we illuminate many different facets, all varied aspects of the "truth." (Bleier, *Science and Gender,* 203)

The generosity of this perspective is reflected in the feminists' ability to see not only multiple truths but changing truths. About the work on early humans, for example, Longino and Doell note that "at this stage the woman-the-gatherer framework offers the more comprehensive and coherent theory [compared with the man-the-hunter], but this may be due to its elaboration after and partially in response to man-the-hunter theories. . . . We suspect that a less gender-biased theory will eventually supersede both currently contending accounts" (175).

This different way of doing science may be reflected in a different way of writing science. For instance, Altmann's feminist study of baboons and their young (*Baboon Mothers and Infants* [1980]) is distinguished by "a particularly tight relation between content and form, in which preferences for multiplicity and complexity, strategic moves to foreground the previously obscure and underrated, and substitution of compelling simplicity for high drama in the writing itself and in its message all converge to produce a powerful effect . . . [allowing] Altmann to move across species and cultures to reconstruct meanings of female and women without seeming to flatten contradictions" (Haraway, *Primate Visions,* 315). Instead of attempting to articulate a single Truth, feminists are more interested in multiple truths "for the sake of the connections and unexpected openings" (Haraway, "Situated Knowledge," 590).

> **2.** *In other systems of thought, what seems peripheral to us becomes central, and distinctions essential to us do not matter.*
>
> (Gould and Vrba, 4)

Like the feminists who argue that we should "think through a particular field and try to understand just what its unstated

Mary Rosner and Georgia Rhoades

and fundamental assumptions are and how they influence the course of inquiry" (Longino, "Can There Be a Feminist Science," 55), Gould and Lewontin think through their field of evolutionary biology to raise questions about some of its fundamental assumptions—in particular, assumptions underlying the adaptationist program,[4] which argues that almost all changes "of form, function, and behaviour" (156) come about only through natural selection as organisms adapt to the prevailing conditions of their environment. This popular interpretative framework, they suggest, denies the complexity of nature, which requires an equally complex and plural response from students of nature: the model may not always account for why changes are the way they are. In asking biologists to evaluate what they have long espoused and to embrace the possibility of alternative theories through which they may better understand their subjects, these scientists implicitly argue for liberation and plurality—for a more feminist approach to science.

Like feminists, Gould and Lewontin explicitly and implicitly base their argument on the assumption that the world is complex; organisms, for example, are "integrated entities, not collections of discrete objects" (151); they are "integrated wholes, fundamentally not decomposable into independent and separately optimized parts" (157); scientifically acceptable explanations for structures and behaviors should be multiple (Gould and Lewontin themselves offer a "partial typology of alternatives to the adaptationist programme" [156]); " 'adaptation' . . . can occur at three hierarchical levels with different causes" (158), though the word "adaptation" unfortunately merges all three levels; "constraints upon evolutionary change may be ordered into at least two categories" (160)—possibly three, if Seilacher is correct. Because of this complexity, reliance on only one strategy of interpretation may distort the organism and produce questionable results; in other words, it may be unscientific.

Gould and Lewontin find the adaptationist program problematic not only because it is represented as *the* single frame of reference for many evolutionary biologists but also because the frame itself may be unnecessarily speculative and narrow. Within the adaptationist program, biologists can account for some contemporary structures—for instance, by measuring traits and environment: "A particular aspect of the organism's life history is isolated as a problem to be solved. By an engineering analysis, the optimal solution is deduced, subject to certain constraints about the nature of the species, and then the species is measured to see whether it has provided the optimal solution. If it has, then a plausible argument is made that the trait examined has in fact arisen as an optimal solution to the posed problem" (Lewontin, 255). But when traits and environment are not measurable, some biologists practice a method that Lewontin identifies as "imaginative reconstruction" (and that Gould and Lewontin call "untestable speculations" [153] and "plausible stories" [154])[5] in which "one simply thinks about a species, past or present, and literally inserts a reason why a certain trait should have been favored by natural selection" (256). Wholesale invention replaces interpretation tied to evidence. Moreover, when the sole criterion of evaluation within the adaptationist program is "consistency with natural selection" (154), interpretations are virtually unconstrained—"plausible stories can always be told" (154)—and science is trivialized.

This type of fictionalizing, "adaptive or speculative story-telling," is reflected in the various "imaginative reconstructions" that open the essay: "The design [of the spandrels] is so elaborate, harmonious and purposeful that we are tempted to view it as the starting point of any analysis" (148); the fan-vaulted ceiling in King's College Chapel in Cambridge exists "because the alternation of rose and portcullis makes so much sense in a Tudor chapel" (149); "Aztec human sacrifice arose as a solution to chronic shortage of meat" (149). In each case, the explanation rests not on facts but on an imaginative tracing back from the present. This procedure has several disadvantages: it results from investigative bias (we follow the most familiar trace, not necessarily the most appropriate one); it distorts the subject by imposing a single explanation on it (rather than looking at alternative explanations); and it thereby does both the subject and the discipline a disservice.[6]

In making their case for alternatives to the adaptationist program, Gould and Lewontin evoke not only the complexity of the natural world but also the complexity reflected in society and in themselves. The essay opens expansively, with "the great central dome of St. Mark's Cathedral in Venice" (147), to remind its intended readers of the possibilities of worlds and truths—"the mainstays of Christian faith" (147)—that are outside their own discipline and time. It moves from Gothic architecture to Aztec culture, from art to anthropology, asking its readers to make connections between the two while again emphasizing multiple worlds and multiple truths. Only after having identified the narrowness of a particular perspective through examples from these two disciplines does the essay argue explicitly for broader meaning-making possibilities than the adaptationist program offers, possibilities that would be in harmony with "Darwin's pluralistic spirit" (156) and his "consistent attitude of pluralism in attempting to explain Nature's complexity" (156).

So the essay *opens* in ways that are more conventional to a feminist than to a patriarchal science. Gould and Lewontin "tell stories" about figures on ceilings and body parts on plates, look for connections among

Mary Rosner and Georgia Rhoades

apparently different spheres, and create a nonaggressive invitation to their readers to reconsider (rather than necessarily reject) science as a distorting mirror, potentially as faulty as the interpreters of the spandrels, ceilings, and Aztec sacrifices: "evolutionary biologists, in their *tendency* to focus exclusively on immediate adaptation to local conditions, do *tend* to ignore architectural constraints and perform just such an inversion of explanation" (149; emphasis added).

Although they raise questions about an ingrained habit of thinking among biologists, Gould and Lewontin do not set themselves up as absolute authorities early in this essay, though they do cast themselves in several roles, offering evidence (again) for complexity. They speak as members of their readers' community—"we are tempted to view [the design] as the starting point of any analysis" (148), "we do not impose our biological biases upon them" (148), "[we have] our biological prejudices" (152). They speak as experts in biology: they know "an old habit" (152) and "an old argument" (152); they recognize "a deeply engrained habit of thinking" (150); they know "evolutionists have often been led astray by inappropriate atomization" (151); they identify problems with Barash's work. As Gragson and Selzer note in chapter 10 of this book, they speak as open-minded, tentative inquirers, not as dogmatists: "We strongly suspect" (150); "We wish to question" (150); "We do not attack these newer interpretations. They may all be right. We wonder, though" (152); "We do not offer a council of despair" (163). As these examples demonstrate, in much of the essay Gould and Lewontin choose to draw attention to themselves by using the active voice; by identifying themselves as interpreters, questioners, advisers; and by making their prejudices and perspectives overt, with phrases like "Panglossian paradigm" (147), "to put it crudely" (150), "beef up the meat supply" (150), "Spandrels do not exist to house the evangelists" (150), "a mighty poor way to run a butchery" (150), "glass beasts as straw men" (153), "our favorite example" (151), "the same phoney he saw before" (154). Facts do not speak for themselves, and scientists who interpret them— even Gould and Lewontin—are not without bias. Instead of adapting the "invisible" role traditionally defined for scientists who write, they adopt the kind of language that reminds us that these two scientists are not always "rational," not always "objective," and that what they speak is not *the* truth but merely one perspective on it, a perspective defined by what they find to be "interesting and fruitful" and rich. And theirs are only two of many interpretations in the essay; others are offered by Harner and Wilson, Sahlins and Ortiz de Montellano, Voltaire and Pangloss, the adaptationist community, Darwin—all of them partial.

In the fruitfulness of alternative approaches, in the plurality of interpretative explanations, in the multiplicity of truths and perspectives, Gould and Lewontin's essay—whatever their intention—is in many ways an implicit argument for a feminist approach to science. Yet paradoxically, the method that forwards this argument is by and large patriarchal and thereby mirrors the behavior Gould and Lewontin explicitly (and many feminists implicitly) condemn: when they argue, they "acknowledge the rival [theory], but circumscribe its domain of action so narrowly [by finding it largely unacceptable] that it cannot have importance in the affairs of nature. Then, [they] . . . congratulate [themselves] for being such . . . undogmatic and ecumenical chap[s]" (151). As traditionally conceived, argument is implicitly hierarchical (with some "chaps" right and some wrong), and Gould and Lewontin repeatedly point out the right from the wrong, the "we" from the "they" (as Gragson and Selzer note in chapter 10). Using "we," they define a community that finds examples of architectural constraints easy to understand "because we do not impose our biological biases upon them" (148). But the large community of biologists is quickly fragmented into the good and the bad when evolutionary biologists "invit[e] the same ridicule that Voltaire heaped on Dr Pangloss": in their tendency to focus exclusively on immediate adaptation to local conditions, they "do tend to ignore architectural constraints and perform . . . an inversion of explanation" (149). The tentativeness of this criticism becomes open mockery by paragraph nine: "The adaptationist programme is truly Panglossian. Our world may not be good in an abstract sense, but it is the very best we could have. Each trait plays its part and must be as it is" (151).

The confident authority of Gould and Lewontin[7] reflected in this ridicule continues throughout the essay. The authors set up a dichotomy between themselves and the adaptationists, the informed and the uninformed, the learned and the naive, through the assertiveness of various claims: "Some evolutionists may regard this [the question of what a trait is] as a trivial, or merely a semantic problem. It is not" (151); "This view is false" (155); "[O]ne cannot claim that, eventually, a new mutation of just the right sort for some adaptive argument will occur and spread" (157); "The mere existence of a good fit between organism and environment is insufficient evidence for inferring the action of natural selection" (159); "The immediate utility of an organic structure often says nothing at all about the reason for its being" (159); "[W]e cannot understand the pattern or its evolutionary meaning by viewing these infrequent and secondary adaptations as a reason for the pattern itself" (162).

Instead of looking for connections between their perspective and

Mary Rosner and Georgia Rhoades

that of Harner, Wilson, and the other scientists they examine, as some feminists might do in order to reduce antagonisms,[8] Gould and Lewontin set up oppositions between themselves and these "rivals," whom they attack for being unscientific and ridiculous. These scientists "invert the proper path of analysis" (148), "impose biological biases" (148), and "have often been led astray through inappropriate atomization" (151); they fail to respect the "key to historical research [that] lies in devising criteria to identify proper explanations among the substantial set of plausible pathways to any modern result" (154). They miss "the notion that suboptimality might represent anything other than the immediate work of natural selection" (151), and they fail to acknowledge "the obvious test of [an] alternative to a conventional adaptive story" (154).

Those who practice the adaptationist program are, like the anthropologists who theorize about "a mighty poor way to run a butchery," victims of a "Panglossian paradigm," oversimplifying what is complex. Their approach is faulted for "its failure to distinguish," "for its unwillingness to consider alternatives," "for its failure to consider adequately . . . competing themes" (147), for "misreading" data (160), for atomizing nature, for failing to explain: "If development occurs in integrated packages, and cannot be pulled apart piece by piece in evolution, then the adaptationist programme cannot explain the alteration of developmental programmes underlying nearly all changes of *Bauplan*" (160). Clearly and repeatedly, Gould and Lewontin find their rivals silly and stupid, so that we imagine these two as members of the community that Tompkins criticizes, those who feel "justified in exposing . . . errors to view, that they may be scourged . . . in the sight of . . . professional peers and superiors. WE feel justified in this because we are right, so right, and they, like the villains in the western, are wrong, so wrong" (588). This hierarchy established by Gould and Lewontin runs counter to their open-minded and open-handed position ("We wish to question a deeply engrained habit of thinking among students of evolution" [150]; "We do not attack these newer interpretations; they may all be right" [152]). They do more than question; they ridicule.

If the adaptationists are the "bad" biologists, who are the good ones? Initially, it might seem that Gould and Lewontin valorize pluralistic approaches to evolutionary biology since they are arguing, after all, for "*alternatives* to immediate adaptation for the explanation of form, function, and behaviour" (156; emphasis added). In fact, however, they privilege a single alternative, a single perspective—Seilacher's, whose work on architectural constraints (epitomized in the examples of spandrels, vaulted ceilings, and Aztec sacrifice) "deserves far more attention than it has received" (161). As Gould and Lewontin explicitly instruct us, this work is "full of potential insight" (161); at least one of his examples is "fascinating" (161); Seilacher's arguments "seem generally sound" (162); and he is "probably right" (162). He is the "good" scientist, the hero of this piece, the one who does not ignore "historic themes of developmental morphology and *Bauplan*" (163). Yet in an essay that ostensibly argues for multiplicity, their privileging this single perspective suggests that Gould and Lewontin are simply replacing one approach (the adaptationists') with another (Seilacher's); they are closing off conversation, questioning, uncertainties. And their assertion of the indisputability of what they claim (there is truth, and they see it) does not acknowledge that they could be limited by their own perspective, even though their perspective may be more satisfying than the adaptationists'. They fail to acknowledge, as a feminist might, that there may be other explanations since science, like Nature, is complex. In sum, they fail to examine their own biases.

It is not only in setting up oppositions and hierarchies that Gould and Lewontin reveal a patriarchal view of science; it is also revealed in the universe they describe. The wrong-headed scientists rejected by Gould and Lewontin are males in a world populated by males (like the discipline of science itself)—from the figures in the spandrels to Aztec kings, from Candide to Pangloss, from Darwin and Wallace, Weisman and Harner, to Seilacher and Riedl and the others. (Since the convention for bibliography in scientific texts identifies the scientists only by the initials of their first names, not by the first names themselves, and since males have conventionally dominated the field, it is also easy to read all the scientists noted in the bibliography[9] as male.) The single female acknowledged (and even then only in the article's abstract) is a dinosaur, one that exists to serve as a passive focus for an active (presumably male) scientist's theory and as a passive sex object for an active male reptile (in either case, she is not treated with much complexity): "male tyrannosaurs may have used their diminutive front legs to titillate female partners, but this will not explain *why* they got so small." (In the paper itself, the dinosaurs become members of the single sex that matters; "glass beasts" are "straw *men*" [153; emphasis added].) In the world of this essay, knowledge and power reside in the males, with Gould and Lewontin as the very Kings of the Hill. True, ignorance resides in the males too—witness Harner—but at least the males are worthy of comment. Females are absent, invisible, insignificant, and their absence is noteworthy since it results from choices made by Gould and Lewontin. If female examples and authorities are available, these choices may reflect a narrow perspective belonging to the two scien-

Mary Rosner and Georgia Rhoades

tists; if unavailable, they may reflect a restricted perspective definitive of the discipline that ought to be acknowledged somehow.

Haraway writes that feminism does not love science "ruled by phallocentrism (nostalgia for the presence of the one true Word) and the disembodied vision"; it loves another science: "the sciences and politics of interpretation, translation, stuttering, and the partly understood . . ., those ruled by partial sight and limited voice—not partiality for its own sake but, rather, for the sake of the connections and unexpected openings situated knowledges make possible" ("Situated Knowledge," 589–90). Gould and Lewontin give a nod to the stutter when they suggest their "partial typology" and a glance to partial sight when they acknowledge alternate disciplines and alternate "truths." But once they enter their own discipline, they define themselves as the "one true Word" through a "disembodied voice" of scientific jargon, the voice of the angry God who knows "the truth" and who uses, like Darwin's opponents, "the common attempt to caricature and trivialize" (155) in order to castigate those who fall away from it. While arguing against constraints, these authors fail to examine the constraints of their own interpretation of science and of Nature. Substituting one perspective for another does not bring "richness" and "complexity" for long. Nor does it bring pluralism.

In the spirit of plurality, and in an effort to reflect the open-endedness of the conversation that we would encourage about Gould and Lewontin, this essay, and feminist science, we offer several versions of a "final" section, which we hope is not "final."

3. *They changed the facts of nature by changing the visions of possible worlds, and it has been hard, complicated work.*

(Haraway, "Primatology," 80)

In LeGuin's "The Eye Altering," Genya, one of the first human exiles born on Mars, is an artist who creates what the Earth-born exiles see as "muddy jumbles of forms and lines all in a dark haze, like words half created" (158). One day, he stops taking the "metabolising pills" that his Earth-born doctor believe guaranteed him survival (though a painful and very limited survival). Soon after, Genya paints a landscape that he hangs in the "Living Room," the refuge for Earth-born colonists who hated their new world; it had "indirect lighting, deliberately reproducing the color and quality of sunlight . . . so that to enter the Living Room was to enter a room in a house on Earth on a warm sunny day of April or early May, to see all things in that clear, clean, lovely light" (162). In this context, the Earth-born see Genya's painting as "a landscape of the Earth: a wide valley, the fields green and green-gold, orchards coming into flower, the sweeping slope of a mountain in the distance. . . . It was a complex and happy painting, a celebration of the spring, an act of praise" (168). Only the doctor is able to recognize that this landscape represents Mars as Genya sees it: " 'Take it outside, into the daylight, and you'll see what we always see, the ugly colors, the ugly planet where we're not at home. But he is at home! . . . It's we . . . who lack the key. . . . We were all perfectly adjusted to Earth, too well, we can't fit anywhere else—he wasn't, wouldn't have been; allergic, a misfit—the pattern a little wrong, see? The pattern. But there are many patterns, infinite patterns, he fits this one a little better than we do—' " (169–70).

Like patriarchal science, which authoritatively imposes an interpretation on nature that deforms her, the Earth-born doctor could see Genya only as a reflection of her pattern, so she gave him medicine that was appropriate to her. Believing that she had found the key to survival on Mars, she did not look for other keys, other kinds of life and other kinds of survival. Her science focussed on the single dominant pattern. Gould and Lewontin offer an alternative to the adaptationist "pattern" in which they invite a richer science, richer not only because they introduce an alternative perspective but also because in choosing to look for a different pattern, they must find different stories, different theories, different facts. If "feminist science is about changing possibilities" (Haraway, "Primatology," 81), then "The Spandrels of San Marco" argues for feminist science. But Gould and Lewontin are not radical enough; they see only one alternative when "there are many patterns, infinite patterns" to see if they would look. They have more work to do.

3.   *I am*
standing . . . somewhere just outside the frame
of all this, trying to see.

(Adrienne Rich, "Frame")

As we suggested, *if* Gould and Lewontin wanted to argue a feminist perspective both explicitly and implicitly, they would need to broaden the world of the essay by including examples drawn from female interests and investigated by female scientists; rather than

Mary Rosner and Georgia Rhoades

alienate the adaptationists, they would look for connections between what they find to be important and what the adaptationists value; and they would revise their role so that it is more in keeping with the assumption that because we cannot know everything, multiple answers are possible.

To privilege multiplicity in their own writing, Gould and Lewontin might look at a range of examples current in some feminist writing. Hubbard, for instance, points out that the Boston Women's Health Book Collective's *The New Our Bodies, Ourselves* includes the individual viewpoints of many women ("Science," 129). Brodkey's "The Discourses of Difference and Consensus" represents a dialogue between the voices of two feminists in an effort to distinguish between collaboration and coauthorship and to show how these feminists define terms and scholarship differently: "the collaboration seems only to have clarified the incompatibility of their rhetorics" (165). Yet Brodkey privileges neither. In *Baboon Mothers and Infants*, Altmann draws on "data-rich sophisticated analytical moves . . . [and] compellingly simple prose"—from narratives of apes like "Alice, who slept in contact with her [dead] mother's associate, Peter (probably not Alice's father)" and "Peter . . . a fully adult male, who was 'particularly gentle with infants' " to discussions of "sociological debates, . . . detailed description and figures summarizing data on demography, ecology and maternal time budgets, social relationship analysis . . . , infants' use of space, maternal style, weaning and independence, dominance position and economics of attention, and evolutionary models of parental investment" (Haraway, *Primate Visions*, 314–315). Haraway describes her own *Primate Visions* as "replete with representations of representations, deliberately mixing genres and contexts to play with scientific and popular accounts. . . . The argument in *Primate Visions* works by telling and retelling stories in the attempt to shift the webs of intertextuality and to facilitate perhaps new possibilities of the meanings of difference, reproduction, and survival" (377). In "Stabat Mater" (66) Kristeva presents two voices on the same page, side by side, as she explores the idea of the Virgin Mary. Her text is said to " '[swallow] goddesses and [strip] them of necessity,' [with] words written down two sections of the page in Derridean fashion like the printing of the synoptic gospels" (Todd, 55):

Head reclining, nape finally relaxed, skin, blood, nerves warmed up, luminous flow: stream of hair made of ebony, of nectar, smooth darkness through her fingers, gleaming honey under the wings of bees, sparkling	Very soon, within the complex relationship between Christ and his Mother where relations of God to mankind, man to woman, son to mother, etc. are hatched, the problematics of time similar to that of

strands burning bright . . . silk, mercury, ductile copper: frozen light warmed under fingers. Mane of beast— squirrel, horse, and the happiness of a faceless head, Narcissus-like touching without eyes, slight dissolving in muscles, hair, deep, smooth, peaceful colors.

cause loomed up. If Mary preceded Christ and he originated in her if only from the standpoint of his humanity, should not the conception of Mary herself have been immaculate? For, if that were not the case, how could a being conceived in sin and harboring it in herself produce a god?

A text like this, one that presents "a kind of incestuous challenge to the symbolic order" (Todd, 61), shows that a single writer can make multiple voices available almost simultaneously. As metaphor, Kristeva's essay invites us to question our conventional ideas of reader, writer, and text, and to consider her (and us) as multiple voices.

Making changes like these means, in effect, redefining both science and argument. By admitting multiple voices, science may recognize not only feminist perspectives but also other ways of doing science that have been considered marginal. Traditionally, argument has depended on oppositions between the writer and the intended audience, between true and false positions. This kind of text is clearly adversarial; opposites fight in a context of ideas, establishing "cognitive authority not only by demonstrating the value of one's own idea but also by demonstrating the weakness or error in the ideas of others" (Frey, 512). This aggressive stance is the accepted form of professional scientific discourse; even when it is sublimated by what Myers in chapter 13 of this book and elsewhere calls "the pragmatics of politeness," the goal of this discourse is to set up oppositions and to uncover weaknesses. An essay that privileges a feminist perspective in both form and content might offer connection as an end rather than repudiation, understanding rather than conflict, dialogue rather than debate; and it would celebrate the personal rather than the objective, plurality rather than singularity, self-conscious exploration of assumptions rather than easy acceptance of conventions.

What would science gain by opening itself up to this kind of science and this kind of writing? The scientific answer to any question has long been considered the "best" answer that current research has to offer. Although "there will always be limits and distortions to a . . . single-dimensional world-view" (Altmann, 193), alternatives are often ignored. If change occurs, it is generally because the "best" answer no longer fits the data, and a crisis results. Science continually revises itself, of course, but typically through conflict and domination. If scientists admit that they make knowledge together, adoption of a more open attitude toward alternatives should produce a healthier climate for revision and a richer science; "alternative versions of long-held scientific 'truths'

Mary Rosner and Georgia Rhoades

can be generated" when scientists are "open to different interpretations of . . . data and . . . [able] to ask questions that would not have
occurred within the traditional context" (The Biology and Gender Study
Group, 178–79). As Kolodny points out, "All the feminist is asserting
. . . is her[/his] own equivalent right to liberate new (and perhaps different) significances from . . . texts: and her[/his] right to choose which
features of a text she[/he] takes as relevant because she[/he] is, after all,
asking new and different questions of it" (185).

Though they still depend on many of the tactics of traditional argument and traditional science, Gould and Lewontin demonstrate a concern for this kind of openness, possibly because of their interest in
multiple disciplines. Similarly, as feminists and rhetoricians, we have
looked for connections between what Gould and Lewontin do as scientists and as writers, and what we know and do. Of course, we who
work outside the frame of science are still "trying to see."

3. *Chorus for Leaves*

James Dickey said it's only common courtesy
To know the names of the others we live among:
Cinquefoil, fire-pink, harbinger of spring.
Our names: we don't know
What they call themselves.

Grandmother Spiderwoman lay herself down on the earth
And chose a star to watch all night,
From rising in the West
To melting with the birdsong into the sun.
She knew as well the stars watched her,
And there are no fixed points how much we want them,
Just dizzy spinning distances that we write stories for,
To make them close.

When I finally caught up with Alice,
She was already way ahead of me,
Sitting in the ivy, polishing mushrooms.
She held one up to the light,
And inside I could see a peopled world.
"Ah," she said to it, "Sister."

Here on my planet
I am trying to listen

What my ears are not attuned to hear,
What lies beyond the veneer
We call significance.

NOTES

1. As Woolf wrote in *A Room of One's Own*, "It is obvious that the values of women differ very often from the values which have been made by the other sex; naturally, this is so. Yet it is the masculine values that prevail. Speaking crudely, football and sport are 'important'; the worship of fashion, the buying of clothes 'trivial.' And these values are inevitably transferred from life to fiction. This is an important book, the critic assumes, because it deals with war. This is an insignificant book because it deals with the feelings of women in a drawing-room" (76–77).

2. Bleier details how difficult it is to believe "that scientists' values, beliefs, and expectations can influence what they are actually able to see or hear with their perfectly functioning senses":

> For example, the leading microscopists of the 17th and 18th centuries, including the great van Leeuwenhoek, claimed that they had seen "exceedingly minute forms of men with arms, heads and legs complete inside sperm" under the microscope. Their observations were constrained not by the limited resolving power of the microscopes of the time, but rather by the 2,000-year-old concept, dating form the time of Aristotle, that women, as totally passive beings, contribute nothing to conception but the womb as incubator. Except for Japanese field workers, primatologists in the 1950s and 1960s could not see what female primates were doing; and even if they could see something, their hypotheses, observations, and interpretations were clearly constrained by the cultural concepts available. Attempts to explain female leadership, or dominance, or sexual aggressivity and initiative had to be accommodated within male-centered explanatory systems. ("Introduction," 3)

3. See also Messing, who explains how science has reflected social prejudices to privilege males in eleven categories: selection of scientists, their access to facilities for their work, choice of research topic, wording of hypotheses, choice of subjects, choice of controls, methods of observation, analysis of data, interpretation of data, publication of results, and popularization of results (75–88).

4. Haraway notes that, in biological discourse, adaptationism has been used to support sexism even though it is "not necessarily sexist": "in combination with assumptions of original female inferiority, it is a powerful tool to reproduce females as resources for male action—symbolically, scientifically, and socially" (*Primate Visions*, 322). We see the dinosaur playing this role in the abstract of Gould and Lewontin's article.

5. Although feminist scientists like Haraway often refer to science as "story-

Mary Rosner and Georgia Rhoades

telling"—"the life and social sciences in general, and primatology in particular, are storyladen; these sciences are composed through complex, historically
specific storytelling practices" ("Primatology," 81)—"stories" is used negatively in "The Spandrels of San Marco." In another context, Gould and Vrba
call these stories "unprovable reveries" (13). The problem lies in the imaginative filling in of a puzzle using one's unacknowledged bias. Multiple stories
(explanations) make accuracy/fairness/plausibility more likely.

6. Kitcher argues that Gould and Lewontin make an unfounded claim when
they suggest "that the unfalsifiability of adaptationist claims is an insuperable
obstacle": "Significant confirmation of adaptationist hypotheses is possible—
but only if biologists are prepared to take seriously all the forms of evolutionary scenarios that they admit as possible and prepared to undertake the
investigations necessary for articulating claims about allometry, pleiotropy,
and so forth" (232). Crediting John Beatty, Kitcher goes on to indicate that
Gould and Lewontin's objections could be met in two ways: "One might
encourage all biologists to develop and consider alternative forms of evolutionary explanation. Or one might foster a pluralistic community, in which
some biologists pursue adaptationist explanations, others concentrate on
developmental constraints, and so forth" (233). In their attention to plurality,
both alternatives seem more feminist than Gould and Lewontin's position.

7. Certainly Gould and Lewontin are recognized authorities in this field,
and we do not mean to suggest that their authority should be denied. But we
do want to make a distinction between "authoritative" and "authoritarian,"
the stance they adopt for much of the essay.

8. Feminist writing also often attacks patriarchal thinking that does not
consider or allow for marginal groups. So while feminism's goal may not
always be the reduction of antagonisms, it often has that effect because of its
respect for the coexistence of many views and its emphasis on how perspective informs those views.

9. Keller tells us that by the 1950s "by their own choice, [women's] tell-tale
first names were withheld from publications"—an act that "failed to protect
women from the effects of an increasingly exclusionary professional policy"
but did help "obscure the effects of that policy" ("Gender/Science," 36).

WORKS CITED

Altmann, Jeanne. *Baboon Mothers and Infants*. Cambridge: Harvard University
Press, 1980.
Belenky, Mary Field, Blythe McVicher Clinchy, Nancy Rule Goldberger, and Jill
Mattuck Tarule. *Women's Way of Knowing: The Development of Self, Voice, and
Mind*. New York: Basic Books, 1986.
Biology and Gender Study Group. "The Importance of Feminist Critique for
Contemporary Cell Biology." In *Feminism and Science*, ed. Nancy Tuana,
172–87. Bloomington: Indiana University Press, 1989.

Bleier, Ruth. "Introduction." In *Feminist Approaches to Science,* ed. Ruth Bleier, 1–17. New York: Pergamon Press, 1986.

Bleier, Ruth. "Lab Coat: Robe of Innocence or Klansman's Sheet?" In *Feminist Studies/Critical Studies,* ed. Teresa de Laurentis, 55–66. New York: Macmillan, 1988.

Bleier, Ruth. *Science and Gender: A Critique of Biology and Its Theories on Women.* New York: Pergamon Press, 1984.

Brodkey, Linda. "The Discourses of Difference and Consensus." In *Academic Writing as Social Practice,* 108–66. Philadelphia: Temple University Press, 1987.

Fausto-Sterling, Anne. "Women and Science." *Women's Studies International Quarterly* 4 (1981): 41–50.

Fee, Elizabeth. "Critiques of Modern Science: The Relationship of Feminism to Other Radical Epistemologies." In *Feminist Approaches to Science,* ed. Ruth Bleier, 42–56. New York: Pergamon Press, 1986.

Fee, Elizabeth. "Is Feminism a Threat to Scientific Objectivity?" *International Journal of Women's Studies* 4 (1981): 378–92.

Flax, Jane. "Postmodernism and Gender Relations in Feminist Theory." *Signs* 12 (1987): 621–43.

Frey, Olivia. "Beyond Literary Darwinism: Women's Voices and Critical Discourse." *College English* 52 (1990): 507–26.

Gilligan, Carol. *In a Different Voice: Psychological Theory and Women's Development.* Cambridge: Harvard University Press, 1982.

Ginzberg, Ruth. "Uncovering Gynocentric Science." In *Feminism and Science,* ed. Nancy Tuana, 69–84. Bloomington: Indiana University Press, 1989.

Gould, Stephen J., and Elisabeth S. Vrba. "Exaptation—A Missing Term in the Science of Form." *Paleobiology* 8 (1982): 4–15.

Haraway, Donna. *Primate Visions: Gender, Race, and Nature in the World of Modern Science.* New York: Routledge, 1989.

Haraway, Donna. "Primatology Is Politics by Other Means." In *Feminist Approaches to Science,* ed. Ruth Bleier, 77–118. New York: Pergamon Press, 1986.

Haraway, Donna. "Situated Knowledge: The Science Question in Feminism and the Privilege of Partial Perspective." *Feminist Studies* 14 (1988): 575–99.

Harding, Sandra. "The Instability of the Analytical Categories of Feminist Theory." *Signs* 11 (1986): 645–64.

Harding, Sandra. "Introduction: Is There a Feminist Method?" In *Feminism and Methodology: Social Science Issues,* ed. Sandra Harding, 1–14. Bloomington: Indiana University Press, 1987.

Hooker, C. A. *A Realistic Theory of Science.* Albany: State University of New York Press, 1987.

Hubbard, Ruth. "The Emperor Doesn't Wear Any Clothes: The Impact of Feminism on Biology." In *Men's Studies Modified: The Impact of Feminism on the Academic Disciplines,* ed. Dale Spender, 213–35. New York: Pergamon Press, 1981.

Hubbard, Ruth. "Science, Facts, and Feminism." In *Feminism and Science,* ed. Nancy Tuana, 109–31. Bloomington: Indiana University Press, 1989.

Keller, Evelyn Fox. *A Feeling for the Organism: The Life and Work of Barbara McClintock.* New York: W. H. Freeman and Co., 1983.

Keller, Evelyn Fox. "The Gender/Science System; or, Is Sex to Gender as Nature Is to Science?" In *Feminism and Science,* ed. Nancy Tuana, 33–44. Bloomington: Indiana University Press, 1989.

Keller, Evelyn Fox. "Making Gender Visible in the Pursuit of Nature's Secrets." In *Feminist Studies/Critical Studies,* ed. Teresa de Laurentis, 66–77. New York: Macmillan, 1988.

Keller, Evelyn Fox. *Reflections on Gender and Science.* New Haven: Yale University Press, 1985.

Keller, Evelyn Fox. "Women Scientists and Feminist Critiques of Science." *Daedalus* 116 (1987): 77–91.

Kitcher, Philip. *Vaulting Ambition: Sociobiology and the Quest for Human Nature.* Cambridge: MIT Press, 1985.

Kolodny, Annette. "Dancing through the Minefield: Some Observations on the Theory, Practice and Politics of a Feminist Literary Criticism." In *Feminist Literary Theory: A Reader,* ed. Mary Eagleton, 184–88. New York: Basil Blackwell, 1986.

Kristeva, Julia. "Stabat Mater." In *The Kristeva Reader,* ed. Toril Moi, 160–86. New York: Columbia University Press, 1986.

LeGuin, Ursula. "The Eye Altering." In *The Compass Rose,* 156–70. New York: Harper and Row, 1982.

Lewontin, Richard C. "Sociobiology: Another Biological Determinism." In *Home and Health,* ed. Elizabeth Fee, 243–59. Farmingdale, N.Y.: Baywood, 1982.

Longino, Helen. "Can There Be a Feminist Science?" In *Feminism and Science,* ed. Nancy Tuana, 45–56. Bloomington: Indiana University Press, 1989.

Longino, Helen. *Science as Social Knowledge.* Princeton: Princeton University Press, 1990.

Longino, Helen, and Ruth Doell. "Body, Bias, and Behavior: A Comparative Analysis of Reasoning in Two Areas of Biological Science." In *Sex and Scientific Inquiry,* eds. Sandra Harding and Jean O'Barr, 165–86. Chicago: University of Chicago Press, 1975.

Messing, Karen. "The Scientific Mystique: Can a White Lab Coat Guarantee Purity in the Search for Knowledge about the Nature of Women?" In *Woman's Nature: Rationalizations of Inequality,* eds. Marian Lowe and Ruth Hubbard, 75–88. New York: Pergamon Press, 1983.

Myers, Greg. "The Pragmatics of Politeness in Scientific Articles." *Applied Linguistics* 10 (1989): 1–35.

Namenwirth, Marion. "Science through a Feminist Prism." In *Feminist Approaches to Science,* ed. Ruth Bleier, 18–41. New York: Pergamon Press, 1986.

Rich, Adrienne. "Frame." In *The Fact of a Doorframe: Poems Selected and New,* 303–5. New York: Norton, 1981.

Rose, Hilary. "Beyond Masculinist Realities: A Feminist Epistemology for the

Sciences." In *Feminist Approaches to Science,* ed. Ruth Bleier, 57–76. New York: Pergamon Press, 1986.

Todd, Janet. *Feminist Literary History.* New York: Routledge, 1988.

Tompkins, Jane. "Fighting Words: Unlearning to Write the Critical Essay." *Georgia Review* 42 (1988): 585–90.

White, Hayden. "The Fictions of Factual Representation." In *Tropics of Discourse: Essays in Cultural Criticism,* 121–34. Baltimore: Johns Hopkins University Press, 1978.

Woolf, Virginia. *A Room of One's Own.* New York: Harcourt Brace Jovanovich, 1929.

6

READING DARWIN,

READING NATURE;

OR, ON THE ETHOS OF

HISTORICAL SCIENCE

CAROLYN R. MILLER AND

S. MICHAEL HALLORAN

In "The Spandrels of San Marco" Charles Darwin plays a central role. He is invoked most vividly as "master," "saint," "divinity" (155), invocations that are self-consciously ironized but nevertheless sincere. Gould and Lewontin are careful to disassociate themselves from certain details of Darwin's position on the mechanisms of evolution ("we do not now regard all of Darwin's subsidiary mechanisms as significant or even valid" [156]), but at the same time they associate themselves closely with his general "attitude" and "spirit" (156). It is, above all, his "pluralism" that is important to them, his refusal to be as rigid as contemporary adaptationists. Darwin is pluralist in part because he, like Gould and Lewontin, is willing to consider explanations for evolution other than natural selection: this pluralism is explicit. But we believe that there is another dimension to Darwin's pluralism that is central to "The Spandrels of San Marco," which is that his work is open, interpretable. Darwin is the prize for which evolutionary biologists compete. The adaptationists claim to be strict Darwinians, but Gould and Lewontin claim to be *more* Darwinian by being truer to the spirit if not the letter of his work. Darwin must be read and reread, interpreted and reinterpreted. We find this attention to a body of work that is well over a hundred years old to be highly unusual and worth investigating.

Work on rhetoric of science has to date paid relatively little attention to the ways in which intellectual predecessors figure in current work. Halloran's study of Watson and Crick remarks incidentally on their treatment of previous work, noting, for example, the excessive and possibly ironic civility in their discussion of Pauling, and the slight they offer to Erwin Chargaff by downplaying his responsibility for establishing the chemical ratios in DNA that form such an important part of their argument. But as a general aspect of scientific rhetoric, the relationship of research and theory to intellectual forebears has been neglected.

This neglect is perhaps understandable, in that the sorts of scientific texts that have been studied most often by rhetoricians—and by philosophers and sociologists of science as well—tend not to dwell on their intellectual debts. Garfield discusses the phenomenon of "obliteration by incorporation," by which a theory or method in relatively short order becomes so much a part of the taken-for-granted that it need not be cited and so is not cited ("Obliteration"), and Kuhn's notion of the dynamic of "revolutionary" and "normal" science offers a rationale for why this should happen. Watson and Crick, for example, make no reference to the crucial study, published just nine years before their own announcement of the structure of DNA, in which Oswald Avery et al. demonstrated that DNA is the substance responsible for transmission of genetic traits from one generation of cells to the next. And while in his popular memoir, *The Double Helix*, Watson acknowledges the crucial importance to their work of Erwin Schroedinger's *What Is Life?* no such acknowledgment appears in any of the papers in which he and Crick present their work to scientific colleagues (Halloran, "Birth"). In both cases, as ideas became "paradigmatic," citing them explicitly became so unnecessary that to do so would have seemed gauche and naive.

Another form that the relationship to intellectual forebears takes in science has been called the "Max Planck Effect" (Harris, 286). This relationship is the Oedipal one, in which the younger generation finds it necessary to kill its fathers by rejecting a prevailing paradigm and replacing it with its own, revolutionary one. Kuhn has remarked on the specifically generational nature of much scientific change (151–52), as has Toulmin: "Each new generation of apprentices, while developing its own intellectual perspectives, is also sharpening up the weapons for an eventual professional takeover" (287). Watson and Crick's treatment of the famous but mistaken Linus Pauling is a case in point, as is their slight of Chargaff.

"The Spandrels of San Marco" presents a striking counterexample to

Carolyn R. Miller and S. Michael Halloran

both "obliteration by incorporation" and the Max Planck effect. The average age of the forty-one works cited by Gould and Lewontin exceeds twenty-one years, and four of them date to the nineteenth century. Darwin, neither obliterated nor ostentatiously killed off, is instead cited frequently in an almost reverential tone. Indeed, a major burden of Gould and Lewontin's argument is simply to establish that their own views are in agreement with those of "the master." They seek to liberate a venerable text from the alleged misinterpretations of the "adaptationist programme" by reasserting the validity of Darwin's "pluralist approach," an approach that, not very surprisingly, turns out to be in agreement with that of Gould and Lewontin.

The distinctive relationship to intellectual forebears established by "Spandrels" is characteristic of a mode of science that has to date not received much attention from rhetoricians, a mode designated by Gould (among many others) as "historical science," in contrast to such "experimental–predictive" sciences as physics, chemistry, and molecular biology. Our purpose in this essay is to examine "Spandrels" as illustrative of this historical mode. Our approach derives from classical rhetoric, with particular attention to the Aristotelian notion of ethos. We assume that the relationship between a particular scientific work and its intellectual forbears is central to the ethos of the work, in that it is an articulation of the relevant intellectual community. We attempt to show that, in contrast to works of the experimental–predictive sciences, "Spandrels" privileges a specific set of texts and claims authority through the process of interpreting them. We argue that, considered as an exemplar of historical science, "Spandrels" establishes an ethos within which conformity with the works of Darwin is crucial, and that this move makes the historical sciences akin to what Foucault calls a "field of discursivity," and in this respect problematizes the very concept of "science," and hence "rhetoric of science."

The historical mode of science as a distinctive enterprise has occupied both Gould and Lewontin, separately, in their other writings. Both are concerned to distinguish methods appropriate for studying organisms as integrated wholes with unique histories from methods used for studying repeatable phenomena that can be decomposed into separately manipulable subunits. Gould calls the method appropriate to evolutionary science "historical," and Lewontin calls it "dialectical." In "Spandrels," they are concerned to show how adaptationist history is inferior to a better kind of history, which somehow must distinguish "proper explanations" from "the substantial set of plausible pathways to any modern result" (154).

Both in a 1986 essay in *American Scientist* and in *Wonderful Life* (1989)

Gould makes a strong case for distinguishing the historical sciences from the experimental–predictive sciences, for history as "the methodology for an entire second style of science" ("Evolution," 60). His argument rests on both definition and authority. The definitional argument proceeds by examining the objects of scientific inquiry and emphasizing that methods of "experiment, quantification, repetition, prediction, and restriction of complexity to a few variables that can be controlled and manipulated" (*Wonderful Life,* 277) cannot "get to the heart of" questions like why the dinosaurs died, a "singular event involving creatures long dead on an earth with climates and continental positions markedly different from today's" (278). Questions concerning the evolution of the geosphere and the biosphere are questions about history—unique, unrepeatable events subject to historical contingency, events that experimental–predictive reasoning cannot account for. For this reason, of course, many philosophers and scientists have believed that history cannot be scientific.

The authority invoked to support this dichotomy of the sciences is Darwin himself, who, Gould claims, "was, above all, a historical methodologist" ("Evolution," 60). It was Darwin, Gould claims, who "made evolution doable" (62) by showing that history is "knowable" through a set of methods that can produce research. "Darwin's singular greatness [lies] in his extended campaign to establish a scientific methodology for history—to make history doable for the zealous researchers of science. Darwin was, more than anything else, a historical methodologist" (61). Similarly, Gould claims that the theory of evolution itself has methodological as well as substantive import: "evolution's essential impact upon the practice of science has been methodological—validating the historical style as equally worthy and developing for it a rigorous methodology" (69).

Darwin's method, according to Gould, rests on two principles: "first, the uniformitarian argument that one should work by extrapolating from small-scale phenomena that can be seen and investigated; second, the establishment of a graded set of methods for inferring history when only large-scale results are available for study" ("Evolution," 61).[1] These methods of inference are three—each, Gould says, exemplified in one or more of Darwin's technical works (on worms, coral reefs, orchids, and the like): first, the scientist can extrapolate from evidence of modern, observable processes of change into the past; second, lacking such evidence, the scientist can arrange existing evidence of results (such as fossils) into hypothesized stages, which may be corroborated by other similar evidence; and third, when all other evidence fails, the scientist can infer history from the imperfections in single objects, such as vestigial organs, which can reveal historical relationships that are obscured by more perfectly functional adaptations (Gould has called this last the "panda principle") ("Evolution," 62–63).

Lewontin's views are consistent with Gould's. In a 1969 essay, "The Bases of Conflict in Biological Explanation," he criticizes as "false" the dialectic established between two kinds of biological explanation taken to be distinct and incompatible, the "molecular" and the "evolutionary." Later, in *The Dialectical Biologist*, written with Richard Levins, Lewontin focuses on the organism understood synchronically as well as diachronically, that is, as part of an evolving system. The principles of dialectical method, as Levins and Lewontin describe it, are that the whole is a relation of parts that have no existence apart from the whole; that subject and object, cause and effect are interchangeable; that apparently mutually exclusive categories interpenetrate; that change is inherent in all systems; that organisms and environments are hierarchically ordered systems (273ff). Levins and Lewontin present their book as a brief for this "epistemological stance," which they see as an alternative intellectual tradition to "mechanistic Cartesian reductionism" (vii).

The status of historical–dialectical science relative to Cartesian–experimental science is an important issue for both Gould and Lewontin: Gould addresses it in a characteristically polemic way, Lewontin more obliquely. Each laments the statement by Luis Alvarez (the physicist who first offered evidence that the dinosaurs could have become extinct as a result of a meteor collision) that paleontologists are more like "stamp collectors" than like real scientists (*Wonderful Life*, 281; Levins and Lewontin, 287).[2] In *Wonderful Life*, Gould offers "a plea for the high status of natural history" (280), urging that we "understand the different forms of historical explanation as activities equal in merit to anything done by physics or chemistry" (281). He complains that "the worst of human narrowness pours forth in the negative assessment of monographic work [publication of detailed taxonomic description]. . . . I can't imagine an activity further from simple description as the reanimation of a [fossil] organism. . . . Why do we downgrade such integrative and qualitative ability, while we exalt analytical and quantitative achievement?" (*Wonderful Life*, 100). In his review of Gould's book for the *New York Review of Books*, Lewontin accuses "biologists, even evolutionary biologists" of "physics envy," of trying "to construct their science as a set of universal 'laws,' acceding to the general intellectual disdain for the merely particular" (3). Elsewhere he says, "We must reject the molecular euphoria that has led many universities to shift biology to the study of the smallest units, dismissing population, organismic, evolutionary, and ecological studies . . . and allowing museum collections to be neglected" (Levins and Lewontin, 287).

Clearly, then, both Gould and Lewontin are persuaded that there are at least two quite distinct forms of science, that one—theirs—has been occluded by a hegemonic science modeled on physics, and that the methods and modes of that other science are inappropriate to their science. The "Spandrels" essay, in fact, focuses on their concern for appropriate method. Its central point is the methodological inadequacy of the adaptationist program (150). They fault it for presenting unfalsifiable explanations (155); for treating organisms as collections of discrete "traits," rather than "integrated entities" (151); for dogmatically refusing to consider any mechanism of evolutionary change other than an omnipotent natural selection (150–51). They present a set of methodological alternatives to rigid adaptationism, advocating forms of explanation that are sensitive to historical contingency, respectful of the complexity of organisms and their environments, and attentive to the interactions between levels of organization (156–59). The general effort is to make evolutionary thinking more flexible and more responsible, more historical and less mechanistic, and thereby more appropriate and distinctive.

But Gould and Lewontin are far from the first to argue that evolutionary biology[3] is closer to history than to physics, and the dissatisfaction with the use of physics as a methodological exemplar for all science is widespread among evolutionary biologists. Ernst Mayr, perhaps the greatest evolutionary biologist of the twentieth century, in *The Growth of Biological Thought* cites the nineteenth-century German naturalist Ernst Haeckel as the first to resist the use of quantitative physics as the normative model for all science and to claim that evolutionary biology is fundamentally historical (70). Evolutionist George Gaylord Simpson complained in 1964 that "considerations of the history, methods, and nature of science have been heavily biased by concentration on physical science and not on science as a whole" (92). Mayr himself refers to the "hubris of the physicists" (*Growth*, 33) and concludes that "physics is no longer the yardstick of science" (42). That there is now a "philosophy of biology" distinct from the general philosophy of science, with a journal and a growing literature (Hull, *Philosophy* and "Darwinism"; Mayr, *Growth*, 73–76), suggests the strength of this movement.

In fact, the major methodological division in the sciences may not really be between the physical and the biological sciences, but between the functional or experimental sciences and the evolutionary or historical sciences: some physical sciences are historical (geology, cosmogony), some biological sciences are functional and nonhistorical (molecular genetics, immunology).[4] As early as 1961 Mayr suggested that there are two kinds of biology: functional biology, concerned with proximate causes, with finding out "how" something happens; and evolutionary biology, concerned with ultimate causes, or "why" something came to happen as it did. In the same year, Goudge's *Ascent of Life* explored the forms of explanation available to evolutionary biology, forms that are, he claimed, both causal and historical without being law-like or predictive and so not consistent with traditional "covering-law" models of scientific explanation. And although the term "natural history" had earlier been discarded as a quaint nineteenth-century artifact in favor of modern disciplinary terms such as cytology, genetics, malacology, paleobotany, and the like, it is making a comeback (Gould, *Wonderful Life*, 280; Lewontin, "The Corpse," 34; Mayr, *Growth*, 67; O'Hara, "Homage to Clio," 153).

The characterization of evolutionary biology as a "historical science" has recently been called a "commonplace" (O'Hara, "Homage to Clio," 142). What are the characteristics of this kind of science? First, as Gould emphasizes, its objects or questions concern unique and unrecoverable events. Second, it uses distinctive types of evidence and modes of inference. For example, it relies heavily on thick description (Gould, *Wonderful Life*, 100; Gross; Mayr, *Growth*, 70), depends on sufficient rather than exact explanation (Lewontin, "Bases of Conflict"), and uses heuristic rules of thumb for inferring the past, rather than covering laws to deduce it (Mayr, *Growth*, 37–42; Simpson, 128; Hull [*Metaphysics*, 191] cites a source listing one hundred such rules for reconstructing phylogeny). Third, it relies on a set of distinctive concepts for interpreting evidence. Some of these are the focus on variation and diversity rather than essence (de Queiroz; Lewontin, "Is Nature Probable" and "Darwin's Revolution"; Mayr, *Growth*; O'Hara, "Homage to Clio") and the related population thinking (Mayr, *Growth*, 45–46), historical-process or "tree thinking" (de Queiroz; O'Hara, "Homage to Clio," 150; Sober, 6), teleology (Hull, *Philosophy*, 101–24; Mayr, *Growth*, 47–51), "emergence" and hierarchy (Hull, *Philosophy*, 125–41; Lewontin, "The Corpse"; Mayr, *Growth*, 63–66), similarity (Dawkins; de Queiroz, 248–50; Gould, *Wonderful Life*, 213; Lewontin, "Fallen Angels"; Sober, 13ff), and contingency (Gould, *Wonderful Life*, 284–85; Simpson, 133).

Historical science offers explanations that are narrative and sufficient rather than predictive and necessary. Goudge made the argument that evolutionary biology does not explain phenomena by reference to "covering laws" that permit prediction but, at least in part, by providing historical narratives: "Narrative explanations enter into evolutionary theory at points where singular events of major importance for the history of life are being discussed" (71); "Narrative explanations are constructed without mentioning any general laws" (75). And recently Hull has argued that historical narratives explain not by fitting an instance under a general law but by integrating an event into an organized whole (*Metaphysics*, 200; also Simpson, 133–34); Mayr holds that narrative explains by providing a possible chain of causes (*Growth*, 72; also O'Hara, "Homage to Clio," 149). Serious attention is being paid to the relevance of the philosophy of history and narrative theory for the reasoning and presentation of work in the field, especially to the need for narratives to have narrators with perspectives and subjects with origins and endings (Hull, *Metaphysics*, 181–204; O'Hara, "Telling the Tree").[5]

We would add to this characterization several additional features specifically pertaining to the rhetoric of evolutionary biology, features that are highlighted by the "Spandrels" essay. First, as we initially noted, the essay bears a distinctive relationship to its own antecedent discourse: that is, it has historical depth, making use of its own discursive history in a way that is not characteristic of experimental science. Second, it is preoccupied with interpreting a set of master texts in a way that seems to us unique among the sciences. And finally, all these features are revealed and held together by a distinctive rhetorical ethos.

We have noted the way that "Spandrels" invokes antecedent discourse in evolutionary biology. We would like to generalize this relationship beyond the single article by Gould and Lewontin. Studies of citation patterns in the general scientific literature have suggested some norms against which to check our impressions about evolutionary biology. Derek de Solla Price suggests that there are two "populations" of citations: one, a uniform raiding of the entire archive, in which one would expect as many citations to a work published two years prior to the citing as to a work published twenty years prior; the other, a "hyperactivity" of the recent literature, showing what he calls the "research front." Thus, purely archival referencing, with no use of a research front, will yield a proportion of references that are five years old or less ("Price's Index") of 22 percent to 39 percent, depending on how fast the literature is growing (22 percent for growth of 5 percent a year [doubling time of fourteen years], which Price finds about average, and 39 percent for the most rapid growth of about 10 percent a year [doubling time of seven years]). A rate greater than 39 percent will probably reflect research front referencing. In a pilot investigation of 154 scientific and nonscientific journals (originally published in 1970, although no date is given for the journals examined), Price found a mean Price's Index of 32 percent. In the upper ranges of the index, higher than 60

Carolyn R. Miller and S. Michael Halloran

percent, were journals in the highly competitive "hard" sciences, such as chemistry, biochemistry, and physics; in the lower ranges, he found some journals in history, philosophy, and criticism with indices of 10 percent or less, indicating a "negative research front," or proportionately greater use of older rather than recent source material. Although his investigation did not include any journals in evolutionary biology, Price suggests: "Among the sciences, I think one would find similarly that the taxonomic sciences . . . would also display an anomalous appearance of a negative amount of research front" (177). Incidentally, in referring to this pattern as "pathological," he confirms the general defamation of the historical sciences.

A citation measure that provides more historical depth than Price's Index is that of A. J. Meadows, who examined the astronomical literature for any difference in the "rate of decay" of citations in two subfields: one subfield emphasizes theory and resembles physical science, while the other emphasizes observation, including historical observation, and, he says, resembles biological science in this respect. Meadows calculated an "immediacy index" by dividing the number of citations to literature of the prior six years by the number of citations to literature more than twenty years old.[6] He found that the physical science subfield had an immediacy index of 44.1, and the observation-based subfield had an index of 1.2.

We were curious to see whether such a distinction could be shown between historical and functional biological sciences. By comparing the citation patterns of selected journals in biochemistry and molecular biology with those in evolutionary biology,[7] we wanted to determine whether the pattern we had noted in "Spandrels" is idiosyncratic or representative of a particular style of biological science that could be distinguished from the more voluminous and better-funded molecular biology; we examined 1990 volumes of each journal (or, in the case of the much larger molecular biology journals, a selection from the 1990 volume to give a total number of citations in the same range as that in the evolutionary biology journals). We found that the Meadows "immediacy index" ranged from 13.05 to 43.5 in molecular biology and from 1.9 to 4.1 in evolutionary biology,[8] suggesting that indeed these two disciplines do have very different relationships to their own prior discourse, with molecular biology operating primarily on a "research front" and evolutionary biology maintaining greater contact with its own history.

One additional feature of the citations in evolutionary biology should be noted, a feature that neither the Price nor the Meadows measure can show. The samples from all four evolutionary journals included citations to nineteenth-century material (the percentage of the total num-

ber of citations ranged from 0.4 percent to 2 percent), and all four included at least one citation to Darwin's original works (0.05 percent to
0.3 percent of the total citations). This pattern is consistent with what
we see in "Spandrels." None of the molecular biology journals, of course,
included any nineteenth-century references. Eugene Garfield has documented a steadily increasing number of citations to Darwin's *Origin of
Species* over the period 1955–84 ("The Articles Most Cited," 192–93).[9]
Some of the use of Darwin is no doubt due to the "Darwin industry"
created by historians of science and Victorianists and to the Darwin
centennials in 1959 and 1982 (see Lenoir). There is a local peak of citations in 1959 and another in 1981, but the greatest increase in Garfield's
graph takes place in the period from 1975 through 1978, and his figures
come from the *Science Citation Index*, not the *Arts and Humanities Citation
Index*.

Although Darwin's work is foundational to evolutionary biology—
definitive of the methods, the "view of life," the objects of investigation—his texts are also understood as inherently problematic. Mayr
once explained that Darwin actually had at least five "theories of evolution"[10] that have been differentially accepted and rejected by evolutionists in the later nineteenth century and even in the twentieth century;
Darwin's thought, he maintains, was flexible, complex, and occasionally
inconsistent ("Darwin's Five Theories," 755–57). Hull also describes
the complexity of "Darwinism" and the contemporary disagreement
about what it is, finding in the biological literature an "amazing variety" of opinions about what the essential concepts of Darwinism are
("Darwinism," 773), in both its original and its twentieth-century (neo-
Darwinist or "new synthesis") forms, and also finding a fundamental
"pluralism" in Darwin's own thought (804). Gould has been one of the
important proponents of this view as well, in other work besides "Spandrels." For example, in the journal *Paleobiology* he has claimed that "Darwinism, as a set of ideas, is sufficiently broad and variously defined to
include a multitude of truths and sins" (119); and in *Science* he has
called Darwin a "pluralist," noting the early twentieth-century "battle
for rights to the name 'Darwinian'" (380). Lewontin has noted that
"Darwin's work is filled with ambiguities, contradictions, and theoretical revisions" ("Darwin's Revolution," 25).

It is in fact not difficult to find even in the titles of the evolutionary
biology literature a fairly large number that illustrate the extent to which
Darwin is a subject of discussion and Darwinism a matter of contention.[11] Quotations from Darwin are often used as epigraphs or within
the first paragraph of an article.[12] The debate over punctuated equilibrium ultimately centered on whether it was a distinctly new idea or

Carolyn R. Miller and S. Michael Halloran

not, that is, on whether it was non-Darwinian or neo-Darwinian.[13] Throughout the literature, positions are routinely characterized as Darwinian, neo-Darwinian, anti-Darwinian, or non-Darwinian. Lewontin has remarked, "We are regularly treated to 'new' theories about evolution that claim in one way or another to renovate Darwinism without actually overturning it" ("Fallen Angels," 6). An important strategy in evolutionary argument might be characterized as the "more Darwinian than thou" *topos*.

"Darwinism" as a term and as a field of concepts is such a presence in this literature that it might seem that Darwinism itself is the discipline, rather than evolutionary biology, or paleontology, or zoology. One observer has noted: "Darwinism is much more than a scientific theory. . . . There are no such things as Maxwellism or Einsteinianism. Only historians speak of Copernicanism or Newtonianism. But there is a Darwinism, in the same way that there is Freudianism or Marxism" (Roger, 814). Lewontin has made a similar observation: "Darwin's writings have a great deal more in common with those other grand theorists of the nineteenth century, Marx and Freud, than with, say, Newton" ("Darwin's Revolution," 25).

The comparison with Marx and Freud is suggestive, since Foucault has characterized both as a distinctive kind of author that appeared in Europe in the nineteenth century. In "What Is an Author?" he calls these two figures unique, in that "they are not just the authors of their own works. They have produced something else: the possibilities and the rules for the formations of other texts" (114). These possibilities include not only analogies and continuities but also divergences that arise from the original works. Foucault calls these authors "founders of discursivities" (114) and he distinguishes such discursive structures from those of science. In science subsequent discourse is "on an equal footing" with the founding texts: the founding act can be superseded by new texts, and the validity of a statement is judged against the current knowledge of the discipline. In contrast, the founding text in a "discursivity" is "heterogeneous" to its subsequent transformations: it has a privileged status. The validity of a statement is defined in relation to the founder, not in relation to the discipline: "one does not declare certain propositions in the work of these founders to be false; instead, . . . one sets aside those statements" in various ways (116).

Foucault does not mention Darwin, and indeed Darwinism has not played the role in continental philosophy and social thought that Marx and Freud have. But we see striking similarities between his description of how Marx and Freud have come to their positions of power in such thought and our description of how Darwin has been rhetorically appropriated in current evolutionary biology. The comparison makes evolutionary biology even more distinctly different from experimental–Cartesian science; it suggests that the better comparison might be with a general model for the human sciences, a model not yet rhetorically articulated, to our knowledge. Foucault's concern was to characterize a particular kind of authorship, or authority, achieved by influential nineteenth-century writers, an authorship that seems to direct, or guide, subsequent writers in an unprecedented way. This being the case, we might expect to find that the image of the author in evolutionary biology will be controlled by Darwin, or at least indebted to him, in a way that is characteristic of a "discursivity." To examine this hypothesis, we look at the rhetorical dimension of ethos in "Spandrels."

The ethos established by Gould and Lewontin is, most obviously, that of the openly argumentative critic. They set up an adversary called "the adaptationist programme," expecting their readers to recognize it as "a deeply engrained habit of thinking among students of evolution" (150). From the opening sentences of their abstract, they adopt a combative tone, using sharply pointed active-voice verbs ("We criticize this approach . . ."; "We fault the adaptationist programme . . . ") to establish a clear dichotomy between a wrongheaded adaptationism that is "based on faith" and their own "pluralistic approach," which, not at all incidentally, is also the approach of Darwin (147). In addition to the accepted "scientific" method of logical argument, they employ wit and ridicule—weapons more often associated with the polemicist—most obviously in the extended architectural analogy, the acerbic citations of Voltaire, and the sarcastic puns ("an unusual way to beef up the meat supply" [150]) of the introductory section, but throughout the rest of the paper as well. (See, for example, the sarcastic use of the imperative mood in the catalogue of styles of adaptationist argument [152–53].) They do not hesitate to make sweeping and dogmatic claims, as for example: "Some evolutionists may regard [the question 'What is a trait?'] as a trivial, or merely a semantic problem. It is not" (151); or, "Darwin has often been depicted as a radical selectionist at heart. . . . This view is false" (155). In these and other examples, they foreground their own role as interpreters of "the master," advocates of their own interpretations, and arbiters of the work of others.

But while consistently harsh, their critical tone retains a high level of civility, owing in part to the humorous effect of their satirical tactics, but also, more importantly, to an asymmetry in the degree to which the argument is personalized. On the side of the Darwinian "pluralists," the argument is identified personally with Gould and Lewontin themselves. As we noted earlier, they use active-voice verbs and the con-

Carolyn R. Miller and S. Michael Halloran

comitant first-person pronouns in stating their view. Indeed, the frequency with which the pronoun "we" appears in this paper is quite striking, and it suggests a strong willingness, even an eagerness, on the part of the authors to place their own professional reputations squarely behind the position they espouse. But on the side of the adaptationists, Gould and Lewontin avoid personalizing the argument, citing relatively few specific instances of fallacious adaptationism and placing these in subordinate positions as illustrations of a habit of thought they consider widespread and even admit to being guilty of themselves ("We all say that not everything is adaptive; yet, faced with an organism, we tend to break it into parts and tell adaptive stories" [152]).

Their strategy is a risky one in that it depends radically on their readers' assent to a highly generalized depiction of something that is itself a high-level abstraction, a "habit of thinking" shared by a large number of evolutionists, presumably a large majority of them, though Gould and Lewontin are imprecise on just how widespread the fallacy of adaptationism is supposed to be. The strategy requires, furthermore, that their readers admit to having indulged in the wrong-headed habit themselves. But by taking this risk and the further one of placing themselves so clearly on the line in this dispute, Gould and Lewontin manage to remain consistently civil while advancing a strenuous polemic. By pointing the accusatory finger not at specific individuals, but at a generalized "us," they avoid seeming overly churlish or mean spirited.

The polemical tone is also tempered throughout by a playful, urbane, and literate quality. By "literate," we mean to invoke the Victorian image of the "man of letters." Like a Ruskin or Carlyle, Gould and Lewontin display an easy and confident familiarity with a range of literature and art, one that rests primarily on a display of their own refined taste rather than on expert testimony or arcane expertise. Thus, in their opening discussion of St. Mark's Cathedral and the King's College Chapel, they cite no authorities on architecture or art history. When they tell us that it was architectural constraint and not the desire for ornamentation that produced the spandrels of San Marco and the interstices of the King's College Chapel ceiling, the strength of their assertion rests upon their own authority as close and informed observers of the architecture itself, not upon the learned authority of experts. In this respect, they are like that other Victorian man of letters, Darwin himself, the consummate close observer of the world around him. Overall, they present themselves as persons of broad experience and learning, and the witty allusions to Voltaire bolster their claim to this sort of generalized cultural authority.[14]

It should be noted, however, that Gould and Lewontin's claim to authority on architecture is not particularly well founded. The notion that architectural constraint is primary and "ornamentation" an epiphenomenon is by no means the universal principle they suppose it to be. Indeed, the understanding of the San Marco mosaics and King's College Chapel designs as "mere" ornaments is the product of a modernist aesthetic that Gould and Lewontin unconsciously universalize. The role of what to the modernist is extraneous decoration in premodern art and architecture is by no means as simple and lowly as they assume. As Frances Yates shows in *The Art of Memory,* the provision of specifically dimensioned spaces for "ornamentation" was often a major consideration in architecture of the classical, medieval, and renaissance periods. The fan vaulting of the King's College Chapel was, to be sure, a means of holding the roof up, but that it provided spaces for "decorations" encoding specific ideological meanings may well have been a consideration of major importance in the selection of that particular means of support.

Even the objection that their extended trope may in the strictest sense be "wrong" does little damage to either the argument or the ethos developed by Gould and Lewontin. Their argument against adaptationism does not rest in any substantive way upon the analogy with architecture, and what matters about it with respect to ethos is not whether it displays accurate expert knowledge. Rather, it shows the authors to be persons of wit and broad learning, willing to venture beyond their narrow specialty and able to make connections in an elegant style. In this sense, as Gragson and Selzer also note in chapter 10 of this book, the very title of their paper, with its balanced syntax and dual alliterations, is a significant contributor to their ethos.

Such considerations as stylistic elegance in "Spandrels" are important because in the essay Gould and Lewontin themselves become something like story-tellers, and a compelling voice is crucial to the success of any narrative. The narrative impulse of the paper is most clearly reflected in the organization of part 2, "The Adaptationist Programme," around a temporal sequence. The "Panglossian Paradigm," we are told, "is rooted in a notion popularized . . . toward the end of the nineteenth century" (150). And further, "studies under the adaptationist programme generally proceed in two steps" (151). An objection is both framed and met in the language of temporally situated and sequenced events:

> *At this point,* some evolutionists will protest that we are caricaturing their view of adaptation. . . . In natural history, *all possible things happen sometimes;* you generally do not support your favoured phenomenon by declaring rivals impossible in theory.

> Rather, you acknowledge the rival, but circumscribe its domain of
> action so narrowly that it cannot have any importance in the affairs
> of nature. *Then,* you often congratulate yourself for being such an
> undogmatic and ecumenical chap. We maintain that alternatives to
> selection for best overall design have generally been relegated to
> unimportance by this mode of argument. *Have we not all heard* the
> catechism about genetic drift. . . ." (151–52, emphasis added)

Here we find both narrative and explicit argument working together;
as Susan Wells demonstrates in her analysis of "Spandrels," similar
language mixing narrative and argument can be found throughout the
paper. Although scientific argument is the primary structure, as we
noted earlier, the habit of thought Gould and Lewontin seek to correct
is sketched as a *story,* as is the "pluralist approach" they attribute to
Darwin.

All of this casts something of a shadow on what appears to be their
chief accusation against adaptationism, that it is a compulsion to "tell
stories" about evolution. If its penchant for telling stories is all that is
wrong with adaptationism, Gould and Lewontin are hoist with their
own petard, for they are as much story-tellers as any adaptationist. In
fact, they do appear in the section headed "Telling Stories" to be applying a criterion of falsifiability to the adaptationists, a move that would
seem to place them in the camp of experimental–predictive science and
deny that they themselves are story-tellers too. Stories, they claim, do
not make good evolutionary explanations because there is no principled way to reject them on the basis of evidence (153) and because the
criteria for acceptance are so "loose" that a plausible story "can always
be told" (154). But of course there is more to the criticism than that. The
real problem with adaptationism is that all its stories have the same
monotonous plot: "the rejection of one adaptive story usually leads to
its replacement by another, rather than to a suspicion that a different
kind of explanation might be required. Since the range of adaptive
stories is as wide as our minds are fertile, new stories can always be
postulated (153)."[15] In their discussion of architecture, Gould and Lewontin themselves commit this very error: they take one interpretive
procedure and absolutize it. They offer a single story that is supposed
to account for the relationship between architectural form and ornamentation in all cases. Had they applied their own principles, they
would have recognized the possibility of different interpretations, different *stories,* different relationships between architecture and ornament.

Nowhere in this essay do Gould and Lewontin define "story" precisely, but the implication is that it is an expedient account of history

that "explains" because it supports the adaptationist dogma of optimal adaptation. Yet in his later work Gould, at least, takes a very different approach to story-telling, making narrative central to historical science as he has come to conceive of it: "History is the domain of narrative— unique, unrepeatable, unobservable, large-scale, singular events" ("Evolution," 61; see also *Wonderful Life*, 277, 283). The accounts that good historical scientists give, then, are better not because they aren't stories but because their stories recognize "the central principle of all history— *contingency*" (*Wonderful Life*, 283), and the dependence of any result on an "unpredictable sequence of antecedent states," not on "direct deductions from laws of nature" (283). And Gould comes to characterize the work of historical narration as interpretation. He presents *Wonderful Life* as an account of the difficulties of interpreting fossils, specifically of the initial "misinterpretation" of a group of fossils and of their subsequent "reinterpretation."

"Spandrels," we suggest, represents a step toward the full articulation of the principles of historical science. It suffers from the ethical contradiction of leveling against adaptationism an accusation to which it is equally vulnerable. The contradiction would eventually be resolved by the realization that historical science must be historical, that is to say, based in narration. But in a not yet self-conscious way, "Spandrels" illustrates the narrative method of historical science, part of which is the ethos adopted by Gould and Lewontin, with its emphasis on the personal role of the speaker as advocate, interpreter, and story-teller.

Ethos is a complex phenomenon that to be understood adequately must be seen from two apparently distinct vantage points. On the one hand, it is the distinctive "voice" of an individual or a tightly knit collaborative team. On the other, it is the spirit or group character of a broader community of speakers. What makes ethos one concept rather than two is the fact that the individual voice is always heard and interpreted against the background of the group character that gives it "authority," while the group character is, conversely, at stake in the performance of the individual. A voice achieves ethos by making present the spirit or character of a community, but it necessarily enacts "the community" in a distinctive way.[16]

"The Spandrels of San Marco" represents an ethos at a moment of inchoateness. Gould and Lewontin seem to be groping toward an ethos for historical science, on the one hand claiming the authority of an experimental–predictive science they would later define as alien to their enterprise, on the other identifying themselves both with and against adaptationism and assuming the narrative voice even as they reject "story-telling"—a strategy Gould would later recognize as central to

Carolyn R. Miller and S. Michael Halloran

historical science. The crucial figure in the full emergence of this ethos is Darwin himself, who described himself as an interpreter in *On the Origin of Species*: "I look at the natural geological record, as a history of the world imperfectly kept, and written in a changing dialect; of this history we possess the last volume alone, relating only to two or three countries. Of this volume, only here and there a short chapter has been preserved; and of each page, only here and there a few lines" (310–11). If, as Gould has claimed, Darwin is the consummate historical methodologist, the ethical task of evolutionary biologists may be simply to become as much like Darwin as possible. He illustrates for Gould and Lewontin both the importance and the difficulty of interpreting "the book of nature," and hence the inadequacy of pure adaptationism with its single plot-line. His work is a model of the sort of pluralistic and flexible interpretation they advocate, but paradoxically it becomes subject to this same process of interpretation.

The contention over Darwinism that is central both to "Spandrels" and to much of the recent literature in evolutionary biology poses a question about the nature of historical method. Is the debate to be understood as moving toward a determinate answer, toward rejecting "misinterpretation," toward settling finally the matter of what Darwin "means"? Or is the debate to be understood as a continuing one that constitutes historical science? If the first understanding is correct, if the debate over Darwin is moving toward closure, then historical science may be nothing more than an immature form of experimental—predictive science. But if, as we believe, the contention over what Darwin means—more broadly, the debate over how to interpret history—is itself constitutive of evolutionary biology, then historical science must indeed be a distinctive form of intellectual inquiry, like the Foucauldian field of discursivity, that places its founding texts on a plane with its objects of inquiry, reading the books of Darwin and of Nature against each other in an endless interpretive chiasmus.

NOTES

We would like to thank Robert J. O'Hara (currently at the Center for Critical Inquiry in the Liberal Arts, University of North Carolina at Greensboro) for his advice about current issues in evolutionary biology and his helpful reading of an earlier draft of this essay.

1. Uniformitarianism, as Gould points out, is both methodological and substantive. He himself believes that substantive uniformitarianism (the belief that all change in fact results from the slow accumulation of small genetic

variations) does not necessarily follow from methodological uniformitarianism (extrapolation from small-scale variations because they are the best historical evidence available); contemporary "strict Darwinism" tends to collapse the distinction, much as Darwin ultimately did. Gould's theory of "punctuated equilibrium" is perhaps the most salient challenge to substantive uniformitarianism (Eldredge and Gould), and it generated much discussion within the literature.

2. Ernst Mayr attributes the characterization of biology as "postage stamp collecting" to Ernest Rutherford ("Is Biology Autonomous," 9).

3. We use the term "evolutionary biology" loosely, to include all biological science with a historical dimension, including paleoanthropology, paleontology, systematics, and even some forms of population genetics.

4. Mayr lists the major historical sciences as cosmogony, geology, paleontology (phylogeny), and biogeography (*Growth*, 72), and Hull lists four historical disciplines: cosmogony, geology, paleontology, and human history (*Metaphysics*, 190). Linguistics might also be added to these lists.

5. These arguments are to be distinguished from the more general one that all scientific knowledge is embedded in narrative understanding; see Rouse.

6. One reader of our manuscript found it suggestive that Meadows didn't call his "immediacy index" an "ephemerality index" or an "immaturity index."

7. The three journals in biochemistry and molecular biology (*Cell, Journal of Biological Chemistry,* and *Journal of Molecular Biology*) were chosen from that category in the *Journal Citation Reports* of 1980; they are among the most highly cited of all journals in the 1980 *Science Citation Index*, ranking thirteenth, first, and thirty-third, respectively, out of 4,398 journals. The four journals in evolutionary biology (*American Naturalist, Evolution, Paleobiology,* and *Systematic Zoology*) were suggested to us by Robert O'Hara as being central to the work of Gould and Lewontin. Indeed, all four contained citations to Gould as primary author, two contained citations to Lewontin as primary author, and three contained citations to the "Spandrels" essay. The four journals ranked 227th, 308th, 1,424th, and 906th for being cited. The total number of citations examined in each journal (with the volume and issue numbers of each) were *Cell* (60:5)—1582 citations; *J. Biol. Chem.* (265:9)—1005; *J. Mol. Biol.* (215:1)—1308; *Am. Nat.* (136:1-6)—2754; *Evolution* (44:1-3)—2503; *Paleobiology* (16:1-4)—1862; and *Syst. Zool.* (39:1-4)—826.

8. Price's Index could not be precisely calculated because our data covered citations within the previous six years rather than within Price's period of five years; on the basis of the six-year figure, the Price's Index was 37 percent to 45 percent for evolutionary biology and 56 percent to 77 percent for molecular biology. The figures will be a bit high because of the inclusion of the extra year's worth of citations in the measure.

9. Garfield's graph also shows a generally flat distribution of citations to Watson and Crick's 1954 *Nature* paper, one of the founding works in molecular biology, and a steadily decreasing number of citations to a 1952 paper on genetic transduction in bacteria, which Garfield identifies as an example of the "obliteration" phenomenon ("The Articles Most Cited," 192).

Carolyn R. Miller and S. Michael Halloran

10. The five theories are evolution "as such," common descent (or common ancestry), gradualism, multiplication of species, and natural selection.

11. For example, "A Neo-Darwinian Commentary on Macroevolution," *Evolution* 36 (1982): 474–98; "Darwin's Gradualism and Empiricism," *Nature* 309 (1984): 116; Darwinism and the Expansion of Evolutionary Theory," *Science* 216 (1982): 380–87; "Darwin and the Definition of Phylogeny," *Systematic Zoology* 34 (1985): 97–98; "Neo-Darwinian Evolution Implies Punctuated Equilibrium," *Nature* 315 (1985): 400–401; "Two Hypotheses on Darwin's Gradualism," *Systematic Zoology* 34 (1985): 201–5; "Non-Darwinian Evolution: A Critique," *Nature,* 225 (14 March 1970): 1025–28; "Species and Neo-Darwinism," *Systematic Zoology* 39 (1990): 399–413.

12. Of the eleven articles in the four evolutionary biology journals that cited Darwin, two used a direct quotation as an epigraph and seven used a direct or indirect quotation in the first paragraph.

13. See Gould and Eldredge's account of this debate, which they summarize in this way: "Punctuated equilibrium—first dismissed as simply false, then rejected as apostasy against Darwinism, and now depicted as the necessary and logical outcome of what the Modern Synthesis always knew—has evidently come of age" (143). Dawkins has also discussed this debate; he concludes: "The theory of punctuated equilibrium is a minor gloss on Darwinism, one which Darwin himself might well have approved if the issue had been discussed in his time" (250).

14. Their treatment of Voltaire as a textual source is Victorian rather than scientific and is different from their treatment of Darwin. Although Gould and Lewontin quote from Voltaire directly (153), they do not cite his work formally or list it in the references, as they do Darwin's.

15. This monotonous plot is articulated by Gould in the British magazine *New Scientist,* in a 1978 essay that criticizes sociobiology for telling adaptationist stories: "The common political character and effect of these stories lies in . . . a defense of existing social arrangements as part of our biology" (532). His criticism of these stories is similar to that in "Spandrels," relying in part on experimental–predictive standards of falsifiability and consistency. For more on this matter, see Susan Wells's discussion in Chapter 3.

16. For more complete discussions of ethos, see Halloran ("Aristotle's Concept") and Johnson.

WORKS CITED

Darwin, Charles. *On the Origin of Species and the Descent of Man.* Facsimile of the first edition, with an introduction by Ernst Mayr. 1859. Cambridge: Harvard University Press, 1964.

Dawkins, Richard. "Puncturing Punctuationism" and "The One True Tree of Life." In *The Blind Watchmaker.* New York: Norton, 1986.

de Queiroz, Kevin. "Systematics and the Darwinian Revolution." *Philosophy of Science* 55 (1988): 238–59.

Eldredge, Niles, and Stephen Jay Gould. "Punctuated Equilibria: An Alternative to Phyletic Gradualism." In *Models in Paleobiology*, ed. T. J. M. Schopf, 82–115. San Francisco: Freeman, Cooper, 1972.

Foucault, Michel. "What Is an Author?" In *The Foucault Reader*, ed. Paul Rabinow and trans. José V. Harari, 101–20. New York: Pantheon, 1984.

Garfield, Eugene. "The Articles Most Cited in the SCI from 1961 to 1982. 7. Another 100 *Citation Classics:* The Watson-Crick Double Helix Has Its Turn." In *Essays of an Information Scientist* 8, 187–96. Philadelphia: Institute for Scientific Information, 1985.

Garfield, Eugene. "The 'Obliteration Phenomenon' in Science—and the Advantage of Being Obliterated!" In *Essays of an Information Scientist* 2, 396–98. Philadelphia: Institute for Scientific Information, 1977.

Goudge, Thomas Anderson. *The Ascent of Life.* Toronto: University of Toronto Press, 1961.

Gould, Stephen Jay. "Darwinism and the Expansion of Evolutionary Theory." *Science* 216 (23 April 1982): 380–87.

Gould, Stephen Jay. "Evolution and the Triumph of Homology, or Why History Matters." *American Scientist* 74 (January–February 1986): 60–69.

Gould, Stephen Jay. "Is a New and General Theory of Evolution Emerging?" *Paleobiology* 6 (1980): 119–30.

Gould, Stephen Jay. "Sociobiology: The Art of Storytelling." *New Scientist* 80 (16 November 1978): 530–33.

Gould, Stephen Jay. *Wonderful Life: The Burgess Shale and the Nature of History.* New York: Norton, 1989.

Gould, Stephen Jay, and Niles Eldredge. "Punctuated Equilibrium at the Third Stage." *Systematic Zoology* 35 (1986): 143–48.

Gross, Alan. "The Origin of Species: Evolutionary Taxonomy as an Example of the Rhetoric of Science." In *The Rhetorical Turn: Invention and Persuasion in the Conduct of Inquiry*, 91–115. Chicago: University of Chicago Press, 1990.

Halloran, S. Michael. "Aristotle's Concept of Ethos, or If Not His, Somebody Else's. *Rhetoric Review* 1 (1982): 58–63.

Halloran, S. Michael. "The Birth of Molecular Biology: An Essay in the Rhetorical Criticism of Scientific Discourse." *Rhetoric Review* 3 (1984): 70–83.

Harris, R. Allen. "Rhetoric of Science." *College English* 53 (1991): 282–307.

Hull, David L. "Darwinism as a Historical Entity: A Historiographic Proposal." In *The Darwinian Heritage*, ed. David Kohn, 773–812. Princeton: Princeton University Press, 1985.

Hull, David L. *The Metaphysics of Evolution.* Albany: State University of New York Press, 1989.

Hull, David L. *Philosophy of Biological Science.* Englewood Cliffs, N.J.: Prentice-Hall, 1974.

Johnson, Nan. "Ethos and the Aims of Rhetoric." In *Essays on Classical Rhetoric and Modern Discourse*, ed. Robert J. Connors, Lisa Ede, and Andrea Lunsford, 98–114. Carbondale: Southern Illinois University Press, 1984.

Carolyn R. Miller and S. Michael Halloran

Kuhn, Thomas J. *The Structure of Scientific Revolutions.* 2nd ed. Chicago: University of Chicago Press, 1970.

Lenoir, Timothy. "The Darwin Industry." *Journal of the History of Biology* 20 (1987): 115–30.

Levins, Richard, and Richard Lewontin. *The Dialectical Biologist.* Cambridge: Harvard University Press, 1985.

Lewontin, Richard. "The Bases of Conflict in Biological Explanation." *Journal of the History of Biology* 2 (1969): 35–45.

Lewontin, Richard. "The Corpse in the Elevator." *New York Review of Books* 29 (20 January 1983): 34–37.

Lewontin, Richard. "Darwin's Revolution." *New York Review of Books* 30 (16 June 1983): 21–27.

Lewontin, Richard. "Fallen Angels" (review of *Wonderful Life*). *New York Review of Books* 37 (14 June 1990): 3–7.

Lewontin, Richard. "Is Nature Probable or Capricious? *BioScience* 16 (1966): 25–27.

Mayr, Ernst. "Cause and Effect in Biology." *Science* 134 (10 November 1961): 1501–6.

Mayr, Ernst. "Darwin's Five Theories of Evolution." *The Darwinian Heritage,* ed. David Kohn, 755–72. Princeton: Princeton University Press, 1985.

Mayr, Ernst. "Is Biology an Autonomous Science?" In *Toward a New Philosophy of Biology: Observations of an Evolutionist,* 8–23. Cambridge: Harvard Belknap, 1988.

Mayr, Ernst. *The Growth of Biological Thought: Diversity, Evolution, and Inheritance.* Cambridge: Harvard Belknap, 1982.

Meadows, Arthur Jack. "The Citation Characteristics of Astronomical Research Literature." *Journal of Documentation* 23 (March 1967): 28–33.

O'Hara, Robert. "Homage to Clio, or, toward an Historical Philosophy for Evolutionary Biology." *Systematic Zoology* 37 (1988): 142–55.

O'Hara, Robert. "Telling the Tree." *Biology and Philosophy* 7(1992): 135–60.

Price, Derek J. de Solla. "Citation Measures of Hard Science, Soft Science, Technology, and Nonscience." In *Little Science, Big Science . . . and Beyond,* 155–79. New York: Columbia University Press, 1986.

Roger, Jacques. "Darwinism Today." In *The Darwinian Heritage,* ed. David Kohn, 813–23. Princeton: Princeton University Press, 1985.

Rouse, Joseph. "The Narrative Reconstruction of Science." *Inquiry* 33 (1990): 179–96.

Simpson, George Gaylord. *This View of Life: The World of an Evolutionist.* New York: Harcourt, 1964.

Sober, Elliott. *Reconstructing the Past: Parsimony, Evolution, and Inference.* Cambridge: MIT Press, 1988.

Toulmin, Stephen. *Human Understanding: The Collective Use and Evolution of Concepts.* Princeton: Princeton University Press, 1972.

Yates, Frances A. *The Art of Memory.* Chicago: University of Chicago Press, 1966.

7

CONSTRUCTING SCIENTIFIC

KNOWLEDGE IN

GOULD AND LEWONTIN'S

"THE SPANDRELS OF SAN MARCO"

DOROTHY A. WINSOR

When people compare scholarly writing in science and in the humanities, they tend to see these two ways of writing knowledge as more different than similar. Most obviously, scientific texts deliberately devalue the individual interpretations and sensitivities that many other kinds of writing exhibit, especially academic writing in the humanities. Thus readers are struck by the highly technical vocabulary, impersonal tone, and heavy emphasis on data found in scientific prose. Indeed, texts are often classified as "scientific" to the extent that they display these qualities (cf. Kinneavy, 88–89, 131–32).

A closer look, however, reveals similarities as well as differences between scientific texts and other kinds of scholarly work. Like other kinds of scholarly writing, for instance, scientific prose must strive to convince its audience of its validity. This rhetorical stance is necessary because knowledge in any field may be defined as the ideas that are shared by a group of the field's recognized practitioners. This aspect of the nature of knowledge is not always easy to remember. People still sometimes speak of knowledge as if it existed somewhere in a Platonic form that we simply "discover." Knowledge, however, does not exist apart from knowers. An idea becomes knowledge only when a significant number of experts accept it as such. Knowledge is, in other words, socially constructed.

At first glance, one might be tempted to see scientific knowledge as an exception to this rule because it is impersonal and empirical. The very impersonality of scientific discourse, however, actually makes it

Dorothy A. Winsor

more socially dependent for validity, not less so, for a field that devalues personal insights and sensitivities necessarily looks for consensus as a sign of validity. In a way, the respect scientists accord replication is a methodological expression of their reliance on consensus. In theory, as more scientists are able to produce similar experimental results, the results assume a more fact-like status. But in practice, scientists seldom replicate the work of others because rewards and recognition for replication are minimal (Collins, 29–49). Instead, scientists form judgments of knowledge claims based on how consistent published results are with their own work. In either case, what counts as knowledge grows out of an achieved consensus.

Scientific knowledge is constructed, then, not just in labs or at field sites, but in arguments that scientists conduct through the medium of scientific papers. Any scientific article is like a freeze frame from a film. It is a moment in a disciplinary discussion that existed before the article was written and will exist after the article is published. Fully understanding a scientific article, then, means examining how it affects that discussion, how it pushes already existing articles closer to or further from status as knowledge, and how it adds its own claims to future episodes in the conversation.

The social construction of scientific knowledge has been an increasingly accepted notion among philosophers, sociologists, historians of science, and scientists themselves at least since Thomas Kuhn's *The Structure of Scientific Revolutions* (1970). Charles Bazerman has reviewed the extensive literature in this area ("Scientific Writing"). But despite the enormous amount of attention social constructionism has received, relatively scant attention has been paid to what it means in practice, in words, in individual scientific texts. (For exceptions, see Myers.) Moreover, much of the interesting work that has been done in this area has been conducted by sociologists, not scholars of writing. Latour, for instance, in his *Science in Action* treats the creation of scientific knowledge as an agonistic activity in which scientists use written versions of reality to negotiate knowledge with other members of their disciplinary communities. Using the term "inscription" to describe the written data, charts, tables and so on that scientists create and then work from, Latour says that the conversion of physical reality into inscriptions is one of science's main activities (cf. Latour and Woolgar). He identifies arguments over the relative merit of various inscriptions as a second main activity, one that locates the creation of scientific knowledge in the realm of the social (cf. Gilbert and Mulkay).

This essay is an attempt to examine the consequences of the social construction of scientific knowledge for a specific article by Stephen Jay

Gould and Richard Lewontin. Its premise is that an article like "The Spandrels of San Marco and the Panglossian Paradigm" is not a simple attempt to inform a waiting public of newly "discovered" scientific knowledge. That information transfer model of scientific prose would imply that Gould and Lewontin's ideas count as knowledge from the moment Gould and Lewontin begin to hold them—clearly, as social constructionism suggests, an impossibility. Rather, "The Spandrels of San Marco" is an attempt to negotiate the existence of that knowledge with other members of Gould and Lewontin's discipline. Specifically, whether Gould and Lewontin's article will count as knowledge will depend on how other scientists treat it in later papers. If it is cited approvingly and its claims are used by others, then it will have succeeded in adding knowledge to its field. If it is cited negatively or, worse yet, ignored, then its claims are only ink on a page and will not become knowledge (cf. Gilbert, 294).

Because Gould and Lewontin, like other scientific authors, are dependent on later writers to turn their claims into knowledge, they use a series of tactics to encourage later readers to treat them as facts. First, they enlist other scientists in their cause by means of references. In citing previous articles approvingly, Gould and Lewontin both bolster their own work and increase the fact-like status of the works cited. Second, they point to evidence from nature and the methodology used to gather it. Using approaches suggested by Latour in *Science in Action: How to Follow Scientists and Engineers Through Society* (see also Latour and Woolgar), this essay will examine the use Gould and Lewontin make of both human and natural backing for their claims. It will then look at the use more recent articles have made of "Spandrels," in an attempt to measure how well Gould and Lewontin succeeded in creating scientific knowledge.

References: The Scholarly Discussion Already Going On

Nearly all scholarly articles, of course, use references. In doing so, they make use of an argument from authority (see Gilbert, "Referencing"). The authority is that of authors with whom the current writers agree, plus the journals that publish the references and the referees of those journals, plus all the references cited by the cited articles and their journals, and so on. This combined authority can make for a powerful force that is difficult to resist. In social constructionist terms, collective authority is knowledge. It is not enough, however, for an article to cite references as authority. Rather, an article must explain to readers how they are to interpret those references and how the references all add up to support the current work. It is false to think of the relationship between an article and its references as flowing in one direction only, with the references influencing the shape of the article. Any article also influences its readers' perception of the shape of the items referenced, for it tells us what is important in them, how much faith should be put in them, and how they fit together to define the field of study (Edge; Small). Gould and Lewontin, like all other scholarly writers, use references to reconstruct their field to make room for their own claims and to demonstrate how their own ideas fit consistently into that field and how they improve it.

The most obvious thing Gould and Lewontin do to array their references for their own support is to cite them with more or less acceptance. Generally speaking, the language in which scientific statements are written reveals the degree of acceptance they are accorded. Latour and Woolgar arrange scientific statements along a scale of five types (75–88). A type 5 statement is of knowledge so thoroughly accepted that it need not even be made explicit. A scientist working today, for instance, would probably refer to "the structure of the DNA molecule" without stopping to explain what the structure is. A type 4 statement, on the other hand, makes accepted knowledge explicit. Most textbook statements would be type 4; the assertion "DNA has a double helix structure" is a type 4 statement.

In contrast to the accepted knowledge carried in type 5 and 4 statements, type 3 statements frame knowledge claims in a way that refers to their creation and thus place them as less factual. At one time, for instance, statements about DNA's structure could be phrased only as type 3 claims, such as: "DNA is *generally accepted* as having a double helix structure," or "*Watson and Crick have demonstrated* the double helix structure of the DNA molecule," or even "The DNA molecule *has been shown* to have a double helix structure *(Watson and Crick, 1953).*" A type 3 statement becomes type 4 when it can be made without modalities or attribution, when it is simply accepted as part of the knowledge of a field and is no longer on the controversial cutting edge.

As one moves down the scale to types 2 and 1 statements, one finds wording that implies less and less certainty about claims. Type 2 statements are phrased more contentiously than type 3, drawing attention to the uncertain nature of current evidence. In their 1953 paper, for instance, Watson and Crick use type 2 statements when discussing a theory competing with their own: "*In our opinion*, this structure is unsatisfactory for two reasons: (1) *We believe* that the material which gives the X-ray diagram is the salt, not the free acid. Without the acidic hydrogen atoms *it is not clear* what force would hold the structure together. . . . (2) Some of the van der Waals distances *appear to be* too small" (737; emphasis added). The modalities here are negative, but they might have been positive. Watson and Crick could, for instance, have said that in their "opinion," the proposed structure was "plausible," although it was "not yet clear" precisely how the molecule held together. The tone would still suggest a claim that had not yet been proven. Finally, type 1 statements are phrased as conjectures or speculations; they are most frequently found at the end of an article in which the author is suggesting future research or tentative connections of other work to that of the author. In practice, it is sometimes difficult to distinguish among types of statements, and of course some overlap is inevitable among the five types. Nevertheless, Latour and Woolgar's typology provides a way of understanding how Gould and Lewontin work to create knowledge in "The Spandrels of San Marco."

When scientists are engaged in controversy, their object is to push their own claims higher up the scale and to push rival claims lower, activities that Gould and Lewontin practice in "The Spandrels of San Marco." Thus, for instance, in the second section of their article Gould and Lewontin phrase adaptationist claims as type 2 or type 1 statements. Biological characteristics that were "once" explained in one way, they say, have "become" something else, an interpretation Gould and Lewontin say they "do not attack" because "they may all be right" (152). But the wording Gould and Lewontin use implies that these claims are without evidence, reducing them thereby to a weak type 2 at best and possibly a speculative type 1. In the next paragraph, Gould and Lewontin refer to other researchers who "hoped to find a correlation" but "failed," and so "assume that another must exist" (152). The researchers' claims are thus reduced to type 1 speculation. In general, the adaptationist theory that rivals Gould and Lewontin's is characterized in a manner that makes it most easily demolishable. Gould and Lewontin themselves recognize to some degree that their description of rival claims might be criticized—"some evolutionists will protest that we are caricaturing their view of adaptation" (151)—but they go on to assert the general fairness of their portrayal.

In contrast, the claims that Gould and Lewontin agree with are phrased more as type 3 statements—a strong placement for claims made in the midst of controversy. Note, for instance, how in their opening section Gould and Lewontin contrast the work of Harner and Wilson, who "ask us" to believe a theory about the adaptive nature of Aztec sacrifices (149), with the work of Sahlins: "as Sahlins argues, it is not even clear that human sacrifice was an adaptation at all" (150). Sahlins makes a statement that is accepted as a type 3 argument. It refers to its originator, Sahlins, but is apparently correct. Sahlins's claim in turn reduces Harner's and Wilson's claims to unproven type 2 statements: their validity is "not clear."

Strongly accepted references are apparently not always available to Gould and Lewontin, so they work to move the claims of their allies further up the scale of types so that they can use those claims to shore up their own. A continental theory, for instance, "has not been appreciated, but deserves to be" (160). Having supported this theory, Gould and Lewontin are then able to quote one of its proponents, Reidl, in support of their own work. On the next page, they prop up "Seilacher, whose work deserves far more attention than it has received" (161). Seilacher's research is then treated in type 3 sentences: he "has shown that the divaricate form of architecture . . . occurs again and again" (161). They then use this work as the basis of a strongly positive type 2 statement of their own: "Seilacher is probably right in representing this case as the spandrels, ceiling holes, and sacrificial bodies of our first section" (162). Seilacher, of course, has made no mention of spandrels, ceilings, or bodies, but Gould and Lewontin make good use of him: they push Seilacher's claims a bit closer to status as knowledge, hoping he will pull their own claims with him.

This validation of previous claims is evident even when Gould and Lewontin cite their own work. Thus at one point they simply assert that "allometric patterns are as subject to selection as status morphology itself (Gould 1966)" (157). This is a strong type 3 claim. Later, a long quote from Lewontin constitutes the entire evidence for "selection without adaptation" (158). Apparently that quote is simply to be taken at face value. In these sections, Gould and Lewontin try to move their own previous work closer to status as knowledge.

While this framing of previous claims is useful in placing Gould and Lewontin's own work, it is not enough. Writers also typically place the rephrased claims in relation to one another so as to control opponents further. Note, for instance, how Sahlins and Ortiz de Montellano are used to neutralize Harner and Wilson in the paper's first section. Harner's claims and Wilson's agreement are first laid out in as weakened a form as possible. Then Sahlins is introduced with a "but": "But Sahlins (1978) has argued that human sacrifice represented just one part of an elaborate cultural fabric" (150). In the next paragraph, Ortiz de Montellano is used to represent "many experts," presumably because of the extensive citing he does in his article: "many experts doubt Harner's premise in the first place (Ortiz de Montellano 1978)" (150). The same technique is used in the second section, where Gould and Lewontin characterize adaptationists' views about genetic drift: "Have we not all heard the catechism about genetic drift: it can only be important in populations so small that they are likely to become extinct before playing any sustained evolutionary role (but see Lande 1976)" (151–52). Lande, presumably, is to be seen as authoritatively countering the opposition's points.

Gould and Lewontin also arrange to have Darwin, the most powerful ally of all, on their side. They call Darwin's support equivalent to "God's" and observe that he "has often been depicted as a radical selectionist at heart" (155). "This view is false," they flatly assert. They then move Darwin into their own corner, first by making a positive reference to a 1900 article by Romanes that, they say, "deserves a resurrection" (155). Such a resurrection would shore up a potential ally for Gould and Lewontin because Romanes apparently viewed Darwin as they do. Then they characterize Darwin's work in a way that allows them to choose the aspects that support their own: "We do not now regard all of Darwin's subsidiary mechanisms as significant. . . . But we should cherish his consistent attitude of pluralism" (156).

Gould and Lewontin, in short, use references in a way common to scientific articles. In the words of Bruno Latour, "The rules are simple enough: weaken your enemies, . . . help your allies if they are attacked, . . . [and] oblige your enemies to fight one another" (37–38). When these tactics are used to modify the existing literature, the effect is striking. The reader is faced not only with a significant number of claims, but also by claims arrayed to form a coherent whole. The reader is thus acted upon by an ordered set of references that Gould and Lewontin modify so that they all point to the same conclusion. They have constructed the field to support their own claim.

Data and Method: "Nature's" Contribution to the Discussion

In addition to drawing on previous scientists, Gould and Lewontin also try to place on their side various natural allies: data and methodologies that are accepted in their field. In the sciences in general, of course, data from nature and from the instruments and methodologies that produce them are highly respected. As a group, scientists see little lasting value in unique, individual perceptions and try to circumvent them. Scientists thus constantly seek experiences

Dorothy A. Winsor

and techniques that will allow nature to shape and restrain their theories. Charles Bazerman calls this phenomenon the active seeking of passive constraints and sees it as one of the hallmarks of science (*Shaping Written Knowledge,* 193, 209). Despite these passive constraints, however, it is also true that scientists negotiate knowledge of the shape of nature in their writing. For the most part, knowledge of the form of nature doesn't settle the argument. Rather, the argument settles what is accepted as knowledge of the form of nature.

In "The Spandrels of San Marco," Gould and Lewontin add no new evidence to the argument about evolution but rather borrow the evidence of others and argue about what that evidence implies for theory about the shape of nature. For instance, they make good use of work by Seilacher on molluscs (161–62), using his evidence to prove their point (which they say is also Seilacher's own) about the nonadaptiveness of some variations. They work with his evidence to prove their point by means of what Latour calls "stacking" (50–52): the extension of bits of evidence to support large theories. Latour uses the example of a biologist who looks at small slices of flesh that are first "three hamster kidneys," then extended to be "hamster kidneys," then "rodent kidneys," and finally "mammal countercurrent structure in the kidney." Such stacking has to occur if theories are to be built, and the tendency is, as Latour says, to "prove as much as you can with as little as you can considering the circumstances" (51). In examining Seilacher's shells, Gould and Lewontin move from shells, to shell patterns, to patterns that are adaptive in only rare cases, to a "fundamental architectural constraint" (162).

Gould and Lewontin draw Seilacher's evidence into their own text not only in words, but also in a figure—which appears in Seilacher's text too—showing various shell patterns. Figure 3 in Gould and Lewontin's article is of a type common to scientific articles: it attempts to show the reader reality itself and thus to allow the reader to see the validity of the writer's claims with his or her own eyes. The chart is not, however, nature itself; rather it is a humanly produced and ordered representation of it. Anyone who has ever looked for, selected, and arrayed pieces of a larger reality knows how much shaping has happened before such a figure is produced. Such a figure is a written trace of reality that in some ways is more valuable for creating scientific knowledge than nature itself. It is, for instance, portable, stable, reproducible, and easy to manipulate (Latour, 219–23). A figure is a piece of writing meant to define reality, not simply reflect it. Moreover, by means of more writing in the figure's caption and the explanation in the text, the writers tell the reader what to see in the figure.

The first two figures in "The Spandrels of San Marco" present other interesting examples of the merging of nature and human creation. They show architectural features of the ceilings of St. Mark's Cathedral and King's College Chapel. Architecture is, of course, a human creation, and a photograph of architecture is even more so. But in Gould and Lewontin's text, the two ceilings are presented as natural facts against which we are to measure our "biases" (148). They are then treated as analogies for evolutionary "architecture" (147), and the analogy is used to argue Gould and Lewontin's point of view. Like the references I discussed earlier, "reality" in the shape of figures is arranged to support the text's claim unmistakably.

Gould and Lewontin's use of Seilacher's shell evidence is not a great coup if Seilacher does, as they claim, agree with them. A more interesting move occurs when Gould and Lewontin borrow evidence from articles they disagree with and turn it to their own account. In section 6 of "The Spandrels of San Marco," for instance, Gould and Lewontin suggest that Haeckel's data support them instead of Haeckel, who "misread" the data (160). In section 3, they make another piece of adaptationist evidence turn traitor and support their own position. In the first of the two adaptationist studies of bluebirds, Barash claimed that male bluebirds attacked rival males more often before females laid eggs then they did afterward. After Gould wrote a 1978 article criticizing this claim, Morton, Geitgey, and McGrath repeated Barash's work and obtained different results, which they nonetheless saw as consistent with Barash's. Gould and Lewontin, on the other hand, describe the observed birds' behavior as possibly patternless and certainly unrelated to the passing on of genes. That is, they appropriate it to their own claims. It would appear that even perceptions of nature, which one would assume to precede the writing of an article, are amenable to shaping in the scholarly discussion of scientists.

Because scientists obtain evidence for discussion by means of techniques and methodologies, they value method as well as data and stress it in their writing. In scientific terms, methods are better as they get "harder"—that is, as they become capable of producing more quantitative evidence that resists attempts to demolish it. Latour calls such evidence "black boxes" (1–17). In contrast to the "plausible stories" of adaptationists, Gould and Lewontin claim to have method on their side. For instance, their "non-adaptative hypothesis can be tested by conventional allometric methods (Gould [1974] in general; Lande [1978] on limb reduction) and seems to us both more interesting and fruitful than untestable speculations" (153). Even though this method is credited to Gould and the identifiable ally Lande (he is cited as offsetting adaptationist "catechisms" at the top of page 152 and as engaging in "personal communication" with the authors on page 157), the claim that method is on the author's side still carries weight. Similarly, Gould and Lewontin say that computer simulation of shell growth supports their idea of the nonadaptiveness of the growth patterns (162), and the evidence of computer simulation seems indisputable.

Parts of scientific articles that refer to method are "technical" both in the sense that they refer to technique and in the sense that they are often difficult for outsiders to read because they tend to be densely packed with the prior knowledge of the field and couched in its specialized vocabulary. Techniques—part of the given knowledge of a field— are thus often referred to rather than explained. Additionally, a field's specialized language is itself often the embodiment of looking at the world in a particular way. The language will, for instance, sort reality into relevant categories by assigning names. The parts of reality worth naming vary from field to field, and the names given have connotations apparent only to a field's insiders. As James Kinneavy says, "the scholar engaged in scientific discourse should strive as assiduously as his literary colleague for *le mot juste* to express his concept with precision, and if this means engaging in jargon or neologisms, so much the worse for the lay and amateur reader" (176).

The most technical part of "The Spandrels of San Marco" is section 5, which contains concentrated, condensed references to method. Latour says that scientific articles become most technical when they are trying hardest to anticipate objections (45–46). In section 5, Gould and Lewontin are proposing their own alternatives to adaptationist theories, so it is not surprising that they fortify themselves most heavily here. In contrast to the accessibility of much of "The Spandrels of San Marco," section 5 contains sentences that to an outsider's eye are highly technical because of both their specialized vocabulary and their references to statistical methodology. Thus we find phrases like "the 0.2 and 0.4 intraspecific scaling among homeothermic adults" and "strongly *r*-selected feeders" and entire passages like the following:

> Even if an allele is favoured by natural selection, some portion of population, depending upon the product of population size N and selection intensity s, will become homozygous for the less fit allele because of genetic drift. If Ns is large this random fixation for unfavourable alleles is a rare phenomenon, but if selection coefficients are on the order of the reciprocal of population size ($Ns = 1$) or smaller, fixation for deleterious alleles is common. (157)

Passages like these are meant to anticipate objections from the most knowledgeable readers and shut out less knowledgeable ones entirely.

In its treatment of nature and method, then, "The Spandrels of San Marco" works at arranging nature to support the claims Gould and Lewontin make, just as it arranges references. Nature is not something in front of which scientists passively sit. It is material whose shape they work together to construct.

The Fate of "The Spandrels of San Marco": The Subsequent Discussion

To what degree, then, did Gould and Lewontin succeed in creating knowledge? This question can only be answered tentatively, and only by looking at the treatment their article received in later articles.

A computerized search of the SciSearch Database using the Dialog System reveals that "The Spandrels of San Marco" certainly escaped the worst of fates: it was not ignored. In English-language articles alone, it was cited 488 times between its publication and the end of 1990. These may be broken down into 287 references (or an average of forty-one per year) for the period 1980 through 1986; eighty-three references (or about fifty-five per year) for the eighteen-month spread between January 1987 and week 28 of 1988; fifty-one references for the fifty-four-week period running from week 29 of 1988 through week 31 of 1989; and sixty-seven references (or about forty-eight per year) for the seventeen months from week 32 of 1989 through the end of 1990. This is an unusually large number of references, made over an unusually long time: in a 1965 study, Derek J. de Solla Price determined that only 1 percent of the articles he studied was cited more than six times per year (511). An overall increase in the number of journals and articles has occurred in the twenty-five years since Price's work, so some inflation in the number of citations might be expected, but "The Spandrels of San Marco" is nevertheless a highly visible article in its field. Price also showed that because of a bias in science in favor of recent research, the average article was no longer useful for citing after ten years (512). The continued frequent citation of Gould and Lewontin's article says both that it is still seen as relevant and that the argument it makes is not yet accepted as common knowledge, for if the article were either dated or accepted as common knowledge, it would no longer be cited.

In an attempt to determine the nature of those citations, I examined a

Dorothy A. Winsor

sample from the lists of citing articles during 1987–88 and 1988–89, selecting these recent lists because articles published at long intervals after "Spandrels" are most likely to reflect the long-term judgment of the scholarly community on Gould and Lewontin's article. The sample consisted of the first article on these two lists and every fifth article thereafter. This process resulted in a sample of twenty-eight articles. (They are listed in an appendix at the end of this essay to illustrate the wide variety of journals in which the citations appeared.)

When "The Spandrels of San Marco" is cited, the two ideas most often drawn on are the primary claim that not all traits result from adaptation and the subsidiary one that when changes are claimed to be adaptive, proof better than "plausible stories" is needed. Gould and Lewontin presumably believe that better proof would often be unavailable because the changes would not in fact be adaptive. In articles that cite it, "The Spandrels of San Marco" is seen as an early entry in a "continuing debate about adaptation" (Knoll and Niklas, 128) in which the "pan-selectionist view of Nature has come under strong attack over the last decade" (Shine, 195). The article appears to be treated as a classic statement of one position in this debate. Sometimes it is the only one cited as representing the antiadaptationist argument (Motta, 60; Bradshaw, "Desert Reptiles," 156; Grime, 7). That Gould and Lewontin's article is still citable suggests that the argument over adaptationism is still going on. If their position had been completely accepted, it would be unnecessary to cite them, just as it is now unnecessary to cite Watson and Crick. If their position had been rejected, their article would be useless for citing.

Despite what Knoll and Niklas call the "continuing debate" (128), however, of the twenty-eight articles I examined only one—Caplan's "Rehabilitating Reductionism" (1988)—is explicitly hostile to Gould and Lewontin's work. According to Caplan, Gould and Lewontin's argument that an organism is more than the sum of its parts is a semimystical position, not justified by science. The other twenty-seven articles examined refer to Gould and Lewontin's position as an accepted one. This is particularly interesting since the authors of seventeen of these twenty-seven articles go on to claim that while it is necessary to avoid "adaptationist abuses" (Barbault, 269), their own work justifies an adaptationist claim. In other words, "The Spandrels of San Marco" is frequently cited approvingly in articles that make adaptationist claims. Cannon and Block, for instance, write that though they want "to avoid the 'Panglossian paradigm' of seeing every feature as a perfect adaptation (Lewontin, 1978; Gould & Lewontin, 1979), it is clear from a wide range of studies that species occupying extreme habitats generally show consistent adaptations to the combined stress of cold, drought and anoxia" (38). Nishikawa cites Gould and Lewontin to the effect that proving adaptiveness is difficult and says that therefore "cases that are amenable to historical analysis of selection pressures should be of special interest." His own study of aggression in salamanders "is one such case" (263). Bernardi and Bernardi provide proof for adaptiveness in genome evolution that, they say, does not depend "on 'adaptive stories,' which can rightly be criticized (Gould and Lewontin 1979 . . .)" (9).

In all these articles, Gould and Lewontin's work has been incorporated as part of the discipline, although that incorporation has taken place in a way they probably would not entirely approve of. Subsequent readers have constructed the meaning of Gould and Lewontin's article, just as they themselves shaped the meaning of the articles they referenced. Those subsequent readers agree with Gould and Lewontin that adaptationist claims need to be supported by evidence, but they then assert that in their own cases the evidence is sufficient.

The ease with which Gould and Lewontin's article has been assimilated by adaptationists is unsurprising since Gould and Lewontin themselves appear to have been content to construct a rather "soft" fact. Latour defines soft facts as those that leave a reader room to negotiate and adapt them to local circumstances (206–10). Soft facts spread more easily than hard ones because readers can alter them to fit their own ideas rather than having to alter their ideas to fit the new facts. In creating soft facts, however, a scientist trades control over what is made of his or her work in return for easier acceptance. Collins discusses this phenomenon as well: A scientist can choose, Collins says, either to make a claim that fits well with current views and is therefore more acceptable or to chance making one that challenges current views. The more radical claim may meet with more resistance, but it may also win the scientist recognition as a more original thinker (129–52; cf. Myers, "Social Construction").

Gould and Lewontin indeed phrase their claims softly. Note, for instance, how they concede that some adaptationist claims "may all be right" (152), although the claims are at present insufficiently supported. Again, at the beginning of section 3 of their article, the two say that they "would not object so strenuously to the adaptationist programme if its invocation . . . could lead in principle to its rejection for want of evidence" (153). Subsequent writers, taking this admission to heart, have hastened to provide what they see as better evidence. "The Spandrels of San Marco" thus is interpreted primarily as an admonishment to use better methodology, to create better black boxes and harder facts. Its

Dorothy A. Winsor

subsidiary argument about method, originally intended to support its overall antiadaptationist stance, is treated instead as the article's main message. A call for better methodology, of course, is easily accepted by the scientific community, since, as I said earlier in this essay, harder methodologies are always preferred in the sciences. Gould and Lewontin apparently believed that better evidence would prove the antiadaptationist position. In many cases, however, their position gets treated as they have said the effects of constraints on evolution are treated: it is "simply acknowledged and then not taken to heart and invoked" (151).

Conclusion

People often think scientific articles are more objective and less socially influenced than other kinds of writing. This is not so. While Gould and Lewontin's paper differs in some ways from most papers appearing in scientific journals, it resembles other scientific papers in that it is written in and to a scientific community, on which it depends for its genesis, its force, and its fate. In fact, as Latour says, "although it sounds counter-intuitive at first, the more technical and specialized a literature is, the more 'social' it becomes, since the *number of associations* necessary to drive readers . . . into accepting a claim as a fact increase" (62). These "associations" may be both human allies and evidence from nature. In seeking to tie itself to "associations," a text tries to point outside itself, to tie itself to reality, to pull reality in, and ultimately to erase itself as text so that the knowledge claim can be accepted as a transparent representation of reality. To construct a joint acceptance of its knowledge claim, "The Spandrels of San Marco and the Panglossian Paradigm" rearranges both the field's practitioners and its perceptions of nature. Only the field can then decide whether the rearrangement has been successful. As Latour has fruitfully shown, knowledge—including scientific knowledge—is always a social construct.

APPENDIX: A SAMPLE OF ARTICLES CITING
"THE SPANDRELS OF SAN MARCO"

Armbruster, W. Scott. "Multilevel Comparative Analysis of the Morphology, Function, and Evolution of *Dalechampia* Blossoms." *Ecology* 69 (1988): 1746–61.
Arnold, Steven J. "Behavior, Energy and Fitness." *American Zoologist* 28 (1988): 815–27.

Auld, Tony D. "Population Dynamics of the Shrub *Acacia suaveolens* (Sm.) Willd.: Survivorship Throughout the Life Cycle, a Synthesis." *Australian Journal of Ecology* 12 (1987): 139–51.

Barbault, R. "Body Size, Ecological Constraints, and the Evolution of Life-History Strategies." *Evolutionary Biology* 22 (1988): 261–86.

Baumgartner, Jeffrey, Michael A. Bell, and Philip H. Weinberg. "Body Form Differences Between the Enos Lake Species Pair of Threespine Stickle-backs (*Gasterosteus aculeatus* Complex)." *Canadian Journal of Zoology* 66 (1988): 467–74.

Bernardi, Giorgio, and Giacomo Bernardi. "Compositional Constraints and Genome Evolution." *Journal of Molecular Evolution* 24 (1986): 1–11.

Bradshaw, A. D. "Comparison—Its Scope and Limits." *New Phytologist* 106 (1987): 3–21.

Bradshaw, S. D. "Desert Reptiles: A Case of Adaptation or Pre-Adaptation?" *Journal of Arid Environments* 14 (1988): 155–74.

Cannon, R. J. C., and W. Block. "Cold Tolerance of Microarthropods." *Biological Reviews of the Cambridge Philosophical Society* 63 (1988): 23–77.

Caplan, Arthur L. "Rehabilitating Reductionism." *American Zoologist* 28 (1988): 193–203.

de Buffrenil, V., J. O. Farlow, and A. de Ricqles. "Growth and Function of *Stegosaurus* Plates: Evidence from Bone Histology." *Paleobiology* 12 (1986): 459–73.

Ferguson, J. W. H. "Dietary Overlap in Plocepasserine Weavers (Aves: Ploceidae)." *South African Journal of Zoology* 23 (1988): 266–71.

Galen, Candace, and Maureen L. Stanton. "Bumble Bee Pollination and Floral Morphology: Factors Influencing Pollen Dispersal in the Alpine Sky Pilot, Polemonium Viscosum (Polemoniaceae)." *American Journal of Botany* 76 (1989): 419–26.

Grime, J. P. "The Stress Debate: Symptom of Impending Synthesis?" *Biological Journal of the Linnean Society* 37 (1989): 3–17.

Heideman, Paul D. "The Timing of Reproduction in the Fruit Bat *Haplonycteris fischeri* (Pteropodidae): Geographic Variation and Delayed Development." *Journal of Zoology* 215 (1988): 577–95.

Jamieson, Ian G. "Behavioral Heterochrony and the Evolution of Birds' Helping at the Nest: An Unselected Consequence of Communal Breeding?" *American Naturalist* 133 (1989): 394–406.

Knoll, Andrew H., and Karl J. Niklas. "Adaptation, Plant Evolution, and the Fossil Record." *Review of Palaeobotany and Palynology* 50 (1987): 127–49.

Moore, Janice, and Daniel R. Brooks. "Asexual Reproduction in Cestodes (Cyclophyllidea: Taeniidae): Ecological and Phylogenetic Influences." *Evolution* 41 (1987): 882–91.

Motta, Philip J. "Functional Morphology of the Feeding Apparatus of Ten Species of Pacific Butterfly Fishes (Perciformes, Chaetodontidae): An Ecomorphological Approach." *Environmental Biology of Fishes* 22 (1988): 39–67.

Nishikawa, Kiisa C. "Interspecific Aggressive Behaviour in Salamanders: Spe-

Dorothy A. Winsor

cies-specific Interference or Misidentification?" *Animal Behaviour* 35 (1987): 263–70.

Pansera, Francesco. "Pathogenesis of Osteoarthritis from an Evolutionary Perspective." *Acta Biotheoretica* 36 (1987): 281–82.

Ross, Caroline. "The Intrinsic Rate of Natural Increase and Reproductive Effort in Primates." *Journal of Zoology* 214 (1988): 199–219.

Shine, Richard. "Constraints, Allometry, and Adaptation: Food Habits and Reproductive Biology of Australian Brownsnakes (*Pseudonaja:* Elapdiae)." *Herpetologica* 45 (1989): 195–207.

Sluys, R. "On Adaptation, the Assessment of Adaptations, and the Value of Adaptive Arguments in Phylogenetic Reconstruction." *Zeitschrift fur Zoologische Systematik und Evolutionsforschung* 26 (1988): 12–26.

Turner, John R. G. "The Evolutionary Dynamics of Batesian and Muellerian Mimicry: Similarities and Differences." *Ecological Entomology* 12 (1987): 81–95.

Webb, Jacqueline F. "Gross Morphology and Evolution of the Mechanoreceptive Lateral-Line System in Teleost Fishes." *Brain Behavior and Evolution* 33 (1989): 34–53.

Young, Craig M. "Direct Observations of Field Swimming Behavior in Larvae of the Colonial Ascidian *Ecteinascidia Turbinata.*" *Bulletin of Marine Science* 39 (1986): 279–89.

Zink, Robert M. "The Study of Geographic Variation." *Auk* 106 (1989): 157–60.

WORKS CITED

Bazerman, Charles. "Scientific Writing as a Social Act: A Review of the Literature." *New Essays in Technical and Scientific Communication: Research, Theory, Practice,* ed. Paul V. Anderson, R. John Brockmann, and Carolyn R. Miller, 156–84. Farmingdale, N.Y.: Baywood, 1983.

Bazerman, Charles. *Shaping Written Knowledge: The Genre and Activity of the Experimental Article in Science.* Madison: University of Wisconsin Press, 1988.

Collins, Harry. *Changing Order: Replication and Induction in Scientific Practice.* Beverly Hills: Sage, 1985.

Edge, David. "Quantitative Measures of Communication in Science: A Critical Review." *History of Science* 17 (1979): 102–34.

Gilbert, G. Nigel. "Referencing as Persuasion." *Social Studies of Science* 7 (1977): 113–22.

Gilbert, G. Nigel. "The Transformation of Research Findings into Scientific Knowledge." *Social Studies of Science* 6 (1979): 281–306.

Gilbert, G. Nigel, and Michael Mulkay. *Opening Pandora's Box: A Sociological Analysis of Scientists' Discourse.* Cambridge: Cambridge University Press, 1984.

Kinneavy, James. *A Theory of Discourse.* Englewood Cliffs, N.J.: Prentice-Hall, 1971. Reprint. New York: Norton, 1980.

Kuhn, J. Thomas. *The Structure of Scientific Revolutions.* Chicago: University of Chicago Press, 1970.

Latour, Bruno. *Science in Action: How to Follow Scientists and Engineers Through Society.* Cambridge: Harvard University Press, 1987.

Latour, Bruno, and Steve Woolgar. *Laboratory Life: The Construction of Scientific Facts.* Princeton: Princeton University Press, 1986.

Myers, Greg. "The Social Construction of Two Biologists' Proposals." *Written Communication* 2 (1985): 219–45.

Myers, Greg. "Text as Knowledge Claims: The Social Construction of Two Biology Articles." *Social Studies of Science* 15 (1985): 593–630.

Price, Derek J. de Solla. "Networks of Scientific Papers: The Pattern of Bibliographic References Indicates the Nature of the Scientific Research Front." *Science* 149 (1965): 510–15.

Small, Harry G. "Cited Documents as Concept Symbols." *Social Studies of Science* 8 (1978): 327–40.

Watson, J. D., and F. H. C. Crick. "Molecular Structure of Nucleic Acids: A Structure for Dioxyribose Nucleic Acid." *Nature* 171 (1953): 737–38.

8

ANGELS IN THE ARCHITECTURE

A BURKEAN INVENTIONAL

PERSPECTIVE ON "SPANDRELS"

JOHN LYNE

Rhetorical invention, concerned with how speakers and writers can determine what is sayable on a given topic, is the most content-centered part of the rhetorical art. The tradition has produced various strategies for helping rhetors to create sayables, including topical lists and common places, "stock" issues, tropes and figures, and the taking of different perspectives. There have been various calls for a renewed attention to rhetorical invention as an engine of argument; more specifically, some scholars have begun to analyze scientific arguments in terms of inventional strategies.[1] I believe it can be productive to approach "The Spandrels of San Marco" with an eye to the inventional strategies it uses and to some that it might have used.

The Gould and Lewontin essay pits a "pluralist" conception of evolution against an "adaptationist" program. Its rhetoric is generated, or at least triggered, by the image of the spandrels. This figure is used to play architectural considerations against a model that, it would appear, reduces all structural aspects to single, atomized "traits." Image and argument work together to underscore the importance of resisting atomization, both in describing natural processes and in the activity of theorizing itself. The antiatomistic, or structuralist, theme recurs at several points in "Spandrels," and at different levels of analysis—for instance, in Gould and Lewontin's lament over the separation of morphology and modern biology (atomized inquiry), in their *Baupläne* arguments, in their movement across various fields of knowledge, and in their professed pluralism. The issues are partly logical ones and partly empirical ones, but the rhetoric of the essay is powerful not because it rushes to us with some new empirical finding or theoretical breakthrough but because it invents powerful arguments from materials already at hand and presents them in a persuasive way. That is what it means to succeed at rhetorical invention.

The object of rhetorical invention, as distinguished from any other form of creation, is to induce movement of discourse through a series of points or perspectives. As distinguished from logic, invention yields consideration of things that do not flow from straightforward deductive or inductive inference. It is generally achieved by acts that I will call *juxtaposition* and *rounding out*. By juxtaposing a familiar or readily understood pattern of any sort—visual, textual, figurative, indexical, tropic, or enumerative—one produces a kind of road map through a less familiar territory. Gould and Lewontin thus impose the image of the spandrel, easily grasped by the nonexpert, upon biological reasoning, giving the reader a new lens through which to see it.

This much is only incipiently an argument, however. To make the new perspective function argumentatively, one must round out the image, so to speak, by inventing architecture within the context of biological theory. Thus the authors generate implications of thinking about structures, relationships—architecture—and these rebound in provocative ways throughout the essay. Once the act of juxtaposition occurs, the architectural figure of the spandrels gets rounded out partly in accordance with the spirit in which it is introduced. But then one suspects it begins to round itself out in partly unforeseen ways as well. There is an active spirit hiding in the apparently static architectural imagery.

Dramatism

Kenneth Burke's "dramatistic" method offers a generally useful strategy that can shed some light on the rhetoric of the Gould and Lewontin essay. The central notion of dramatism is that well-rounded accounts of human action have certain constituents that are presupposed by the very grammar of action.[2] Thus, for instance, one cannot logically have an action without an actor or agent. Moreover, actors do not act without purposes, without means, or outside a context. These commonsense notions are all a part of the chemistry of action, but they are configured in different molecular arrangements in accounts of action. For instance, when people are portrayed as being compelled to act in certain ways by pressure from their environment (Burke's "scene" of the action), the "scenic" component may dominate the account, almost to the exclusion of references to the purposive agent. The familiar courtroom strategy of blaming a bad environment for the actions of the accused competes with the equally familiar strategy of asserting indi-

John Lyne

vidual responsibility for action. The same action can thus be cast up and contested in different ways, depending on which of the grammatical constituents is "featured" and how the "ratios" between terms are configured.

Burke, in short, works through what he calls a "dramatistic pentad," five terms that constitute the grammar of action: the *act* itself; the *agent* performing it; the means or *agency* used; the *scene,* or context; and the *purpose* of the action. If one speaks as a "pragmatist," one might give more emphasis to means (Burke's "agency") than to purposes; accounts of action emphasizing character feature "agents" more than environmental factors ("scene"); and so on. The kind of relationship posited between any two terms is a "ratio," which may be variously adjusted. Disputes about "free will" versus determinism, for instance, play upon the problematic scene–agent ratio; that is, they wage a rhetorical contest over how much weight should be given to circumstance and how much to personal choices in accounts of action. Arguments about the relative priority of means versus ends are a play upon the agency–purpose ratio; and so on. Between any two terms in the pentad, there will always be a tension; and among the five there will be a dynamic interplay. These tensions and dynamics are creatively exploited by rhetoric.[3]

One may speak without direct reference to any of the grammatical components of action, of course. The point is that the relationships of the grammatical constituents are always lurking as rhetorical possibilities, as ways to alternatively configure and round out the analysis. One can usually show that the pragmatist is in fact deeply committed to certain purposes, not merely expediency, thus setting the stage for playing the one "motive" off against the other, or for exploring their tensions. In that sense the grammatical terms for action are inventional as well as critical resources.

The dynamic tension between physical motion and motivated action is a pervasive theme in Western thinking, and its rhetorical overtones can be picked up in scientific writing as in virtually any other. Although Burke contrasts meaningful human action with natural processes, many of his examples reveal how, perhaps unavoidably, anthropocentric understanding tends to conflate the two, producing depictions of nature as "motivated" in ways analogous to how humans are motivated. Historically, the two have been intertwined. In prescientific times, the doctrine of animism, which held the natural world to be moved by the various spirits inhabiting it, provided an obvious continuity between human action and natural processes. In modern science, one has to work a bit harder to find the points of continuity, because science does not usually ascribe motives or purposes to nature.

The general culture, on the other hand, seems to mix scientific and purposive notions of nature in a variety of ways, ranging from intellectually sophisticated religions that see God working through nature to the familiar pastiche of science and the paranormal that is the stuff of supermarket tabloids. The point to be made across the board, from the dramatistic perspective, is that it is very difficult, if not impossible, to understand things in categories that escape anthropomorphism, which in turn suggests that the language and imagery of human motivation will also influence discourse about nature. Modern science, in its methods and theories, struggles to escape the limitations of anthropocentric explanation. But science cannot be utterly discontinuous with culture and its rhetorics. It seems reasonable to expect, therefore, that a dramatistic perspective can open up additional dimensions of the adaptationist–pluralist controversy represented in the Gould and Lewontin essay.

It may be easy enough to make the world of physics seem only a world of "motion," not action (at least one has to dig beneath the surface of physics to find the signifiers of action), but in biology the telltale signs of action crop up everywhere. Forms of life are said to "struggle" in their battle to survive. Nature, the all-powerful arbiter, is said to "select" those that are to stay on the lifeboat and those that are not. Even genes are now said to apply "strategies" to secure the continuation of their progeny. Perhaps this is all no more than figurative language, just enough personification to make complex processes comprehensible. Perhaps we as humans cannot think in any other way. Or perhaps we are just repeating some age-old habits, good or bad. In any case, the language of action in writings on biology invites the Burkean probe to see how the grammar is being used and how it is being deployed rhetorically.

What, for instance, is the *scene* in which the story takes place? Gould and Lewontin in "Spandrels" describe a setting in which inquiry has been boxed in by restrictive academic practices. At least part of the problem, they contend, is that the modern academic landscape has left the traditions and insights of morphology virtually cut off from modern biology (160). A more pluralistic approach would presumably require changes in that general academic environment. It might increase the variety of applicable methods *(agencies)* for investigating these complex issues. Every element of an action has ramifications for every other: a better intellectual environment would produce better intellectual action.

John Lyne

Who's On First?

Who or what is *acting* in the account given by Gould and Lewontin? This is a good place to start our probe. At one level, we have the actions of the scientists whose theories are in contention. The matter of who is contesting these theories, and in the names of which patrons and principles, is very much a part of the drama enfolding in the essay. It appears that those scientists with whose program Gould and Lewontin disagree have acted not altogether wisely. The problem has not been one of particular errors so much as one of a bad pattern of thinking. (The anthropological controversy over Aztec cannibalism makes the point that this pattern is not confined to biology.) It appears that the adaptationists have bought into a faith that has failed them. Gould and Lewontin assign them a patron saint in the figure of Dr. Pangloss, and they coin a name for the faith (obviously, not a name adaptationists would have chosen). Pangloss was no ignoramus; in fact, he knew quite a lot—which only shows that very bright minds can be trapped by doctrinal prejudice and platitudes. And so it is with the adaptationists, who accord to natural selection something near "omnipotence," as if the object of religious awe.

If Gould and Lewontin are iconoclasts, they are no agnostics, however. They too act with piety, perhaps more self-consciously than the actors they contest. Their doctrinal correction is offered in the name of the deified Darwin, in an explicit attempt to "invoke God's allegiance" (155). They comb through the sacred texts with care to produce the necessary authority for their position. Avowing that they are no fundamentalists ("We do not now regard all of Darwin's subsidiary mechanisms as significant or even valid. . . ." [156]), the authors endorse the great message of pluralism. And in the spirit (if not the letter) of Darwin, they present their partial hierarchy of alternatives to extreme adaptationism. They show the single-minded adaptationists a fuller menu of possibilities, each supported with plausible examples, creating an inventional checklist—a series of slots that might be investigated whenever the question of adaptation and selection arises. The argument of this text might therefore be reenacted in response to any further instance of adaptationist research.

At one level, then, Gould and Lewontin are the protagonists of the drama, and hence agents. At another level, they are telling the story of evolution itself. Dramatism would predict that this story of Nature itself will produce actors, as indeed it does. Traditionally, Nature is indeed an actor—wise, cruel, inscrutable—whose principal handiwork is to make the all-important selections in evolution. Nature determines which creatures will be admitted on the planetary lifeboat, not arbitrarily but in accordance with the Law. But organisms also have their roles in the drama. Their lot is to adapt, to struggle, and to succeed or fail. The stories switch back and forth between the two kinds of agent, one of them local, the other omnipresent.

A rhetoric that combines these two kinds of agents has been worked out fairly thoroughly in contexts other than biology. In his economic writings, for instance, Adam Smith stressed the role of the individual entrepreneur, but also said the market was guided as if by an "unseen hand," controlling the whole process for the general good. And to a large extent American public philosophy upholds a picture of self-determining individuals who belong to a single, purposeful collectivity. So we are rather accustomed to discourses that configure agency dually in this way. The adaptationist program has given us an image of Nature as the free market economist, a rational optimizing agent dealing in trade-offs and cost–benefit analysis (151).

Gould and Lewontin are reacting to a recent twist in the story. As Barry Schwartz has noted, in its extreme form adaptationism has arguably become a new kind of Social Darwinism, kept alive and well by a symbiotic relationship among major strands of economic, psychological, and sociobiological theory. Equally important, it has shifted the story from one about organisms and species to one about traits. This change in the dramatis personae sets traits free to determine their own course in the natural economy. Their connection with organisms and communities of organisms becomes only an accidental feature of their lives as traits seek their place in natural order. The atomization Gould and Lewontin speak of is, from a dramatistic standpoint, a displacement of organisms by traits on the stage of life.

One might not think at first that a trait would make a very good candidate for an agent. It would be as if the parts of a personality were themselves characterized as dramatic actors. In fact, however, the adaptationist program produces such actors, and they populate the literature of sociobiology, the study of the biological bases of social behavior. Sociobiology solves the dramatistic awkwardness by giving the active role to genes, which are the invisible correlates to traits. The genes act as "survival machines" programmed to perpetuate themselves, even to the extent that they will favor self-perpetuation at the expense of the host organism in those cases where the trade-off is necessary.[4] Genes are seen as acting "as if" strategically, to assure their own survival, and they enlist organisms and social techniques toward this end. E. O. Wilson, the best-known advocate of human sociobiology, summarizes one of the essential arguments of his book *On Human Nature* as follows:

John Lyne

"The brain exists because it promotes the survival and multiplication of the genes that direct its assembly" (4–5). It is not coincidental that Gould and Lewontin criticize Wilson's interpretation of the Aztec sacrificial practices, which they see as linked with the genetic version of the adaptationist program.

If Agents, Then Purposes

"The fact that, prior to the theory of natural selection, the term adaptation had teleological connotations should be completely irrelevant for a Darwinian," writes eminent biologist Ernst Mayr (116). Mayr may be right in saying that teleological connotations *should* be irrelevant for modern Darwinism, but there seems ample evidence that their presence lingers, mainly at the level of connotation and the rhetorical operations they permit. In fact, Darwin himself solved part of his own rhetorical problem by invoking teleological imagery. In constructing a plausible case for his theory of evolution, it was a rhetorical problem for Darwin that his audience could not readily accept the notion of nature without purpose. As Campbell has shown, Darwin found in the figure of the domestic animal breeder, gradually changing the traits of his livestock, one key to a successful rhetoric of evolution by natural selection.

Like Darwin, Gould and Lewontin use an example of striking purposefulness to illustrate what the theory claims not to be purposeful at all. Deliberately or not, "Spandrels" seems to reflect the irrepressibility of teleological thought and expression. It is a good case to illustrate Kenneth Burke's claim that the suggestion of *purpose* will insinuate itself between the lines even in those places where it is most expressly prohibited. One might even say that it is part of the basic architecture in accounts of action. In this instance, it supports a version of the agent–purpose ratio that competes with the one used by the adaptationist program.

Teleology figures into biology whenever biological entities are described as pursuing an end. In Aristotle's biology, "final causes," or purposes, operated explicitly in biological explanation. Latter-day critics of evolutionary theory have thrived on supposed analogies between cultural artifacts such as clocks and the "creations" of nature, used with the intent of showing how life could not have developed without a purposeful creator.[5] In modern biology, certain "teleonomic" characteristics of organisms are posited, but characterization of natural evolution as behaving purposefully will be introduced mainly for stylistic reasons, to help us conceptualize unities within complex processes.

In principle, accounts of evolution do not require references to *purpose*. In fact, the Aristotelian notion of *telos*, applied to biology, is considered laughable. But if one can locate action within the story, then purpose cannot be far behind. It is, in fact, very difficult to tell the story of organisms changing over time without using teleological, or purpose-laden, language. The grammar itself seems to militate against excluding it. Our language and culture seem to incline us to connect functions with purposes. What were the dinosaur's short front legs *for*? This is the question that adaptationists must resist asking, as natural as it might seem. Stories that feature organisms solving adaptation problems can produce the Lamarckian impression that changes in animals are the result of adaptive strivings by those animals. Judging by the way evolution is represented in science fiction movies and popular discourse, it would appear that the popular conception of evolution is still more Lamarckian than Darwinian. If legs have functions (as tools do), then it is difficult not to think of what their purpose is (as we do with tools). Gould and Lewontin remind the reader that acquired traits are not heritable, despite the tenacity of that view in the public mind.

Locating a purposive agent within a general life force has had philosophical proponents, most notably among French intellectuals from Bergson to Teilhard de Chardin. In fact, cosmological notions of higher-order mechanisms that regulate the general development of life enjoy considerable current popularity.[6] In evolutionary theory something like this apparently presents one of the conceptual possibilities. Speculation about higher-level mechanisms of evolution certainly does not require an appeal to cosmological principles, however. "Spandrels" itself speaks of hierarchic levels, and Gould has elsewhere advocated the view that macroevolutionary changes occur at the group or species level.[7]

Yet Gould and Lewontin are careful in walking the fine line between various sorts of reduction—the action must be carefully distributed over a complex assembly of players.[8] In pulling theory back from the "parts and genes" model, for instance, they do not want to go too far in the direction of greater wholes (mega-agents, so to speak), any more than they want to give too much agency to individual organisms. Their alternative to atomism is not the "strong" form advanced by some European evolutionists, who have posited an unknown mechanism that constructs the *Bauplan* itself. This, Gould and Lewontin feel, takes things too close to mysticism (159). They want to tell a story in which organisms, genes, and environment interact, thus avoiding a reduction to something like a *telos* within either type of "agent."

If we follow Burke's argument, merely banning the teleological in principle does not eliminate its rhetorical undercurrents. It may simply relocate them. The architecture of a cathedral, after all, represents human creation, not a natural one, and such creations are deliberately constructed according to preconceived plans, even if bent to material and structural necessities. It is a bit ironic, after all, that this powerful critique of the Panglossian view of nature, and of the general tendency to read purposes into nature, should take as its rhetorical peg such a monument to human purposefulness as a cathedral. Perhaps this says something about the limits of figurative comparison, or perhaps it illustrates the deconstructionists' point that all imagery can be inverted. In any case, one might turn the authors' own logic back on them and ask in what ways the case they construct has, itself, unplanned byproducts shaped by the very materials used. The reference to architectural blueprints, for instance, might open some unanticipated implications of design and planning, sending readers in directions quite at odds with Darwinian thinking.

Shifting the Scene

From a dramatistic perspective, one might say that Gould and Lewontin counter one agent–purpose ratio with another, and that it serves their purposes well. The dramatistic pentad may suggest additional strategies, however. If they had chosen an explicitly dramatistic strategy for criticizing the adaptationist program, one might ask, what other possibilities might have been available? The dramatistic perspective considers how the terms of the pentad are configured in relation to each other, and these configurations can to some degree be abstracted from particular settings. Indeed, by experimentally shifting the "scene," one can sometimes get a clearer sense of the contours of the drama. As a kind of experiment, and for fun, let us imagine how the adaptationist program might have been held up for examination within a completely different setting.

Suppose one were to apply the adaptationist scenario to the very human activity of engaging an audience rhetorically. Arguably, rhetorical practices call for some degree of "adaptationist" theory, insofar as one must adapt to an audience and environment. Put simply, rhetoric that does not make the necessary adaptation can be selected out. The speaker who does not find common ground with the audience is doomed to failure. In disciplinary writing, one must know how to command the established forms and formats, or one does not get a forum at all.[9] Thus we seem to have a good case for rhetorical performance as a kind of adaptation to an environment, which includes, among other things, an audience and its expectations. But suppose there developed a program of extreme adaptationism—one that reduced complex rhetorical actions to atomized traits, then proposed an adaptive story for each acting separately? What would such a program look like?

It might look a good deal like the program of political campaign consultants, or at least a certain stereotypical version of it, according to which their trade rests on a certain view of the relation of speaker to environment—a relationship that might be characterized as adaptation of messages and selection by public opinion. Rather than leave message design to random mutation, political consultants make an effort to foresee selectional forces and respond to them. They survey popular tastes and opinions to locate those presentational "traits"—a popular issue position, a personal characteristic, use of certain loaded words— to which the public responds. Test audiences are sometimes monitored for their emotional responses at intervals as brief as one second or less. A negative or positive response is taken to represent a strong "selection" principle, and what emerges is an "optimal" marketing strategy. (I think I have not caricatured unduly here.)

Proceeding further in this heuristic spirit, we might use Gould and Lewontin's pattern to critique these adaptationist assumptions. (Compare this characterization with the one offered on page 151 of "Spandrels.") Postmortem studies of successful political campaigns, according to the adaptationist paradigm, generally proceed in two steps: 1. campaign texts are atomized into elements, which are explained as optimally designed for their functions; and 2. when the claim to optimality becomes unconvincing, the experts point out that trade-offs were involved. (For instance, holding the line on taxes is a good campaign pitch only to the degree that it does not create the perception of a threat to popular social programs.) What is selected is seen as a trade-off between competing demands, the best possible outcome under the circumstances. Under fire, the political consultants evidence common styles of argument, including shifting accounts of why things went as they did, citing unforeseen factors that were not figured into the equation, and so on. Rather than look to different modes of explanation, they try to fine-tune their adaptationist stories. The Panglossian view remains intact.

Certain styles of argument predictably occur (cf. pages 152–53 of "Spandrels"). When adaptationists consider rhetorical form, they see it only in terms of immediate utility. This limited perspective would incline them to ask what President Reagan's characteristic bob of the head when he spoke was *for.* Given his rhetorical success, it would never occur to them to see any of his traits in other than functional terms. Nor would they consider that political behavior is filled with nonadaptive features, or that certain forms of political speech are simply passed down from one generation to the next because they are part of, as Wittgenstein might say, a form of life. No, in the words of Gould and Lewontin, if one adaptive argument failed, they would look for another or attribute the failure to an imperfect data set.

In contrast, the story of adaptation becomes much more complicated, and more interesting, if one follows the spirit of pluralism shown in "The Spandrels of San Marco." What determines campaign success? How do some strategies get weeded out and others perpetuated? Here is a range of alternative stories parallel to the alternatives offered by Gould and Lewontin (156–59):

1. A campaign strategy might reflect *no real adaptation to or selection by the public at all*. In the U.S. presidential election of 1988, for instance, it was widely observed that virtually no one, from the public to the candidates, was satisfied with the character of the "negative" campaign advertising, yet no amount of dissatisfaction seemed to affect the course of it. Moreover, the major consulting firms roll on, regardless of failures.

2. Particular features of a politician's campaign *may not be objects of adaptation or selection at all*. Some traits may persist simply because they are tolerated, not approved, by the public. Voters understand that they cannot pick and choose among the traits of a campaigner—they have to choose a whole package. Hence it is risky to speak as if specific traits were selected in elections.

3. *Selection and adaptation are sometimes decoupled*. An effective campaign strategy might be dropped or added simply because of the candidate's disposition or for any of a variety of other reasons unrelated to its "optimal fit" to the political environment.

4. *Adaptations and selection take place, but there is no selective basis for differences among adaptations*. There may be several viable but alternative campaign themes that can each adapt to a given environment. If so, then one could not reach any conclusion about the selective basis for one over the others.

5. *Adaptation and selection take place, but the adaptation is a secondary utilization of the parts present for reasons of architecture, development, or history*. The general "architecture" of the political culture and its mechanisms are put to various secondary uses. If there is a two-party system, the structural logic of a two-party election will dictate certain kinds of compromises, appeals to the middle, coalitions, and so forth.

To a superficial view, many of these outcomes will appear to have been specifically "chosen" in a campaign. But the pluralist does not assume that every political ritual has an adaptive, still less an optimally adaptive, function.

Caveats

As Gould and Lewontin show with their spandrels analogy, and as this last little experiment on political campaigns attempts to show, rhetorical resources can be borrowed from one context to produce insights and invention in another, almost regardless of material content. The dramatistic perspective can give one form of guidance in this process, but there are others. Juxtapositions of forms, images, and lines of argument are generative within scientific discourse as they are elsewhere. This is not to say that doing science is just a matter of creating plausible dramas or telling useful stories, however. For just as politicians should in the end do more than tell plausible stories, so too must scientists be held to account for additional responsibilities. Scientific talk moves in and out of a variety of rhetorical contexts, including politics and the justice system; but scientific discourse must eventually square itself with theory and observation.

This is not to suggest an acultural conception of science. To most who would pick up this book, perhaps, the very notion of a science purified of cultural influences would seem a contradiction in terms. All forms of representation are, after all, rooted in culture. Moreover, the idea that there must be a single correct description of natural phenomena seems to have been abandoned even by much of realist philosophy of science.[10] In an eagerness to show the cultural influences on the discourses of knowledge, however, we must be careful not to rush to reductive formulas of our own (or, indeed, to brandish the term "positivist" as a way of limiting inquiry). In that connection I hasten to observe that the relationship between dramatism and the scientific content of the Gould and Lewontin essay is a *rhetorical* relationship, not strictly speaking a scientific one. Rhetorical analysis can help us better understand how scientific processes, as well as the rhetoric that attends them, are embedded within a multileveled web of cultural and linguistic practices.

To the degree that the social sciences or even the humanities borrow the terminology or concepts of the natural sciences—and the terminology of evolution pervades both—they need to be aware of the limitations that some of these methods may hold. Gould and Lewontin do a service to a broader audience than those expert in the study of evolu-

John Lyne

tion. They have also sent up a warning flag to those in the humanities and social sciences who might too confidently latch onto what is perceived to be a proven theoretical program in a more prestigious discipline. If the apparent strength of the adaptationist program is taken as warrant for applying a similarly one-sided mode of explanation in other disciplinary contexts, knowing how to "round out" the alternative possibilities can at least be a counterbalance to reductive approaches, even those that seem to speak with the authority of science. Burke's dramatistic pentad offers a way of scouting out the repressed constituents of action, thereby rounding out the story in terms of its plural elements.

NOTES

1. See Lloyd F. Bitzer and Edwin Black, *The Prospect of Rhetoric;* Michael Billig, *Arguing and Thinking: A Rhetorical Approach to Social Psychology;* Lawrence J. Prelli, *A Rhetoric of Science: Inventing Scientific Discourse;* and Herbert W. Simons, *The Rhetorical Turn.*

2. See Kenneth Burke, *A Grammar of Motives.* For helpful glosses on dramatism, see Hugh Dalziel Duncan, *Communication and Social Order,* and Herbert W. Simons and Trevor Melia, eds., *The Legacy of Kenneth Burke.*

3. For a good example of the use of the pentad to understand the rhetoric of social science, see Gusfield.

4. The term "selfish gene," along with some of the imagery discussed here, has been introduced to a popular audience by Richard Dawkins.

5. See Richard Dawkins, *The Blind Watchmaker.*

6. The resurgent popularity of this kind of cosmology is discussed by Stephen Toulmin in *The Return to Cosmology.*

7. See S. J. Gould, "Is a New and General Theory of Evolution Emerging?"

8. Compare this to the "dialectical materialist" perspective taken by Levins and Lewontin (104ff.), which sees the organism and the environment as codetermining. The organism is described as both subject and object of its own evolution—not simply adapting to its environment but actually playing a significant role in constructing that environment.

9. See Charles Bazerman, *Shaping Written Knowledge: The Genre and Activity of the Experimental Article in Science.*

10. See, for instance, Rom Harré, *Varieties of Realism: A Rationale for the Natural Sciences.*

WORKS CITED

Bazerman, Charles. *Shaping Written Knowledge: The Genre and Activity of the Experimental Article in Science.* Madison: University of Wisconsin Press, 1988.

Billig, Michael. *Arguing and Thinking: A Rhetorical Approach to Social Psychology.* Cambridge: Cambridge University Press, 1987.

Bitzer, Lloyd, and Edwin Black. *The Prospect of Rhetoric.* Englewood Cliffs, N.J.: Prentice-Hall, 1971.

Burke, Kenneth. *A Grammar of Motives.* 3rd ed. Berkeley: University of California Press, 1969.

Campbell, John A. "Scientific Discovery and Rhetorical Invention: The Path to Darwin's *Origin.*" In *The Rhetorical Turn: Invention and Persuasion in the Conduct of Inquiry,* ed. Herbert W. Simon, 58–90. Chicago: University of Chicago Press, 1990.

Dawkins, Richard. *The Blind Watchmaker.* New York: Norton, 1986.

Dawkins, Richard. *The Selfish Gene.* New York: Oxford University Press, 1976.

Duncan, Hugh Dalziel. *Communication and Social Order.* New York: Bedminster, 1962. Reprint. New York: Oxford University Press, 1968.

Gould, Stephen Jay. "Is a New and General Theory of Evolution Emerging?" *Paleobiology* 6 (1980): 119–30.

Gusfield, Joseph. "The Literary Rhetoric of Science." *American Sociological Review* 3 (1984): 70–83.

Harré, Rom. *Varieties of Realism: A Rationale for the Natural Sciences.* New York: Basil Blackwell, 1986.

Levins, Richard, and Richard C. Lewontin. *The Dialectical Biologist.* Cambridge: Harvard University Press, 1985.

Mayr, Ernst. *Toward a New Philosophy of Biology: Observations of an Evolutionist.* Cambridge: Harvard Belknap, 1988.

Prelli, Lawrence. *A Rhetoric of Science: Inventing Scientific Discourse.* Columbia: University of South Carolina Press, 1989.

Schwartz, Barry. *The Battle for Human Nature.* New York: Norton, 1986.

Simons, Herbert W., ed. *The Rhetorical Turn: Invention and Persuasion in the Conduct of Inquiry.* Chicago: University of Chicago Press, 1990.

Simons, Herbert W., and Trevor Melia, eds. *The Legacy of Kenneth Burke.* Madison: University of Wisconsin Press, 1988.

Toulmin, Stephen. *The Return to Cosmology.* Berkeley: University of California Press, 1982.

Wilson, Edward O. *On Human Nature.* Cambridge: Harvard University Press, 1978.

9 TACTICS OF EVALUATION IN

GOULD AND LEWONTIN'S

"THE SPANDRELS OF SAN MARCO"

JEANNE FAHNESTOCK

No easy generalizations about "the rhetoric of science" can come from Stephen Jay Gould and Richard Lewontin's "The Spandrels of San Marco and the Panglossian Paradigm." The unusual essay is not a report or research article; it does not argue for the existence of a phenomenon or experimental result or for a causal factor, nor does it debate issues of definition and classification, though it builds on many such subsidiary arguments. Instead it attempts, overall, to devalue one method of evolutionary interpretation in favor of another, or rather of various others. In that sense, "Spandrels" comes closest in kind to the review article, that genre of academic discourse that surveys and assesses the state of a field. Every discipline creates occasions for these, and in many fields "annual reviews" are vehicles for articles that measure progress in a subdiscipline (they usually find progress) or that report new discoveries and techniques. Gould and Lewontin's piece, however, is more overtly polemical and deliberative in its ends than the typical summarizing review article.[1]

Special features of the discipline and of these two authors may explain the uniqueness of "Spandrels" in terms of the norm of the review article. Several disciplines pursue questions of adaptation and evolution, seeking to "explain" the existence of physical or behavioral traits chosen for scrutiny. But while genetics and molecular biology may pursue these questions through experimental means, evolutionary biology, like paleontology and archaeology, comes closer to being an *interpretive* discipline. Elsewhere Gould coined the term "historical science" as opposed to "experimental science" for disciplines that do not act directly on what they study (*Hen's Teeth*, 122, 257; see Miller and Halloran in chapter 6 of this book). Some evolutionary biologists certainly

158

conduct experiments; "Spandrels" refutes the design of one such experiment. But experiments of necessity, designed according to a priori hypotheses, produce results that must be explained by some interpretive routine. The process is not as easy as looking for tritium in a cold fusion jar or insulin in genetically engineered *E. coli.*

Thus, while their objects of inquiry are living and fossil organisms rather than texts, evolutionary biologists must nevertheless approach their objects hermeneutically, and the correct hermeneutics becomes a matter of argument. In one sense, what we have in "Spandrels" is an argument favoring one theory of interpretation over another. There is no obvious test by which competing theories of interpretation or explanations of a trait can be measured, no simple prediction that can lead unambiguously to the falsification of an interpretive theory. In a sense, what is at stake is the kind of explanation an audience will find satisfying.

Another complication in this argument comes from the fame of its two authors, a fame that may either help or hinder its persuasiveness. A critique like Gould and Lewontin's "Spandrels" might be aimed ideally at uncommitted readers, neophytes in a discipline; a critique normally amounts to a contest between two views for the allegiance of the spectators. But, as Davida Charney shows in detail in chapter 11, readers familiar with the authors and the issues respond idiosyncratically. Gould and Lewontin have associated themselves so polemically with the critique they defend in this article that their position and many of their arguments could be anticipated by prepared readers from the moment they read the title and the authors' names. If they continue reading anyway, readers familiar with the authors' reputations will find it easier to dissociate the manner of the argument from the points being made. Any such separation between technique and content is potentially damaging to an argument. However, Gould and Lewontin's fame may also win attention despite a reader's disposition to disagree, particularly in the case of Gould, whose flamboyance in arguing is well known from his column in *Natural History.* Some readers may simply wish to be entertained by a certain style of argument. They dissociate to appreciate.

Though "Spandrels" is unusual for the review genre and though the discipline and the reputations of the authors are special constraints on analysis, much in this argument will nevertheless be familiar to rhetoricians, especially rhetoricians with a classical bent. Indeed, while one chapter in this book characterizes "Spandrels" as a "discourse of modernity," I would call it a "discourse of antiquity" since many of the argumentative strategies described in classical rhetorics, particularly the Aristotelian common topics, can be identified in this late twentieth-

Jeanne Fahnestock

century argument in evolutionary biology. Specifically, the topical analysis of an argument characterizes basic strategies in a way so general that it is possible for rhetoricians to see similar patterns even in arguments on vastly different subjects. Aristotle described these generic moves in the *Rhetoric* (see especially book 2, chapters 18–23) and in the *Topics*. A modern work that synthesizes both these less-than-accessible sources is Chaim Perelman and Lucie Olbrechts-Tyteca's *The New Rhetoric*, the immediate source for much of the analysis which follows. Aristotle, for example, talked about "size," the bigger and the smaller, the more and the less, as a topic common to all oratory (*Rhetoric*, 129). *The New Rhetoric* rechristens these "loci of quantity," value warrants that say, for example, that more of a good thing is better than less of it (85).

From the perspective of classical rhetoric, reincarnated in *The New Rhetoric*, the only analytical challenge lies in seeing how features of the discourse that seem referential and informational, or even accidental and trivial, embody certain standard moves and hence serve the argument—or, to put this point in the lexicon of the discipline under discussion, how the epiphenomenal is adaptive. Exploring some of the connections between stylistic choices in Gould and Lewontin's text and argumentative strategies—by means of the well-established analytical tactics of classical rhetoric—is the aim of the following analysis.

Overall Goals and Strategies

Gould and Lewontin have three interconnected goals in "Spandrels." The first, very clearly articulated goal is, as I have stated, the demotion of a single kind of evolutionary interpretation in favor of several kinds of explanations.[2] The overall scheme that accomplishes this goal seems a clear and common one: discredit one view by showing its flaws and inconsistencies and recommend another that presumably avoids those flaws. But instead of replacing one kind of interpretive routine with another fully elaborated and consistent explanatory theory, a fairly tough thing to do, Gould and Lewontin simply want adaptationist explanations to move over and make room for a pluralism of interpretive strategies. They want, in other words, to replace the one with the many, the less with the more. Once I recast the argument in these very general terms, the pattern becomes obvious.

Since Gould and Lewontin want to persuade their readers that multiple interpretive strategies are better than a single approach, their second goal, in effect, is arguing, ideally to an uncommitted audience, about what standards of evidence it should find convincing, about what patterns of thinking it should follow—about, in short, what kind of audience it should be. To accomplish this goal of audience construction, "Spandrels" creates roles that readers are to identify with or to reject. (See also Gragson and Selzer, "Fictionalizing," and chapter 10 of this book.)

But if they are to accomplish the first and second goals together, Gould and Lewontin must, in effect, accomplish a third: recasting a disciplinary issue as essentially a moral one. In other words, the choice offered readers becomes not only one between better or worse interpretive routines but even between a better or worse character as an evolutionary biologist and a moral being. This third goal is accomplished primarily by the extensive analogies in part 1 of "Spandrels" and by various syntactic gestures that transfer value from this illustrative material to key disciplinary terms.

Rather than emphasize a one-to-one fit between device and goal, I will discuss several strategies that accomplish these three goals simultaneously. I take the position that the precise wording of an argument is a determining aspect of its persuasiveness so that, although it is certainly possible to paraphrase various arguments (indeed, teaching topical invention would be impossible without paraphrases at higher and higher levels of generality), specific argumentative moves are often achieved by precise linguistic choices that are better repeated than summarized. Specifically, I am going to look at certain argumentative and stylistic maneuvers that are prominent in the classical tradition and in the recent reformulation of that tradition by Perelman and Olbrechts-Tyteca: how the article creates an opposition; how certain syntactic devices are used to transfer value; how series, register shifts, and sentence length are exploited to continue the characterization of two sides; and how the argument relies, basically, on two argumentative strategies identified by Aristotle.

Creating an Opposition

Perhaps the least obvious tactic in this argument is its creation of an opposition to point to and push against. To belabor the obvious, in order to oppose a view, one has to have a view to oppose. One has to encapsulate it, distance it, be able, as it were, to compact it into a sphere, attach a string, and dangle it at arm's length for an audience's contemplation. And the string connecting the sphere to the hand should be invisible. For if Gould and Lewontin have no opposition, they have no exigence; and if they have no exigence, they have no argument. (One can easily see how this process would work in reverse; the need to create an argument, say because of publishing pressures in academia, requires creating an exigence, which requires the creation of an opposition.)

The easiest way to create something in a text is to give it a name. Thus one of Gould and Lewontin's techniques is, simply, labeling and thus bringing into existence an approach called the "adaptationist programme." Though when Gould and Lewontin introduce this term in the text they call it a "habit of thinking" (150), the label in the title and elsewhere in the text nevertheless suggests the existence of an organized body of evolutionary biologists pursuing a unified agenda with a consistent body of methods.[3] Both authors used the label "adaptationist program" in articles written separately but published around the same time. Lewontin's article in a 1978 issue of *Scientific American* speaks of an adaptationist program as something evolutionary biologists work out of (216) or that informs their work (217). It is a program they might abandon (228), were it not that adaptation is a real phenomenon. In this less polemical article (which nevertheless contains the alternatives to adaptationist explanations listed in "Spandrels"), the "adaptationist program" remains more an approach independent of a group than a group defined by its approach. In a 1982 *Science* article, Gould lists the "adaptationist program" as an additional but unnecessary consequence of the Darwinian worldview (381), a postulate vulnerable to a reprise of the critique in "Spandrels" (383). And in the collection of *Natural History* articles from the early eighties, *Hen's Teeth and Horse's Toes*, Gould makes "the adaptationist program" a subject that takes a verb, like a human agent who does things, but he puts the term in quotation marks, perhaps indicating a slight uneasiness with this usage (13). Repetition of "adaptationist program" in several forums, over a period of years, might have brought it into currency as a label for a particular approach and those who practice it.

If the title embodies one technique for creating an opposition, the first line of the abstract continues with two others: "An adaptationist theory has dominated evolutionary thought in England and the United States in the last forty years" (147). Having created an adaptationist program and therefore, implicitly, a something else, an approach or approaches other than the adaptationist, Gould and Lewontin now map it onto a geographical dissociation, making explicit the British allusion. A reader's mental map now has a location for an unnamed view or approach. The second tactic in this sentence seems somewhat surprising, since characterizing one's opposition as the dominant party in the field, even if a geographically circumscribed field, amounts to suggesting that one's own view is in the minority. But the tactic of limiting the initial adherence to one's views, as counterproductive as that seems, is certainly not an unheard-of device in scientific controversies (Fahnestock). It is a way of suggesting that these less widely held views are new views, up-and-coming views, the latest and best-informed views; and in science, presumption usually favors the latest rather than the established view. It is no accidental observation in Gould and Lewontin's sentence that the adaptationist view has been around for forty years. Later in the argument, the adaptationist thought style will be dismissed as an "old habit" and an "old argument" (152).

Despite the fact-like claims of the title and the first sentence, there was no discoverable external group in the late seventies that called itself "the adaptationist programme" and that, in the process, demoted itself from representing the entire field to just part of it. Insiders, particularly readers of a British journal, would probably read in the allusions to British evolutionary thought references to John Maynard Smith, the foremost British evolutionary biologist, especially since Maynard Smith arranged and attended the symposium that included "Spandrels" and contributed to it one of his own essays. Maynard Smith, originally trained as an engineer, practices a certain style of mathematical argument in evolutionary biology that involves identifying specific traits, like foraging behaviors, as functions that are ideally optimized under the pressure of natural selection. Maynard Smith refers to his own approach as "optimization theory." (For the opposition between Gould and Lewontin and Maynard Smith, see Gragson and Selzer in chapter 10 and in "Fictionalizing"). Maynard Smith is not, however, mentioned anywhere in the article or the references, perhaps since mentioning him could give "presence" to his writing (see *The New Rhetoric*, 115–20), where refutations of Gould and Lewontin's characterization could be found.

Not only do Gould and Lewontin have to bring an "adaptationist programme" into being, they must also make it an approach recognizably different in kind from their own. If explanations of traits differ only in the degree to which they depend on adaptationist explanations, then there are not two sides but only two "more or less" positions on a continuum. Only adjustments become necessary, not recantations, and the force of Gould and Lewontin's critique—chiding people who basically don't disagree with them—would seem out of place.

Darwin himself was adept at arguments that reconceptualize differences in degree and kind. In chapter 2 of the *Origin*, for example, he attacks the notion that individual species are radically different in kind from one another. The basic tactic for such reconceptualization is described in *The New Rhetoric:* "When we are confronted with two realms of a different order, the establishment of degrees within one of them is often for the purpose of diminishing the break between them" (347). A stark distinction between the rich and the poor, for example, is bridged by talking about "levels of income" in each group; the gulf between "certainty" and "uncertainty" (word opposites are a fertile source of differences in kind in the first place) shrinks when we talk about "degrees of certainty." Darwin, in a magnificent *gradatio,* divides species into subspecies, varieties, and individual living forms exhibiting individual differences. "These differences blend into each other by an insensible series; and a series impresses the mind with the idea of actual passage," so evolution from one species to another by the accumulation of small differences becomes possible, the difference in kind bridged by internal differences of degree (*Origin,* 45).

Gould and Lewontin want to do the opposite; they want to erase the gradations and create mutually exclusive categories, so that instead of conceptualizing their field as a spectrum of individuals holding more or less the same views but emphasizing them differently, they want opposing camps separated by an either/or test. This categorical difference cannot be the issue of whether adaptationists do or do not take "constraints" into account because Gould and Lewontin admit that adaptationists do so. Furthermore, it is easy to pull concessionary quotes from those who emphasize adaptationist explanations. In the third edition of *The Theory of Evolution,* for example, Maynard Smith says explicitly that "not all differences between animals can be explained as adaptive" (18). Nevertheless, a difference in kind has to be mapped over this difference in degree, and the difference chosen is a difference in seriousness. The mutually exclusive notions of taking something seriously or being dismissive provide two categories, an either/or test, to push the two views apart:

> Constraints upon the pervasive power of natural selection are recognized of course (phyletic inertia primarily among them, although immediate architectural constraints, as discussed in the last section, are rarely acknowledged). But they are usually dismissed as unimportant or else, and more frustratingly, simply acknowledged and then not taken to heart and invoked. (151)

So the sincerity with which one takes constraints to heart becomes the difference in kind. This difference in kind is not a particularly compelling one because "seriousness" can be easily reconfigured as something that comes in degrees. But it is a useful moral difference. Gould and Lewontin are, after all, up against a tough argumentative challenge, namely how to handle an opponent's trivialization of one's own main point—the put-down that occurs when you reveal your treasure and your opponent yawns and says, "Oh, that. We know all about that." Aristotle, citing Gorgias, advises killing jesting with earnestness (216); Gould and Lewontin kill trivialization with moral seriousness.

Transfer of Values

Part 1 of "Spandrels" opens with four analogies. The first two, detailing spandrels in St. Mark's Cathedral in Venice and bosses in the fan-vaulted ceiling of King's College Chapel, derive from a pun, for the term "architecture" is used in evolutionary biology in the sense of overall morphology, and Gould and Lewontin maintain that architecture, an organism's *Bauplan*, is a forceful restraint on adaptation. If one wishes to discredit a view of evolution that fails to consider an organism's architecture, one can scarcely do better than choose analogies from architecture in its common sense.

But while the first two architectural comparisons naturally transfer to the subject because the same word is used in both fields, the third analogy, drawn from Voltaire's Dr. Pangloss, does much of the work of the explicit evaluation. "Anyone who tried to argue that the structure exists because the alteration of rose and portcullis makes so much sense in a Tudor chapel would be inviting the same ridicule that Voltaire heaped on Dr. Pangloss" (149), and that Gould and Lewontin heap on the opposition. The Pangloss figure, apparently derived from J. B. S. Haldane (Maynard Smith, "Optimization," 42), condenses two complaints, for, first, Pangloss justified the world as it is as the best of all possible worlds, and, second, he explained every detail as serving some greater good (though his goods were all reductions to absurdity: noses were made for eyeglasses). Thus the Pangloss comparison connotes both insensitive meliorism and unimaginative reductionism (one whose over-simple scheme of interpretation drives him into unethical corners such as justifying venereal disease because of the products derived from New World trade). Furthermore, Panglossian simplification is a kind of moral weakness that seeks an overall purpose in the world as it is. (If you don't like simplicity, you call it reductionism; if you do like it, you call it elegance, the application of Occam's razor.)

Once this set of Pangloss terms is on the table, various stylistic techniques perform the work of transferring their negative connotations to disciplinary terms. Bringing two terms into juxtaposition in various syntactic structures suggests their evaluative equivalence. Thus part 2 of "Spandrels," the first move away from the colorful opening comparisons, establishes the synonymy of the descriptive and evaluative terms. "We wish to question a deeply engrained habit of thinking among students of evolution. We call it the adaptationist programme, or the Panglossian paradigm" (150). If the same thing can be named by two terms, then the two terms are equivalent, and the negative connotations of the one (the Panglossian paradigm) pass to the other (the adaptationist program), assisted by some heavy alliteration.

The *New Rhetoric* calls this technique "bracketing," presenting two terms as syntactically interchangeable to suggest their equivalence. The bracketing continues in the next sentence: "It [the adaptationist programme] is rooted in a notion popularized toward the end of the nineteenth century: the near omnipotence of natural selection in forging organic design and fashioning the best among possible worlds" (150). Here coordination (around the "and") accomplishes the bracketing. The next page gives us predicated equivalence: "The adaptationist programme is truly Panglossian" (151). To do the one is to do the other and invite ridicule à la Voltaire on Dr. Pangloss. The direct evaluation argument is so emphatic after this categorical proposition that Gould and Lewontin pause to acknowledge what their opposition might think: "At this point some evolutionists will protest that we are caricaturing their view of adaptation." But the Panglossian term is not abandoned so long as the tactic of the argument is the characterization of the opposed theory of interpretation. Indeed part 3 opens with a fresh epigraph from Dr. Pangloss on the Lisbon earthquake: "For it is impossible for things not to be where they are, because everything is for the best" (153). The constant foregrounding of this term of ridicule could easily offend rather than persuade those not convinced that a distinctly separate opposition exists. Or readers with a traditional sense of the conventions of scientific discourse might ask why Gould and Lewontin use value-laden comparisons at all when they have substantive points to raise. Gould and Lewontin actually answer this question: they resort to shock value to break the force of habitual thinking (150). But they choose to present strong evaluative appeals before raising problems with adaptationist explanations.

It is tempting to equate the Panglossian put-down with a consistent theme in other writings by Gould alone. Anyone who has read a selection of Gould's *Natural History* articles knows that he is a vociferous foe of teleology in evolutionary explanation. Again and again he sets himself against any belief that describes evolution as progress or that sees pattern or design in changes over time. Nature, in Gould's view, is not moving toward any preordained state; all is randomness and accident.

Furthermore, to believe otherwise is to be morally weak, not macho; it is like believing in something because, sentimentally, you want it to be that way. Interpreting too much pattern into nature is moral weakness because it demonstrates fear of facing the absolute uncompromising randomness and meaninglessness of the cosmos. Within the discipline, the adaptationist program is a version of this weakness because it apparently espouses a teleology and sees fitness as the result of evolution, opening the door for perfectionist (or, as this article would have it, Panglossian) arguments.

Register Shifts and Sentence Length

Once Gould and Lewontin have created an opposition and evaluated it negatively, indeed ridiculed it, they must sustain its isolation from and contrast to their own view. Each view will have, in effect, a stylistic domain in the argument, although of course the opposition's stylistic domain is "spoken" by the authors. The classical rhetorical doctrine of levels of style, each appropriate to a different occasion and effect, provides a way of understanding some of the language practices serving this feature of the argument in "Spandrels." The first-century B.C. *Rhetorica ad Herennium,* for example, defines three levels of style, the Grand, Middle, and Simple: "The Grand type consists of a smooth and ornate arrangement of impressive words. The Middle type consists of words of a lower, yet not the lowest and most colloquial, class of words. The Simple type is brought down even to the most current idiom of standard speech" ([Cicero], 253). Few texts of any length maintain a single level of style throughout. Far more typically, an extended text exhibits different styles from section to section, and just what these differences are and how they map over, or even create, other divisions of the text—say, major moves in the argument—are of particular interest. Readers often intuitively sense these subtleties of style and sometimes refer to the "voice" of the text. But the voice or, as I maintain, voices of a text, are usually resolvable into the co-occurrence of several features.

To create different "levels" of style, the most effective feature singled out by the anonymous author of the *ad Herennium* and by Cicero is word choice. ("Although a word has no force apart from the thing, yet the same thing is often either approved or rejected according as it is expressed in one way or another" [*Orator,* 359].) Speakers, for example, recognize different words and phrases as appropriate to different occasions or levels of conversation. Academics would rarely address a dean

Jeanne Fahnestock

as "buddy," though they may use that term for someone who has just cut them out of a parking space (not the dean, we hope). Switching from one level of formality to another in a conversation or text, the practice recognized by classical rhetoricians, is called a register shift or register variation by contemporary linguists (Lyons, 292). A register shift is a word or phrasing choice that becomes noticeable against a local norm; such shifts occur when an author suddenly borrows a term from a semantic field other than that of the subject or when the author, who has been writing formally, suddenly inserts a word, a phrase, a sentence that represents an informal choice. Admittedly, labeling some usage a "register shift" is often an interpretive call.

Gould and Lewontin occasionally use words and phrases that are surprisingly informal in the setting of the *Proceedings of the Royal Society.* For example, in part 1, in their fourth analogy, Gould and Lewontin criticize the sociobiological explanation of human sacrifice among the Aztecs as a way to supplement the supply of protein in the diet. The phrases italicized in the following excerpts from page 150 represent the register shifts:

> Why invert the whole system in such a curious fashion and view an entire culture as the epiphenomenon of an unusual way *to beef up the meat supply?*

> Since each new monarch had to outdo his predecessor in even more elaborate and copious sacrifice, the practice was beginning to stretch resources *to the breaking point.* It would not have been the first time that a human culture *did itself in.*

> . . . a practice awarding meat only to privileged people who had enough anyway . . . represents *a mighty poor way to run a butchery.*

Similar shifts occur in the discussion of flaws in an experiment testing the response of bluebirds to the model of a male bird placed in the vicinity of a nesting pair: "Male returns at times two and three, approaches the model, tests it *a bit,* recognizes it as *the same phoney he saw before, and doesn't bother his female*" (154; emphasis added). Other researchers repeating the experiment in a different species observe a different response but incorrectly reason, according to Gould and Lewontin, that the different behavior is also an adaptation, though to different circumstances: "Perhaps, they conjecture, replacement females are scarce in their species and abundant in Barash's [the first experimenter]. . . . Eastern bluebird males are *stuck with uncommon mates and had best be respectful*" (155; emphasis added).

Register shifts change the personality of the implied author (the ethos in classical rhetoric) and therefore momentarily change the relationship between author and audience. If a speaker changes register with a listener, suddenly becoming informal, the two achieve the sudden mutual recognition of insiders, a change in footing. Gould and Lewontin use register shifts when they characterize and refute arguments they find not only fallacious but foolish. They suddenly appear at the reader's elbow and together author and reader share the joke.

A register shift of another kind occurs in part 5 when Gould and Lewontin offer explanatory routines other than the adaptationist. Here, as in the following example, the lexicon becomes considerably more formal and disciplinary as Gould and Lewontin now speak for the multiple views they favor:

> Even if an allele is favoured by natural selection, some proportion of the population, depending upon the product of population size N and selection intensity s, will become homozygous for the less fit allele because of genetic drift. (156)

> At least three times in the evolution of arthropods (mites, flies and beetles), the same complex adaptation has evolved, apparently for rapid turnover of generations in strongly r-selected feeders on superabundant but ephemeral fungal resources. (157)

> Sweeney & Vannote (1978), for example, showed that many hemimetabolous aquatic insects reach smaller adult size with reduced fecundity when they grow at temperatures above and below their optima. Coherent, climatically correlated patterns in geographic distribution for these insects—so often taken as *a priori* signs of genetic adaptation—may simply reflect this phenotypic plasticity. (158)

There is another difference between the parts of the essay where Gould and Lewontin discredit exclusively adaptationist views and section 5, where they list their own. In the refutation of the (straw man?) argument by Barash, the average sentence length (leaving out quotations from others) is about eighteen words per sentence. In section 5, the average sentence length is about twenty-five words per sentence. Averages are less significant, however, than the simple observation that there are more long sentences when Gould and Lewontin positively promote alternatives to adaptationist explanations. Longer sentences mean (not inevitably but usually) greater syntactic complexity, more embedding, more modification, and thus denser and more complex thought, just what one expects from an approach that is more sensitive

Jeanne Fahnestock

to the complexities of nature and from authors speaking authoritatively within a discipline.

Enumeration

Enumeration, making lists, has two basic purposes or effects: series either open up or shut down possibilities. On the one hand, a rhetor may list things to give a complete counting or accounting of them, often with the added purpose of achieving climax in a list (i.e., ending it with a most important point or a final extreme along a continuum). On the other hand, a list may be given in such a way as to suggest that the enumerated set is large or incomplete, and the list itself a mere random handful scooped from an unnamed many. The purpose in the second kind of list making is as much to create the possibility of other items as it is to accumulate the items it names. Various devices of syntactic presentation produce an impression of complete or incomplete series, and Gould and Lewontin show an unerring instinct for creating series with appropriate effects at the appropriate points.

We need go no further than the abstract for an example. In a summary of the main faults of the adaptationist program (a summary that is itself a list of four items, each introduced with "for" for parallelism), the last item reads:

> . . . and for its failure to consider adequately such competing themes as random fixation of alleles, production of non-adaptive structures by developmental correlation with selected features (allometry, pleiotropy, material compensation, mechanically forced correlation), the separability of adaptation and selection, multiple adaptive peaks, and current utility as an epiphenomenon of non-adaptive structures. (147)

How many competing themes have the adaptationists failed to consider? The reader is less likely to come away with an exact count of those cited than with an impression that the adaptationists have forgotten or ignored many possibilities. It is obviously in Gould and Lewontin's interest to present the adaptationists as opaque to as many alternative explanations as possible.

The length of the series is one factor creating this impression of muchness or incompleteness, which we can call *copia*. After about three items (the number that suggests completeness in a series), readers tend to lose count of how long a series is unless there is some compelling pattern or numerical template imposed on it (e.g., the four seasons, the seven dwarfs). The series quoted above, a series within a series, totals six items, some of considerable length and syntactic complexity in themselves, which prolong the length and deepen the impression of "muchness," the more rather than the less. Furthermore, this series within a series has a series within it—"(allometry, pleiotropy, material compensation, mechanically forced correlation)"—and this third series enhances the impression of openendedness or copia in two ways. First, the elements in the parenthetical series are considerably shorter than those in the containing series, producing an impression of rapid ticking off from a list; such quickening of a series also creates an impression of random selection from many possibilities, a kind of onomatopoeia in the sense of pace. Second, the last two items in this series are not separated by a conjunction. The *Rhetorica ad Herennium* labeled this figure two thousand years ago as *asyndeton,* the suppression of conjunctions (331), and Aristotle noted that "the omission of conjunctions acts in the reverse way and makes a single one into many" (197).

When arguers wish to prolong a series, they can use the opposite figure, *polysyndeton,* an excessive use of conjunctions. When conjunctions are placed between each item in a series, the reader is naturally slowed down, and each item seems separately emphasized. With polysyndeton, the parenthetical series would read as follows: "allometry and pleiotropy and material compensation and mechanically forced correlation." When a list is emphasized in this way, it may convey an impression of completeness, of the weightiness of each item. By comparison with the opposite technique, we can see the force of the asyndeton in this series, the potential impression of a hasty tally of some of the things forgotten by the adaptationists in just one of many categories.

The means of presentation chosen in this series simply reinforces the explicit gesture that prefaces it; the introductory phrase, "such competing themes as," suggests a selection among a host of ignored alternatives. The reader is to imagine that other items might be forthcoming if the writer had the space to list them; furthermore, the unspecified quantity of evidence also functions as a kind of typicality warrant, much like the explicit claim "My examples are representative." How much readers trust that unnamed evidence exists, and that the named evidence represents it, is a function of their background knowledge and the credibility of the arguers. Pointing out this technique does not amount to a charge that the unnamed items do not exist; I am merely drawing attention to the rhetorical advantage of the device of incomplete presentation in suggesting quantity; "all of us," said Aristotle, "have to argue that things are bigger or smaller than they seem" (219).

Gould and Lewontin construct open-ended series at two other criti-

cal points. Section 5 is itself a list titled "A Partial Typology of Alternatives to the Adaptationist Programme" and introduced by the phrase
"we present an incomplete hierarchy of alternatives to immediate adaptation for the explanation of form, function, and behavior." Five numbered sections follow, and again, for the sake of copia, it's important
that there are more than three; the third section is subdivided further by
lower-case Roman numerals, but each section is extensively explained.
At this point in the argument Gould and Lewontin attack the dominance of adaptationist explanations by offering alternatives. The colloquial paraphrase of the argument here is "Look at all the things you
guys are forgetting." Clearly it is to their advantage to have as many
alternatives as possible; thus the explicit and implicit gestures of incompleteness and fullness.[4]

Section 2 of the article, labeled "The Adaptationist Programme," includes a list of faulty argumentative strategies used by the adaptationists,
introduced by the following sentences: "The adaptationist programme
can be traced through common styles of argument. We illustrate just a
few; we trust they will be recognized by all" (152).[5] The four styles of
faulty arguing that follow are, then, only a sampling of evidently many?
some? other errors in logic that could be cited. Giving four instead of
two or three also reinforces copia. But enumerating also requires separating, convincing an audience that each item in a list does have an
independent existence, and some readers may have difficulty separating items one and two on Gould and Lewontin's list: "(1) If one adaptive argument fails, try another"; and "(2) If one adaptive argument
fails, assume that another must exist; a weaker version of the first argument." Since the authors admit that (2) looks like a version of (1), another
list maker might not have generated two items out of this failing.

What about using lists in the opposite way, to shut down a series and
suggest complete enumeration, rather than incomplete selection from a
fuller set? First of all, we have to ask what the opposite techniques in
series construction are. Obviously, generalizing across possible contexts might be fairly foolhardy, since all kinds of gestures can accom-
pany series. But in general, naming three items (the last two separated
by a conjunction) can give the impression of an adequate series. (Naming only two items does something quite different.) The items in a
series can also suggest a progression along some gradient the audience
recognizes. In the middle of section 2, in which Gould and Lewontin
characterize their "opposition," and after specifying the two steps that
adaptationists follow, they suddenly interrupt themselves and allow
their opponents to speak: "At this point, some evolutionists will protest
that we are caricaturing their view of adaptation. After all, do they not

admit genetic drift, allometry, and a variety of reasons for non-adaptive evolution?" I would call this an almost-*prosopopoeia*, a half-way ventriloquism in the voice of the opponent. (Specifically, the "after all" is the informal concession marker, but the pronoun choice "they" continues to separate this group from authors and readers.) More to the current point, notice that the opponents make their case in a three-item series in which the last, catch-all term is a generalization that of necessity makes the series complete. While technically this series could incorporate every item that Gould and Lewontin accuse the adaptationists of forgetting, its construction as three items with a predictable conjunction conveys a minimal counterawareness.[6]

A three-item series also carries the negative comparisons of part 1 into later parts of the argument. The fifth alternative to adaptation (adaptation as secondary utilization) listed in part 5 refers the reader to the opening: "We have already discussed this neglected subject in the first section on spandrels, spaces and cannibalism" (159). In the final part, where Gould and Lewontin present in some detail a study of divaricate patterning in clam shells as an architectural constraint, the triad appears again: "But Seilacher is probably right in representing this case as the spandrels, ceiling holes and sacrificed bodies of our first section" (162). In both cases, but especially the first, the authors want to invoke with a kind of quick finality the argumentative point they made earlier. So these three analogies travel together in a complete little series.

Most of the lists in "Spandrels" (and there are enumerations within enumerations) serve copia. Gould and Lewontin are especially adept at suggesting a multiplicity of problems and pitfalls in the approach being discredited, and, by a kind of inherent logic of parallelism, they also manage to suggest that the only way to overcome these many problems is with many interpretive strategies, the pluralism that is the goal of this argument.

Part/Whole Arguments

The strategies involved in list making bring us closest to a critical aspect of Gould and Lewontin's value argument, an aspect that depends on a redefinition of the object of study in evolutionary biology. Again and again, Gould and Lewontin repeat the key identification: "Organisms are integrated entities, not collections of discrete objects" (151; the choice of "object" with its "inorganic overtones" is not accidental). Organisms are "integrated wholes, fundamentally not decomposable" (157); they cannot and should not be atomized into

separate traits or "pulled apart piece by piece" (160), wording that suggests violent dismemberment.

Once Gould and Lewontin have stressed the unity of an organic
entity, the antithesis of the adaptationist view as they have characterized it, they can exploit a common ordering of values that says the
whole has more value than any part. In the words of *The New Rhetoric*,
"The whole includes the part and is consequently more important"
(231), or, "the quasi-logical scheme has no trouble in ascribing greater
value to the whole, to that which includes, to that which explains the
part" (233). This appeal is most often used to order values, not necessarily to establish them, and it is ubiquitous: the welfare of the whole
team is more important than the well-being of a single player; the economic health of the entire country takes precedence over the prosperity
of a single section; a theory that explains all the data supersedes a
partial explanation; and so on. The greater value that attaches to the
organism as a whole is transferred to the interpretive strategy that has
the whole as its object. Quite simply, any approach that can explain a
whole organism is better than an approach that can explain only the
parts. In other words, a value ordering in one sphere is transferred to
another. To select again from the compendium of tactics in *The New
Rhetoric*, what we have here is a double hierarchy argument (377). Perelman and Olbrechts-Tyteca provide as an example of the double hierarchy strategy the theological argument that since God is greater than
humanity, God's laws are greater than human law.

In still another way Gould and Lewontin develop a classical argument that exploits what *The New Rhetoric* calls loci of quantity (elaborating on Aristotle's identification of arguments having to do with the less
and the more). The logic inherent in this strategy is simple: one can
frequently present a greater amount of some things as being better
than a lesser amount. However, the value attached to the numbers in
this argumentative move in "Spandrels" is the inverse of the part/whole
strategy based on the characterization of organisms as integrated entities. In section 5, the section we have singled out for its greater sentence length, its register shift to the disciplinary lexicon, and its strategy of open-ended series, Gould and Lewontin attempt to sketch in the
notion of a large set of possible interpretive strategies. Adaptationism
is then devalued because it is only one of many. The whole in this case
is not a unity but a collection greater by sheer numbers than any one of its
members—greater in quantity and, again, therefore, presumably better. The only way for adaptationists, if they existed, to resist this kind of
"swallowing up" argument would be to claim that adaptationism is the
ground or underlying assumption behind all the other strategies rather

than an equal sibling in a whole family of explanations. When, for instance, Gould and Lewontin cite the arthropod reproductive strategy in which females are eaten from the inside by their own progeny, they claim that "paedomorphic morphology *per se*" is not adaptive because it is merely "a by-product of selection for rapid cycling of generations" (157). But then, as they say, the rapid cycling is the adaptation.

Now we have "the many better than the one," especially after part 5, but we still need to connect this hierarchy to another to anchor it. We need, in short, another double hierarchy to order pairs of values and assure us that pluralism is an inherently better strategy than the guiding principle of adaptation, the central explanatory agency in evolutionary biology. The new hierarchy of choice is complexity versus simplicity. And we may be at the end of the pairs now, for the direct linking of complexity with pluralism occurs at two strategic places in the argument.

The first is at the end of section 4, which contains an extended argument from authority, notably Darwin's (for commentary on the uses of Darwin in "Spandrels," see Miller and Halloran in chapter 6 of this collection). The purpose of this section, conducted in the voices of earlier commentators on Darwin, is to show that Darwin did not consider natural selection the only explanatory agency in evolutionary biology. In the final sentence, after conceding that many of Darwin's other interpretive strategies are no longer believed, Gould and Lewontin write, "But we should cherish his consistent attitude of pluralism in attempting to explain Nature's complexity" (156). The second linking of these two key terms occurs in the very last sentence of the article: "A pluralistic view could put organisms, with all their recalcitrant, yet intelligible, complexity, back into evolutionary theory" (163).

The value pairs that link the unified organism with a pluralistic approach and a recognition of complexity clearly connect with what I called earlier the moral dimension of the issue, the argument with the audience over what kind of audience it should be. "Spandrels," modeling the character of the ideal evolutionary biologist, suggests that it is somehow unsophisticated to rely on only one interpretive schema rather than to practice the subtlety that draws on several competing schema. Anglo-American monism looks like a naive, rigid reductionism next to the playful, tolerant pluralism of the sophisticated continental theorists.

Dissociations

When Gould and Lewontin offer the continental alternative in their final section, they use still another device, especially

Jeanne Fahnestock

common in philosophical argument, a dissociation. A dissociation, according to *The New Rhetoric,* attempts to divide an entity in some way so that its negative aspects are carried away in one half of the resulting pair. The archetypal dissociation distinguishes the "real" or "true" from the apparent, borrowing a "philosophical pair" that has done yeoman's work in western philosophy (Perelman and Olbrechts-Tyteca, 411–19).

Perhaps anticipating assumptions about continental evolutionary thought, Gould and Lewontin attempt to defuse any preexisting negative stereotypes by dissociating this school into two approaches. The dissociation is accomplished, as so many dissociations are, with the assistance of an antithetical pair (*The New Rhetoric,* 422), in this case "strong" and "weak," frequently favored in science to distinguish forces and theories. The "strong" form might be predicted to be the preferred part, but in a clever move Gould and Lewontin make the weaker—the one that inherently should raise less opposition—"paradoxically" the more powerful (159–60). The strong form cannot construct transitions between *Baupläne* and so seems to resort to mysticism. The weak form focuses on the constraints, phyletic and developmental: "It does not deny that change, when it occurs, may be mediated by natural selection, but it holds that constraints restrict possible paths and modes of change so strongly that constraints themselves become much the most interesting aspect of evolution" (160). One might, however, question whether Gould and Lewontin's extended example of the "weak" continental theorizer, A. Seilacher, is really free of areas of "non-explanation" that opponents might label "mysticism." Seilacher has created a unity, a universal pattern, out of features that other evolutionary biologists might not categorize together: divaricate patterning in raised lines, in colors, in internal mineralization, and in incised grooves on mollusc shells. Since Seilacher is sure that many of these characteristics are nonadaptive, his work is a concluding showpiece for Gould and Lewontin. But since he does not know what generates this "universal" pattern (162), he too might be accused of "mysticism," and the strong and weak versions of continental evolutionary biology, which the argument attempts to dissociate, melt together again.[7]

Caveat and Conclusion

Analyses of stylistic techniques and argumentative strategies often appear to be refutations. To cite the compendious *New Rhetoric* one more time, dissociations of manner and content, a specific type of means/end dissociations, always result in a loss of value (436). But I intend no position here on matters of evolutionary interpretation. For readers in the discipline, the degree of adherence invoked by "Spandrels" would probably depend largely on the repertoire of cases they could call up to reinforce or contradict the illustrative cases with which Gould and Lewontin back up their points.

Gould and Lewontin could easily fault the preceding analysis as, well, frankly adaptationist. I have atomized their argument into separate traits and have assumed that these separate traits are "in" the text for reasons of utility, in this case persuasiveness rather than survival. I have also assumed that Gould and Lewontin have made the best possible argument they could summon, thereby optimizing their strategies. The only parts I have called attention to are those that have been defined by rhetorical analysts for the last 2,500 years; I am like the evolutionary biologists who tend to fixate on certain features, like beak shape and size or interspecies aggression. Furthermore, I have not been concerned with unifying Gould and Lewontin's strategies, in either a laudatory or a critical way. I do not consider it a sign of an underlying fracture, for example, that Gould and Lewontin in one case constitute the whole as a unity and in another case as a collection of parts. Instead, that strikes me as rather clever because I see the art of argument, practiced so skillfully by Gould and Lewontin, as essentially an art of improvization from a repertoire of strategies built into any language in use.

To go beyond the list of special features at critical points in "Spandrels" would mean to climb at this point to some unifying theme illustrated by the techniques of rhetoric and their uses. To do so would certainly give me a better essay, rounding off with a suitably general and, one would hope, grand-sounding moral of my own. I would be satisfying the conventions of my own discipline, which prefers essayistic studies that reach for and often concoct unifying themes. But mine is an approach that prefers the parts.

NOTES

1. Applying classical stasis theory, which categorizes arguments according to the kind of issue at stake, one would call "Spandrels" a third-stasis argument; scientific reports and reviews are first- and second-stasis arguments, respectively (Fahnestock and Secor).

2. The overall structure of the argument, divided into six parts, does the basic work of devaluation. Part 2 criticizes the attempt to rationalize each morphological or behavioral trait as having evolved "for" some specific pur-

pose. Part 3 criticizes the standard of "mere plausibility" in adaptationist explanations of traits. Part 4 appeals to authority by arguing, somewhat tongue-in-cheek, that a pluralist approach is truly Darwinian. Part 5 provides alternative descriptions illustrated by some specific cases of natural selection. And part 6 features the continental (German–Austrian) school of evolutionary explanation according to *Baupläne*.

3. In their naming, Gould and Lewontin have also made this entity strange for some readers with a visual, orthographic figure of speech, spelling "program" as "programme" in the title and throughout the article. Doubling the "m" and adding "e" represents, of course, British usage and so satisfies the stylistic conventions of the British journal that published this article. Nevertheless this spelling serves a continental versus Anglo-American dissociation of the contrasting approaches and hence, potentially, of the audience.

4. Another interesting feature of this list is that its headings and categories are what I call figure-driven. That is, the arrangement of words in the title itself suggests other possibilities in the series. So item 3i is "Selection without adaptation," and item 3ii is the *antimetabole* "Adaptation without selection."

5. A rhetorician with a penchant for figural analysis cannot pass up the chance of remarking on the antithesis in the sentence prefacing this series: "We illustrate just a few; we trust they will be recognized by all." The antithesis is, of course, the contrast between "few" and "all." These all but contraries are positioned at the ends of successive clauses for maximum if ersatz contrast. The antithesis is merely verbal, for if we complete the elliptical phrases we see that each of the two words modifies something different, not contrasting traits of the same entity: "We illustrate just a few *of these styles of argument;* we trust they will be recognized by all *who have sufficient background knowledge.*" But in its elliptical form it's a nicely turned phrase.

6. Another example of asyndeton used for copia comes from the final section on the continental alternative to Anglo-American adaptationism. Gould and Lewontin sample Seilacher's arguments for nonadaptiveness in patterns on clam shells: "Others [of Seilacher's reasons] rely more on general principles: presence only in odd and pathological individuals, rarity as a developmental anomaly, excessive variability compared with much reduced variability when the same general structure assumes a form judged functional on engineering grounds" (162).

7. The text does contain a sentence that is supposed to "count as" an explanation of divaricate patterning, forestalling the charge of mysticism: "It [divaricate patterning] must arise from some characteristic pattern of inhomogeneity in the growing mantle, probably from the generation of interference patterns around regularly spaced centres" (162). Seldom does a sentence so successfully restate an effect as a cause. One might paraphrase it as follows: this patterning (detectable only because of differences) must come from some pattern of differences (inhomogeneities) that is caused by "interference patterns around regularly spaced centres," that is, by a repeating pattern, only detectable as a pattern because of repeating differences. (Admittedly, Gould

and Lewontin have in mind something like interference between advancing wave fronts, but the whole sentence still circles back on itself rather than forming a causal chain.)

WORKS CITED

Aristotle. *The Rhetoric and Poetics of Aristotle.* Trans. W. R. Roberts and I. Bywater. New York: Modern Library, 1984.

Bolinger, Dwight. *Aspects of Language.* 2nd ed. New York: Harcourt Brace, 1973.

Cicero. *Brutus, Orator.* Trans. H. M. Hubbell. Cambridge: Harvard University Press, 1952.

[Cicero]. *Rhetorica ad Herennium.* Trans. H. Caplan. Cambridge: Harvard University Press, 1954.

Darwin, Charles. *On The Origin of Species and the Descent of Man.* Facsimile of the first edition, with an introduction by Ernst Mayr. 1859. Cambridge: Harvard University Press, 1964.

Fahnestock, Jeanne. "Arguing in Different Forums: The Bering Crossover Controversy." *Science, Technology, and Human Values* 14 (1989): 26–42.

Fahnestock, Jeanne, and Marie Secor. "The Stases in Scientific and Literary Argument." *Written Communication* 5 (1988): 427–43.

Gould, Stephen Jay. *Hen's Teeth and Horse's Toes.* New York: Norton, 1983.

Gragson, Gay, and Jack Selzer. "Fictionalizing the Readers of Scholarly Articles in Biology." *Written Communication* 7 (1990): 25–58.

Lyons, John. *Language and Linguistics: An Introduction.* Cambridge: Cambridge University Press, 1981.

Maynard Smith, John. "Optimization Theory in Evolution." *Annual Review of Ecology and Systematics* 9 (1978): 31–56.

Maynard Smith, John. *The Theory of Evolution.* 3rd ed. Harmondsworth, England: Penguin, 1975.

Perelman, Chaim, and Lucie Olbrechts-Tyteca. *The New Rhetoric.* Trans. J. Wilkinson and P. Weaver. Notre Dame, Ind.: University of Notre Dame Press, 1969.

10 THE READER IN THE TEXT OF

"THE SPANDRELS OF SAN MARCO"

GAY GRAGSON AND JACK SELZER

The whole thing has gotten more complicated.

It used to be that the parties constituting a rhetorical exchange, whether that exchange was literary or scientific or anything else, could be pretty easily identified and subdued. Under the influence of formalist criticism, texts until the 1970s were usually regarded as more or less unmediated and timeless objects with discrete boundaries; they were "verbal icons" standing apart from their settings and from the people who produced and consumed them. The "writers" and "authors" and "readers" and "audiences" of written discourse weren't considered very often; when they were, they were living, breathing, anything-but-anonymous individuals, responsible for bring texts into existence and for decoding and processing the information in those verbal icons. Now, of course, our notions of *text, author,* and *reader* are much less settled. Under the aspect of Mikhail Bakhtin, for instance, any *text*—any poem or story or scientific article or whatever—becomes not so much an autonomous unit as an entry in a larger conversation; texts are events, not objects— unbounded, dynamic collaborations among seen and unseen writers, present and past and future readers, and prior and subsequent discourses, all cooperating in the creation of meaning. *Authors* too have come to be seen as corporate, communal, less autonomous; a generation after Roland Barthes declared them to be dead, authors have come to be regarded as part of a broad social environment that generates and shapes discourse. The *author,* that is, has become *authors:* on the one hand, indistinguishable from their settings and situations and collaborators (those acknowledged and not); and on the other, broken into identifiable but multiple characters—the authors, coauthors, editors, sponsors (and so forth) living outside the text, and the authors and narrators implied within it.

And so it is with *audiences* and *readers.* Once stable referents, *audience* and *reader* have come to denote a sometimes bewildering array of characters within and without the text—implied readers, narratees, inscribed readers, ideal readers, informed readers, resisting readers, real readers, real audiences, audiences addressed, audiences invoked, multiple audiences.[1] All of these characters have attracted considerable interest, often under the general rubric of "reader-response criticism."

Although there is plenty of variety and overlap among them, reader-response critics belong generally to two main schools: those primarily interested in observing the responses of real readers to written discourse and those whose interest centers on understanding readers—fictional characters, so to speak—within the text. Those in the former group, including Louise Rosenblatt, Wolfgang Iser, Norman Holland, David Bleich, and the "early" Stanley Fish, have taken seriously the pragmatic responses of real readers.[2] In contrast to formalist critics, who dismissed attention to readers' responses as the "affective fallacy," these critics see reading as an event or a transaction that involves fundamentally both texts and readers; the task of these critics has been to uncover the dynamics of that transaction in order to understand better the texts or the readers or both. Fish and Iser have tended to be interested in how texts arouse and fulfill (or disappoint) expectations—and how that process establishes a background or context that influences subsequent reading. Rosenblatt, Holland, and Bleich seem less interested in how textual conventions shape the experience of reading and more in how a reader's personal experiences and psychic associations shape response. But all of these critics are more or less interested in how the subjective dispositions of real readers affect the act of reading and interpretation. All of them have written about how monitoring the activities of readers can contribute new insights into the dynamics of texts—seventeenth-century poetry and prose in Fish's case, the novel in Iser's, just about everything in the case of Holland. And all of them have pioneered new methods of analysis—Fish's microanalyses, Holland's psychoanalytical efforts, Bleich's innovative pedagogies—that are calculated to turn up new information about readers and texts. In the past decade or so, in fact, a number of methods for understanding reading and interpretation have been developed by literary theorists, reading specialists, and cognitive psychologists. In the essay that follows this one, for example, Davida Charney explains and demonstrates how the technique of protocol analysis can be used to understand more completely real readers and their readings. It is a notable instance of one standard approach to reader response—the one that focuses on the actual responses of flesh-and-blood readers.

The approach of this essay, by contrast, is more text-based—more in tune with an earlier generation of reader-response critics, who were working under more formalist assumptions. In the 1950s and 1960s, Walker Gibson and Wayne Booth noticed in fiction what they called "the mock reader" or "the implied reader," in-the-text characters analogous to the narrator or implied author and known through the background knowledge, assumptions, and other human values implied by the language of the text. In Booth's words, "The author creates . . . an image of himself [i.e., the narrator, or implied author] and another image of his reader [i.e., the implied reader]; he makes his reader as he makes his second self" (138). Just as a writer like Fitzgerald creates a fictional narrator separate from himself (Nick Carraway), so too he creates an implied listener or reader within the text who is separate from the real reader.

Readers in the text are quite different from text to text. Sometimes they are quite vivid: Gerald Prince has focused not on "implied" readers but on explicit creations that he calls *narratees* (by analogy with narrators); Prince cites as examples of these explicit creations the caliph in *A Thousand and One Nights,* whose threat to execute Scheherazade shapes the stories-within-the-larger-story that we listen in on, and the comrades aboard the *Nellie* who listen to Marlowe tell the events of *Heart of Darkness.* Other times the readers in the text are less overt: they are "implied"—implicit—indeed. But in every case "the writer's audience is always a fiction," in Walter Ong's well-known words: "the historian, the scholar or scientist, and the simple letter writer all fictionalize their audiences, casting them in a made-up role and calling upon them to play the role assigned" (17). The real audience may or may not accept the cues, may or may not take on the role of a narratee or implied reader—as the efforts of Holland, Bleich, Fish, Charney, and others have shown, real readers are anything but passive—but it is certainly well established that such roles are created in the text.[3]

Most analyses of readers in texts have been performed on belletristic texts, particularly fiction.[4] As Lisa Ede and Andrea Lunsford have noted, "little scholarship [outside literary studies] takes this perspective" (160). But Ong's comments about "the historian, the scholar or scientist, and the simple letter writer" are an explicit invitation to test how other writers—including scientists—create readers in texts: how scientists observe conventions and provide cues that create particular kinds of readers. Through a discussion of "The Spandrels of San Marco and the Panglossian Paradigm: A Critique of the Adaptationist Programme," we hope to demonstrate in this chapter the utility of a reader-response methodology for understanding scientific prose. Our method will be somewhat comparative: we will compare the reader in the text of "Spandrels" with the reader in John Maynard Smith's "Game Theory and the

Evolution of Behaviour"—another essay from the same symposium (and journal) that produced "Spandrels." By analyzing two pieces of scientific prose that were written on the same general topic (evolutionary theory) for precisely the same occasion and community but that nevertheless construct radically different audiences, we will show how a reader-response approach can clarify the rhetoric of science. In passing, we will also demonstrate something of the range of rhetoric possible in a discipline in which the writing is too frequently dismissed as purely conventional.

Background

Part of the fun of comparing "Game Theory" with "Spandrels" is that the authors of the two articles are respectful rivals. The full details of that rivalry are spelled out in the introductory chapter of this book and in the first section of Gould's chapter 15, but a quick accounting of their differences is worth rehearsing again here. Maynard Smith, probably the preeminent evolutionary biologist in Great Britain, is a committed "hardline" adaptationist who is convinced of the centrality, even the virtual exclusivity, of adaptation via the natural selection of an organism's features as an explanation for the diversity of nature. "The hypothesis of adaptation is virtually irrefutable," he has argued confidently in an essay appropriately entitled "Optimization Theory in Evolution"; "there is no reason to doubt the adequacy of the concepts of optimization and evolutionary stability for studying . . . evolution" (53). For Maynard Smith, natural selection is the mechanism that optimizes a species's chances of surviving and flourishing; in his teleology, maladaptive or neutral changes are of no significance. The biological term "microevolutionist" might offer a means of describing broadly Maynard Smith's position. Gould and Lewontin, equally prestigious American scientists, also acknowledge adaptation and optimization as fundamental processes. But they are much more concerned with the effects of natural selection on the individual as a whole than they are with its effects on an individual's features; their approach can thus be broadly labeled "macroevolutionist." Gould and Lewontin also offer alternatives besides natural selection to explain the diversity of nature. Thus, if Maynard Smith "rules out neutrality as a plausible explanation" (36) for adaptations, if he recognizes few constraints on natural selection, Gould and Lewontin argue instead that variations within a species may indeed be neutral or maladaptive. They therefore argue for a pluralistic approach to understanding the agents of evolutionary change, an approach that emphasizes alternatives to directed evolution. Among the alternatives are random processes (such as mutations) and constraints on evolution that prohibit the optimization of every individual trait.

"Game Theory" and "Spandrels" can be seen, therefore, as rival arguments that were offered on December 7, 1978, to the Symposium of the Royal Society of London on "The Evolution of Adaptation by Natural Selection." Maynard Smith and Gould had the roles of ministering to the disciples and of converting the pagans, respectively. Maynard Smith's paper, the first one offered at the symposium (which Maynard Smith, a Fellow of the Royal Society, in fact arranged), is characteristically, even classically, microevolutionary in identifying a single feature (fighting behavior in animals) for discussion. The paper develops his theory that such behavior can be explained by natural selection, and it critically assesses the theory's methodology, applications, and opponents. Then, at the end of the symposium, with the very last paper, Gould and Lewontin argue explicitly against the microevolutionary approach and implicitly against Maynard Smith's own position on adaptation. Though Maynard Smith is never mentioned explicitly in "Spandrels" (nor Gould or Lewontin in "Game Theory"), it is clear from the context, from the entire thrust of their arguments, from the subsequent response by A. J. Cain, from Gould's testimony in chapter 15 of this book, and from one of our own personal conversations with Maynard Smith, that Maynard Smith and Gould and Lewontin are in effect debating the merits of two different schools of thought concerning the role of adaptation in biology.

More to the point here, the essays differ not only in point of view: they differ radically in the rhetorical strategies they employ—particularly in the ways their authors choose to constitute their audiences. A comparison of the two articles is telling: it reveals how self-consciously rhetorical are both performances (including the one that seems most "conventional");[5] how even within the exacting constraints of the scientific journal, writers have considerable freedom to exercise stylistic choices; and how the writers of transactional prose, to enhance their persuasiveness, do indeed give cues that direct implied readers into specific roles. Ong, Booth, Iser, Prince, and others have demonstrated that the authors of novels and short stories fictionalize their audiences; Maynard Smith and Gould and Lewontin, through the choices they make in two essays on the same topic written for the same group of readers, show that there is a good deal of fictionalizing in scientific nonfiction as well.

The Reader in the Text of "The Spandrels of San Marco"

Maynard Smith's Implied Reader

It is not difficult to understand how Maynard Smith constitutes the audience in "Game Theory and the Evolution of Behaviour" because he relies on highly conventional cues that create for his readers a highly conventional role. In a recent article ("Fictionalizing") we attempted to demonstrate how Maynard Smith in another, similarly titled essay ("Optimization Theory in Evolution"), written in the same year, created a conventional, almost stereotypically scientific reader—an objective, fair-minded, sober, impersonal, careful, and reasonable pursuer of factual scientific truth. We further indicated that Maynard Smith's reader in that circumstance was asked to assume a position of respectful inferiority to him: Maynard Smith was in effect an implied schoolmaster lecturing to willing students—graduate students, as it were, who were invited to take notes on information presented by an authority, rather than professional colleagues invited to take issue with an argument. The reader in "Game Theory" is not quite so subordinated; the rhetoric of this essay is not as overtly coercive as it is in "Optimization Theory." But it still creates through recognizable cues a reader in the text who is just as conventional—the objective, fair-minded, sober, reasonable biological scientist—and nearly as subordinate in relation to the implied author.

Certainly with the first two paragraphs (one an abstract, the other introductory) the reader of "Game Theory" is situated in the role of the typical evolutionary biologist:

Game theory and the evolution of behaviour

By J. Maynard Smith, F. R. S.

*School of Biological Sciences, University of Sussex,
Falmer, Brighton BN1 9QG, U.K.*

How far can game theory account for the evolution of contest behaviour in animals? The first qualitative prediction of the theory was that symmetric contests in which escalation is expensive should lead to mixed strategies. As yet it is hard to say how far this is borne out, because of the difficulty of distinguishing a 'mixed evolutionarily stable strategy' maintained by frequency-dependent selection from a 'pure conditional strategy'; the distinction is discussed in relation to several field studies. The second prediction was that if a contest is asymmetric (e.g. in ownership) then the asymmetry will be used as a conventional cue to settle it. This prediction has been well supported by observation. A third important issue is whether or not information about intentions is exchanged during contests. The significance of 'assessment' strategies is discussed.

Introduction

The theory of games was applied to the evolution of fighting behaviour (Maynard Smith & Price 1973) in the hope of explaining the conventional aspects of this behav-

iour (Lorenz 1966 and earlier) in terms of individual selection. Before this there had
been a tendency to account for such behaviour in terms of the good of the species. This
presented an obvious difficulty to a neo-Darwinist, in view of the weakness of group
compared with individual selection as an evolutionary force (Williams 1966; Maynard
Smith 1964, 1976).

After all, the mere act of placing a contribution into the *Proceedings of the
Royal Society of London* generally disposes the reader in the text toward a
scientific role and away from other common roles (e.g., citizen, parent);
in that sense, the genre and medium of a publication probably do much
to create a given role for a reader. The role of scientist is quickly reinforced when Maynard Smith and his editors make the article conform
visually to the conventions of its genre, which the reader is expected to
take in: the familiarly nominalized title, the author's identification of
himself as an academician within a School of Biological Sciences, and
the adherence to the reference system and other iconic dictates of the
Council of Biology Editors. As Prince says, the implied reader always
"knows the tongue and the languages of the narrator" (180).

And in fact the superficial formal cues of "Game Theory" are reinforced by the subject matter, vocabulary, and syntax of the two paragraphs we just quoted. If they have not already become full-fledged
evolutionary biologists, fictional readers become so when the author
invokes explicitly the shared information and common interests of that
field. When Maynard Smith says, "The theory of games was applied to
the evolution of fighting behaviour . . . in the hope of explaining the
conventional aspects of this behaviour in terms of individual selection,"
and "Before this there had been a tendency to account for such behaviour in terms of the good of the species," he is not so much reminding
his readers of information they might need or might have forgotten as
he is asking them to place themselves within the community of shared
beliefs that constitute evolutionary biology in general and microevolutionary biology in particular. When he cites as common knowledge that
"this presented an obvious difficulty to a neo-Darwinist," he secures
the reader in a particular discipline, especially since the nominalized
diction of the abstract and the first paragraph creates a reader comfortable with the insider's language of biology—"evolution," and "individual selection" and "neo-Darwinist," to be sure, but even more particularly "asymmetry," "assessment strategies," "frequency-dependent
selection," "contest behaviour," "qualitative prediction," "symmetric
contests," and "mixed evolutionarily stable strategy" (not to mention
the highly technical language of the remainder of the essay). Even the
syntax of Maynard Smith's sentences locates the essay and its readers

in the domain of conventional science: note the ten passives in these roughly two hundred words and the numerous impersonal constructions (e.g., "it is hard to say," "the difficulty of distinguishing," "the second prediction was," "this prediction has been well supported by observation"—that is, saying without sayers, distinguishing without distinguishers, predictions without predictors, observation without observers). Very early in "Game Theory," then, the fictional reader has taken on a conventional disciplinary role. Nor does anything in the remainder of the essay do much to disturb that circumstance; Maynard Smith continues to invoke the assumptions, shared beliefs, and common knowledge and vocabulary of the biological sciences.

But the essay is also filled with other kinds of directives, other kinds of techniques for ensuring that readers become not just evolutionary biologists but idealized, almost stereotypical models of their kind: objective, impersonal, fair-minded, careful, reserved, and reasonable pursuers of factual truths. For example, in keeping with the norms of the objective scientist, Maynard Smith prefers nominalizations, couches explanations relentlessly in the passive, and resorts to awkwardly impersonal constructions (e.g., "it can be shown" [42]; "what this matrix says" [42]; "to demonstrate, . . . it is sufficient to show" [45]; "it is puzzling that" [46]; "as mentioned earlier" [50]). He treats the work of colleagues with fullness (more than fifty works are cited) and respect and generosity. And he consistently gives the appearance of scientific detachment and impartiality: the only work he leaves open to explicit criticism is his own (52); the only implicit criticism of another is indeed indirect (when he acknowledges a possible counterinstance only with a dismissive "although no data are given" [46]); and in his references to his main critic, Geist, he is most circumspect, first citing Geist approvingly in several contexts (e.g., "Geist makes the interesting suggestion that . . ." and "Geist is surely correct when . . ." [48]), then acknowledging "that there is, I think, some justice in [his] complaints" (51), and in the final paragraphs of "Game Theory" seeking areas of agreement:

> In fact, the disagreement between Geist (1974) and Maynard Smith & Price was more apparent than real. Part of the difficulty arose because of our failure to notice that Geist had earlier put forth part of our argument. In addition to attempting a general formulation for the analysis of contests, we proposed that a specific strategy, namely 'retaliation', was likely to be a component of evolutionarily stable behaviour. I think that this conclusion is correct (despite some mathematical difficulties; see Gale & Eaves 1975), but it had been stated quite explicitly by Geist (1966). In a

later analysis of contest behaviour in ungulates, Geist has
given the concept of retaliation a more central position and has
extended his earlier argument in an attempt to relate the types of
weapons and behaviour to the ecology of different species. Had
there been space, the role of retaliation would have been treated
here as a fourth prediction of game theory; it was omitted only
because it is observationally less controversial. (52–53)

In all these ways Maynard Smith schools his readers in the ethos of
science. Toward the same end he often supplies careful, conventional
qualifiers; within the space of two paragraphs on page 44, for instance,
he uses the words "suggested," "if," "however," "although" (twice),
"may" (four times), "would be," "sometimes," and "fairly." He refers to
his own work in the third person (citing "Maynard Smith" or "Maynard Smith & Price" or "Maynard Smith & Parker" when "Price and I"
would be easier), making himself seem objective, impartial, and fair
and valorizing those traits in general. Then—and this is important—he
explicitly but subtly projects those same qualities onto his fictional readers by serving up an occasional "we" or "our" or "us" without clear
antecedents. The personal pronouns are particularly effective in creating an audience in the image of the author when they come at crucial,
emphatic junctures (as at the beginning of sections 1 and 2).

Just as important, Maynard Smith is careful to keep his language
and style scientifically neutral and dispassionate even when he is being
most pointed. Like the grant proposals analyzed by Greg Myers, a
review essay like this one "must persuade without seeming to persuade" ("Social Construction," 220), not only by observing the conventions of impartiality and impersonality but also by fulfilling its responsibility to map the field with care and completeness. Thus, as he says in
the paragraphs quoted, Maynard Smith's overall aim is not to argue but
"to account for" the dynamics of fighting behavior in terms of the good
of the species, for to argue is seemingly to locate one's discourse somewhat outside the bounds of the review essay in particular and conventional science in general. Maynard Smith simply promises to "discuss"
the significance of assessment strategies, not to argue for them himself;
to "support" predictions and observations, not to make them himself;
to answer questions, not to debate them. He announces that his aim is
not to argue for game theory as an explanation but merely "to discuss
how far the application of game theory . . . is useful in interpreting
animal contests" (41). He merely purports to "seek explanations" and
"to ask" questions and to provide "facts," to "explain" and to "suggest" and "to demonstrate," not to argue. After all, as Kinneavy notes,

"a discourse which becomes [we would say "seems"] . . . directly persuasive . . . is a discourse which is in danger of being unscientific" (88)—and Maynard Smith is determined to keep his readers in a scientific, referential frame of mind. He and his fictional readers remain well within a community of disinterested scientists who merely describe the diversity of nature according to the mandates of established law and who eschew the partisan, sometimes emotional rough-and-tumble of argument. Despite all of these disclaimers, of course, "Game Theory and the Evolution of Behaviour" is a very argumentative essay, particularly in championing game theory as a powerful explanation of adaptation and survival strategies. That becomes quite clear in the final "Discussion" section, when criticisms of game theory are taken up rather formally. But by consciously avoiding the language of argument, substituting the neutral "describe" and "show" and "explain" and all the rest, and then using "we" and "us," Maynard Smith creates the persona of a conventional biologist and then asks his implied readers to become like him.

That is not to say that "Game Theory and the Evolution of Behaviour" is completely conventional in its presentation or in its creation of the in-text reader, however. In our previously published analysis of Maynard Smith's "Optimization Theory in Evolution," we showed how he departed from the most conventional aspects of academic science writing to create a reader who was not only "a typical biologist" but a rather subordinate one: by means of emphatic sentence patterns, hyperconfident diction (e.g., "undoubtedly," "it is clearly impossible"), heavy metadiscourse (especially forecasting), and a decisive personal "I" in his forecasting statements, Maynard Smith established his implied author as an authority and his implied reader as a novice in search of direction. The implied author in "Game Theory" also departs from the conventional scientific role; he may be an objective, fair-minded, reasonable pursuer of facts, but he is also genial. On occasion he uses wit and extrabiological language and comparisons, for example:

> Now it is inconceivable that it would be selectively advantageous for a fish to tell its opponents at the start of a fight that it intended to retreat, and then continue the fight; it is as if a trade union negotiator were to say, 'I demand a 50% increase for my members; furthermore, I will settle for 5%.' (49)

> The argument was put in more mathematical language, but it simply amounts to saying that you should not believe what an opponent at poker tells you. (48)

Gay Gragson and Jack Selzer

> An entertaining example is provided by the study by Alcock (1977) of the bee *Centris pallida*. Male bees adopt one of two strategies. 'Patrollers' search for sites where buried virgin females are about to emerge, dig for them, and attempt mating. Fights over digging sites are common, and are usually won by the larger male. 'Hoverers' wait for females that have not been mated by patrollers, perhaps because a fight was in progress when they emerged. Hoverers are smaller than patrollers. Although measurements of mating success are not available, it seems certain that the larger patrollers are more fit. To a small male, hovering is making the best of a bad job. But the size of a male is determined by the amount of food brought by his mother. Small males are cheaper to produce, both because they need less food and because they are less likely to be parasitized. It is hard to see how producing small males could be forced on a female by circumstances. (45)

In so doing, Maynard Smith implies a reader who is likewise genial, likewise confident, and comfortable enough in the disciplinary role to enjoy an in-house laugh.

But Maynard Smith remains the authority, the person in command; and the implied reader is there to be willingly commanded. Maynard Smith is the active entertainer and teacher, and the reader is there to be entertained and led. In the previous paragraph we quoted the third example at some length because it illustrates not only Maynard Smith's wit but also his use of the same emphatic sentences and hyperconfident language that we observed in "Optimization Theory." The quotation about the hoverer bees and patroller bees contains eleven sentences but only 167 words—a mere fifteen words per sentence, on the average. Those emphatic, declarative sentences (there are only nine subordinate clauses in the entire passage), when combined with confident diction such as "it seems certain" and "it is hard to see," create an authoritative voice that by implication dominates the reader in the text and admits a polite margin of doubt that only those who wish to seem ignorant can actually question. It makes for a very effective strategy, especially when it recurs in paragraph after paragraph and particularly when it is reinforced by other similarly calculated tactics.

In noting that Maynard Smith refers to himself in the third person and in speaking of his use of passive and impersonal constructions, we have implied that the personal "I" is absent from the essay. But that is not so; it is absent only in the "factual" sections of the article. When Maynard Smith engages in metadiscourse to forecast his intentions or to sum up his conclusions, a decisive personal "I" is prominent. The abrupt shift to the personal pronoun begins in paragraph 2 of the article, which amounts to a lengthy thesis and forecasting statement. Then at certain other forecasting spots (e.g., "I shall return to this point") and at several other key junctions, most notably in his concluding "Discussion," Maynard Smith shifts again to the first person, a first person both dominant and directive. Moreover, to clarify many of his points Maynard Smith asks his readers to imagine concrete (or sometimes less concrete) scenarios—and does so in a rather professorial-sounding imperative mood ("suppose" this, "imagine" that) that patronizes readers further into a passive role.

The cumulative effect of Maynard Smith's rhetorical choices—the declarative sentencing, the confident diction, the insistent "I," the heavy forecasting, and the imperatives—is to establish him as masterfully in control of the situation. In the face of such a persona and in response to such cues, readers are guided into becoming patient and passive, if attentive and entertained, bystanders. These fictional readers may be mainstream professional biologists, but they are put into a quiescent role by Maynard Smith's performance. Though they are in fact faced with an argument, not a presentation of facts, they are directed to take notes on information presented by an authority rather than invited to take issue with an argument.

The Implied Reader in "The Spandrels of San Marco"

In characterizing Maynard Smith as a persona who employs rhetorical tactics conventional to scientific writing in the academy and who chooses to fictionalize his readers into a passive role, we mean nothing pejorative. Maynard Smith's scientific expertise is apparent in his mastery of the nuances of game theory and in his easy command of an amazing array of published information; his voice, if unremarkable in some places, is always confident and authoritative and, as we have shown, is often quite graceful and urbane and witty. Furthermore, Maynard Smith is a most resourceful writer who is quite capable of adopting a range of voices in the service of a range of circumstances.[6] But we do wish to emphasize that his approach in "Game Theory," his mostly conventional way of fictionalizing a reader in this instance, is not a scientist's only choice. For in writing to exactly the same group of academic biologists on the same general issue in exactly

Gay Gragson and Jack Selzer

the same environment, Gould and Lewontin fictionalize their audience into a very different role.

The first few pages of "The Spandrels of San Marco" present a striking contrast to Maynard Smith's more conventional opening. On the one hand it is apparent that Gould and Lewontin address their readers as biological scientists, just as Maynard Smith did. They too appear in a mainstream publication that in general disposes a reader to "become" a scientist. They too cue their readers toward a disciplinary role by identifying themselves as scientists and by observing the formal, generic conventions of the scientific article: an abstract is provided, for instance, and the conventional documentation system assumes and creates disciplinary expectations. They too speak the language of the evolutionary biologist, depending on the insider's vocabulary, shared knowledge, and common assumptions of evolutionary biology, which place implied readers within that specific scientific community. The abstract, for instance, mentions phenomena such as "allometry," "pleiotropy," "material compensation," "mechanically forced correlation," and "adaptive peaks" without worrying about definitions or explanations, and the text proper without a blink uses the language of biology ("orthogenetic," "phyletic inertia") and assumes a knowledge of biological concepts: "If we regard the chin as a 'thing' rather than as a product of interaction between two growth fields (alveolar and mandibular), then we are led to an interpretation of its origin (recapitulatory) exactly opposite to the one now generally favoured (neotenic)" (151).

On the other hand, however, Gould and Lewontin constitute their readers as more than members of a particular scientific discipline. Though they locate their readers within the field of evolutionary biology, they also take pains to direct those readers to become more than "mere" biologists. The title of the essay, for example, is most unusual: prefixed to the perfectly conventional "A Critique of the Adaptationist Programme" are the alliterative (and nearly dactylic) "Panglossian Paradigm,"[7] the rather mysterious and decidedly undisciplinary "Spandrels of San Marco," and a colon that has become stereotypical in titles found in the humanities. It is also common enough to use illustrations and figures in scientific texts, but not figures like those on pages 148–49. If Maynard Smith's conventional tables and mathematical formulas in "Game Theory" confirm a disciplinary presence, Gould and Lewontin's photographs of spandrels and cathedrals, because they assume broad, cultural knowledge more than disciplinary background, situate fictional readers outside the conventional limits of stereotypical science.

The opening paragraphs of "Spandrels" also keep the reader in the text from becoming too much the evolutionary biologist. Though Gould and Lewontin in the abstract and later in their essay address their readers as people who share the conceptual and linguistic heritage of evolution, they are careful *not* to point readers toward that role in their first four paragraphs. By beginning with St. Mark's Cathedral and King's College Chapel, Venice and Cambridge, Christian doctrine and Tudor iconography, Biblical tradition and art history, Voltaire and Dr. Pangloss—and the elaborate architectural trappings of quadrants, arches, fans, bosses, and vaults—the authors invite their reader away from disciplinary specialization. Even though the subject of the essay, adaptation, is mentioned in these paragraphs and though the point of the elaborate analogy is anything but hidden, the entire performance calls forth the reader's cosmopolitan self. It is as if the reader is being consciously removed from a scientific role and transformed instead into the broadly urbane reader implied in *Smithsonian* or *Natural History.* Where Maynard Smith begins with Darwinian fitness and assessment strategies, adopts a nominalized vocabulary, and employs a series of scientific passives, Gould and Lewontin reference angels and icons, architecture and Voltaire; employ active verbs and a vocabulary congenial to and reminiscent of the humanities; and luxuriate in their opening figure. While metaphor and analogy are hardly unique to the humanities—scientists trade quite heavily in models and metaphors, in fact, as Maynard Smith's "Game Theory" indicates—this kind of analogy in a scientific essay is unusual and undisciplinary in its elaboration, its self-consciousness, its overtly argumentative edge, and its artistic allusions and literary denouement. It is the kind of analogy that Prince described as being among the "possibilities for mediation between authors and readers": "Dialogues, metaphors, symbolic situations, allusions to a particular system of thought or to a certain work of art are some ways of manipulating the reader, guiding [his or her] judgment, and controlling [his or her] reactions" (192).

It is not insignificant that the same humanities professor who struggles with Maynard Smith can read the opening of "Spandrels" with ease and appreciation: Gould and Lewontin are creating their in-text scientists in a more humanistic image.[8] If Maynard Smith through a variety of techniques chooses to constitute his fictional readers as stereotypical scientists—as objective, impersonal, reserved, and impartial observers—Gould and Lewontin make their readers in another image: not objective but interested; not impersonal but flesh and blood; not reserved but involved; not dispassionate observers of reality but enthusiastic partisans; not blinded by narrow specialization but enlightened by broad experience. Quotations from Voltaire are not confined to the introduction, for instance; they reappear throughout the essay. In addition, instead of devising impersonal constructions, numbered references, and passives, Gould and Lewontin present a conspicuous "we," plenty of other breathing humans, and the active voice. Instead of keeping to current professional literature, they put their implied reader into the Boston Museum of Science (153) and, with the perspective of historians, quote the work of biologists who worked a century ago. Instead of treating their adversaries and allies as Maynard Smith treated Geist—with the impartial detachment and respectful indirection customary in science, where explicit criticisms are usually frowned on (Myers, "Social Construction," 238)—Gould and Lewontin catalogue and anatomize their villains (Harner, Wilson, Wallace, Barash, Morton et al.), venerate explicitly their saints (Romanes, Galton, and of course Darwin), and salute among the living those few who are yet in the state of grace (e.g., Lande, Riedl, Seilacher, and themselves). Instead of reifying the conventional scientist's qualified, guarded conditional language and careful subordinations, Gould and Lewontin prize strong, emphatic generalizations: "Spandrels do not exist to house the evangelists" (150); "Some evolutionists may regard this as a trivial, or merely a semantic problem. It is not" (151); "The adaptationist programme is truly Panglossian" (151); "It is an old habit" (152); "This view is false" (155). If Maynard Smith typically avoids histrionic sentence patterns to respect his implied reader's sober ethos of reserved and dispassionate science, Gould and Lewontin choose rhetorical questions (150–51; 154–55; 158), imperatives (152–53), dashes (147, 158–59), even exclamations ("Under these windows dwell endosymbiotic algae!" [152]) to lend personality and passion to their implied readers and to their larger enterprise. If most scientists imply their in-text readers through a neutral phraseology and the voiceless ideal of objectivity, Gould and Lewontin flout that ideal with contractions, spontaneous "ands" and "buts" at the opening of sentences, and unusual—even outrageous—wit and sarcasm:

> To put it crudely: a system developed for other reasons generated an increasing number of fresh bodies; use might as well be made of them. Why invert the whole system in such a curious fashion and view an entire culture as the epiphenomenon of an unusual way to beef up the meat supply. (150)

> And finally, many experts doubt Harner's premise in the first place (Ortiz de Montellano 1978). They argue that other sources of protein were not in short supply, and that a practice awarding meat only to privileged people who had enough anyway, and who

used bodies so inefficiently (only the limbs were consumed, and partially at that) represents a mighty poor way to run a butchery. (150)

> In natural history, all possible things happen sometimes; you generally do not support your favoured phenomenon by declaring rivals impossible in theory. Rather, you acknowledge the rival, but circumscribe its domain of action so narrowly that it cannot have any importance in the affairs of nature. Then, you often congratulate yourself for being such an undogmatic and ecumenical chap. (151)

Passages like these are worth quoting not only because they are so common in Gould and Lewontin and so striking (cannibalism is described as a way to "beef up the meat supply"!) but also because they do so much to define the role of the reader. By asking rhetorical questions, by creating a voice and a historical awareness for themselves, by risking puns, humor, irony, sarcasm, wit, and literary allusions, Gould and Lewontin imply a reader who eschews traditional and stereotypical roles and who has adopted a broader, more cosmopolitan, more interdisciplinary identity.

And, as Maynard Smith has done for his implied reader, they provide an ideal model for that implied role, too: themselves. In part the persona of "Spandrels" derives from the stylistic choices we have just mentioned. The wit and wordplay, the irony and sarcasm, the informal and emphatic sentences of the speakers all suggest urbane self-confidence and a healthy cynicism toward the circumscribed "certainties" of the biological sciences. Gould and Lewontin cite their own work, just as Maynard Smith does, to establish their professional credibility and authority; but they seem at home not just in the lab or in the field or among scientists, but in museums and cathedrals, in libraries and bookstores, among historians and anthropologists and writers as well. The persona of the essay is relevant to this study of the reader in the text because Gould and Lewontin not only create an ethos that seems at odds with the role of the stereotypical scientist, but they also invite the reader in the text to become like them. In part that invitation is implicit, of course; it depends on the reader's admiration for the essay's persona.

But the invitation is also explicit, particularly in the way the authors manipulate personal pronouns. For who exactly is the "we" who dominate so many sentences in "Spandrels"? Quite often, of course, the "we" refers to Gould and Lewontin; take, for example, the pronouns in the abstract ("we criticize"; "we fault"; "we support"), or the concluding sentences of the introduction, or this passage from the end of section 2:

> (We purposely choose an example based on public impact of science to show how widely habits of the adaptationist programme extend. We are not using glass beasts as straw men; similar arguments and relative emphases, framed in different words, appear regularly in the professional literature.) We don't doubt that Tyrannosaurus used its diminutive front legs for something. (153)

But many times the "we" includes not only the authors but also the reader in the text—that member of the skeptical, unconventional scientific community that is in the process of being created by the authors. The first uses of "we" in the essay, in fact, refer to the implied readers, not to the authors: "The design is so elaborate, harmonious, and purposeful that we are tempted to view it as the starting point of any analysis. . . . Such architectural constraints abound and we find them easy to understand" (148). The same community "we" recurs throughout the essay: "we all say that not everything is adaptive; yet faced with an organism, we tend to break it into parts" (152); "we may be sure of this in numerous cases" (158); "we do not now regard all of Darwin's subsidiary mechanisms as significant or even valid, though many, including direct modification and correlation of growth, are very important. But we should cherish his consistent attitude of pluralism" (158). As some of these quotations imply, it is sometimes quite difficult—purposefully difficult, we suspect—to tell whether the "we" refers to the authors or includes the implied reader as well. Note, for example, how the authorial "we" becomes an "our" that includes the in-text reader in this passage: "We deliberately chose non-biological examples in a sequence running from remote to more familiar: architecture to anthropology. We did this because the primacy of architectural constraint and the epiphenomenal nature of adaptation are not obscured by our biological prejudices" (150). Or note how the referent for "we" shifts within the space of two sentences at the beginning of part 2: "We [Gould and Lewontin] call it the adaptationist programme, or the Panglossian paradigm. It is rooted in a notion popularized by A. R. Wallace and A. Weismann (but not, as we [i.e., including the implied reader] shall see, by Darwin)" (150). The effect is unmistakable. It is not so much that Gould and Lewontin are identifying themselves with their readers as that the implied reader is being asked to identify with—to become like—Gould and Lewontin: disciplinary nonconformists whose irreverence and progressivism (and implied youth) stand so attractively against the established order.

That is particularly true since Gould and Lewontin also serve up a conventional establishment "them" for their fictional readers to measure themselves against. If the "we" of "The Spandrels of San Marco" becomes an interchangeable code denoting both "Gould and Lewontin" and "other people like Gould and Lewontin," then "they" and "them" often denote the stereotypical scientists—scientists like Harner and Wilson who bring narrow disciplinary preconceptions to their work. "Harner and Wilson ask us to view an elaborate social system and a complex set of explicit justifications involving myth, symbol, and tradition as mere epiphenomena generated by the Aztecs as an unconscious rationalization masking the 'real' reason for it all: the need for protein" (150): note the "them versus us" formulation; and note the implication that disciplinary prisoners like Harner and Wilson are hardly as capable of understanding Aztec myths and tradition as the worldly wise Gould, Lewontin, and their implied reader. Later Barash and Morton et al. are identified as other victims of narrow perspectives, as another "they" so blinded by membership in a particular disciplinary community that "we" may dismiss them with irony and sarcasm:

> Since we criticized Barash's work, Morton et al. (1978) repeated it
> [without being able to confirm his conclusions]. . . . Yet instead
> of calling Barash's selected story into question, they merely devise
> one of their own to render both results in the adaptationist mode.
> Perhaps, they conjecture, replacement females are scarce in their
> species and abundant in Barash's. Since Barash's males can
> replace a potentially 'unfaithful' female, they can afford to be
> choosy and possessive. Eastern bluebird males are stuck with
> uncommon mates and had best be respectful. (154–55)

The prominent first-person plural in "Spandrels" also points to one last feature of the role readers are asked to play in the essay. Gould and Lewontin cast their readers as equals, as colleagues. Though their reputations might well permit them to address readers as subordinates, Gould and Lewontin address their implied readers as insiders, initiates, full-fledged colleagues. The heavy-handed forecasting and pedagogical "I" so distinctive in "Game Theory" are missing from "Spandrels," replaced by the inclusive "we" and by architectural and literary allusions that put readers on a common ground with the authors. Toward the same ends, the argumentative language and stance that Maynard Smith avoided with such calculation are brandished with delight by Gould and Lewontin: "describe" and "discuss" and "outline" and "show" are discarded in favor of "we criticize," "we fault," "we contend," "we wish to question," "we object strenuously," and so forth;

Gay Gragson and Jack Selzer

and the expository summaries that put Maynard Smith's implied reader into a role of passivity are replaced by charged language and rhetorical questions designed to insubordinate the implied reader and involve that reader as an active participant. Whereas Maynard Smith directs his implied reader to take notes, Gould and Lewontin invite their reader to become a colleague taking issue—especially taking issue with the advocates of "the adaptationist programme." Kinneavy may be right that a discourse that becomes directly persuasive is in danger of seeming unscientific, but Gould and Lewontin are ready to take the risk because they want their readers to become less conventionally scientific, more like they themselves seem to be. Gould and Lewontin, after all, are challenging conventional "knowledge" in evolutionary biology. In part, therefore, the stance they take is authorized by the nature of the ongoing discussion of which "Spandrels" is a part: to challenge the established concepts of "the adaptationist program," the authors assume for themselves and project onto their reader the role of disciplinary nonconformist. In the same way, for that matter, Maynard Smith's choices are also a response to his own circumstance: his articulation and defense of the scientific status quo imply for him and his reader a more traditional role.

Conclusion

We want to be careful not to overstate. Gould and Lewontin use many of the same rhetorical choices as Maynard Smith. They too write as scientists, adopt a recognizably scientific persona, and constitute their implied reader as a scientist. Indeed, they could hardly afford to do otherwise because their real readers—members of the scientific community—are most likely to respond to a reader in the text who is created in the image of scientist, at least in this circumstance. But while Maynard Smith through conventional means places his reader firmly in the role of conventional scientist, Gould and Lewontin have chosen to address their reader not merely as a scientist but also as an inquisitive skeptic, as a member of an academic and social community whose breadth of human experience gives the implied reader the perspectives and attitudes required for understanding the limits of scientific thought.

Moreover, we should emphasize once more that we are dealing here with the textual features of the two articles and not with the actual reception of those textual features. We are dealing with fictional readers, the "readers in the text," not with human beings engaged in the process of reading. In practice, of course, real readers might not play along with the roles prescribed for fictional readers; in the process of actual reading, real scientists might become neither the rather conventional one implied by Maynard Smith nor the skeptical colleague created by Gould and Lewontin. In fact, they might well resist such roles completely. Thus we are not contending that one article is written more effectively than the other. Those who would favor one method of presentation over the other should consider whether they are bringing the biases of their own disciplinary community to bear on the discourse of another community; moreover, the response of real readers to any text is highly idiosyncratic (as Gould's own record of responses to "Spandrels" indicates and as Davida Charney's analysis of "Spandrels" demonstrates in the next essay in this book).

Nor do we pretend that we have catalogued the complete list of rhetorical tactics used in both articles. Illustrative and suggestive only, our discussion has focused solely on the creation of the reader in the text; we simply explicate two of the many possible ways of proceeding in scholarly articles on science, with the goal of establishing the accuracy and the potency of Ong's assertion that "the historian, the scholar or scientist, and the simple letter writer all fictionalize their audiences." Those who wish to continue this line of inquiry might consider some of the same questions we have puzzled over: How are roles created for implied readers? What specific signals tip readers off about the roles they are expected to play? And how can an emphasis on the active writer as fictionalizer be reconciled with the fact that real readers are active themselves—that they might resist a particular role instead of assuming it? In other words, if reading is a process of negotiation, what are the specific terms of that negotiation? The answers to such questions could lead to improved understanding of everything from *The Origin of Species* and *Walden* to computer manuals, scientific articles, and reports—and, in turn, to the empowerment of inexperienced writers who would imitate the tactics of Maynard Smith, Gould, Lewontin, and other scientific or technical writers who fashion language as masterfully.

NOTES

1. For an attempt at a complete lineup of these characters, see Selzer, "More Meanings of *Audience.*"

2. For reviews of this school of reader response (and a bibliography), see Nan Johnson's "Reader-Response and the *Pathos* Principle." The "early" Fish refers to the work in *Surprised by Sin: The Reader in "Paradise Lost"* (1967) and *Self-Consuming Artifacts* (1972). More recently, as is evident from *Is There A Text*

Gay Gragson and Jack Selzer

in This Class? (1980), he has argued that reading is a personal act but not a subjective one: people can arrive at similar readings because they belong to "interpretive communities" whose assumptions shape the act of reading. In this sense, he now occupies a middle position between reader-based and text-based critics.

3. For a fuller review of work on "the reader in the text" (with bibliographies), see Jane Tompkins, *Reader-Response Criticism* (1980); Susan Suleiman and Inge Crosman, *The Reader in the Text* (1980); and Daniel Wilson, "Readers in Texts" (1981). The journal *Reader* prints studies of both real readers and the readers inscribed in texts.

4. For notable exceptions, see Black and Dillon.

5. For authoritative summaries of the stylistic characteristics of conventional scientific prose, see Kinneavy and Gross.

6. See, for example, the exchange between Maynard Smith and Gould (with Niles Eldredge) in the pages of *Nature,* 10 December 1987 ("Darwinism Stays Unpunctuated"); 17 March 1988 ("Punctuated Equilibrium Prevails"); and 24 March 1988 ("Punctuation in Perspective"). In that exchange it is Gould who adopts the more "conventional" tone and Maynard Smith who writes more provocatively.

7. The phrase "Panglossian Paradigm" actually assumes more disciplinary background than one might imagine. The phrase has been used in discussions of adaptation in biology since at least 1949, as Maynard Smith has noted in "Optimization Theory" (49).

8. Gould commonly adopts this as a conscious strategy, of course. For an account of his desire to help link the sciences and the humanities, see the middle portion of his essay in chapter 15.

WORKS CITED

Bakhtin, M. M. *The Dialogic Imagination.* Austin: University of Texas Press, 1981.
Barthes, Roland. *Image, Music, Text.* New York: Hill and Wang, 1977.
Black, Edwin. "The Second Persona." *Quarterly Journal of Speech* 56 (1970): 109–19.
Bleich, David. *Subjective Criticism.* Baltimore: Johns Hopkins University Press, 1978.
Booth, Wayne. *The Rhetoric of Fiction.* Chicago: University of Chicago Press, 1961.
Council of Biology Editors Style Manual Committee. *Council of Biology Editors Style Manual.* 5th ed. Bethesda, Md.: Council of Biology Editors, Inc., 1983.
Darwin, Charles. *On The Origin of Species and the Descent of Man.* Facsimile of the first edition, with an introduction by Ernst Mayr. 1859. Cambridge: Harvard University Press, 1964.
Dillon, G. *Constructing Texts: Elements of a Theory of Composition and Style.* Bloomington: Indiana University Press, 1981.

Ede, Lisa, and Andrea Lunsford. "Audience Addressed/Audience Invoked: The Role of Audience in Composition Theory and Pedagogy." *College Composition and Communication* 35 (1984): 155–71.

Eldredge, Niles, and Stephen Jay Gould. "Punctuated Equilibrium Prevails." *Nature* 332 (1988): 211–12.

Fish, Stanley. *Is There a Text in This Class? The Authority of Interpretive Communities.* Cambridge: Harvard University Press, 1980.

Fish, Stanley. *Self-Consuming Artifacts.* Berkeley: University of California Press, 1972.

Fish, Stanley. *Surprised by Sin: The Reader in "Paradise Lost."* New York: St. Martin's Press, 1967.

Gibson, Walker. "Authors, Speakers, Readers, and Mock Readers." *College English* 11 (1950): 265–69.

Gragson, Gay, and Jack Selzer. "Fictionalizing the Readers of Scholarly Articles in Biology." *Written Communication* 7 (1990): 25–53.

Gross, Alan. *The Rhetoric of Science.* Cambridge: Harvard University Press, 1989.

Holland, Norman. *The Dynamics of Literary Response.* New York: Oxford University Press, 1968.

Holland, Norman. *Five Readers Reading.* New Haven: Yale University Press, 1975.

Iser, Wolfgang. *The Act of Reading: A Theory of Aesthetic Response.* Baltimore: Johns Hopkins University Press, 1974.

Johnson, Nan. "Reader-Response and the *Pathos* Principle." *Rhetoric Review* 6 (1988): 152–66.

Kinneavy, James. *A Theory of Discourse.* Englewood Cliffs, N.J.: Prentice-Hall, 1971. Reprint. New York: Norton, 1980.

Maynard Smith, John. "Darwinism Stays Unpunctuated." *Nature* 330 (1987): 516.

Maynard Smith, John. "Game Theory and the Evolution of Behaviour." *Proceedings of the Royal Society of London, B: Biological Sciences* 205 (1979): 41–54.

Maynard Smith, John. "Optimization Theory in Evolution." *Annual Review of Ecology and Systematics* 9 (1978): 31–56.

Maynard Smith, John. "Punctuation in Perspective." *Nature* 332 (1988): 311–12.

Myers, Greg. "The Social Construction of Two Biologists' Proposals." *Written Communication* 2 (1985): 219–45.

Ong, Walter. "The Writer's Audience Is Always a Fiction." *Publications of the Modern Language Association* 90 (1975): 9–12.

Prince, Gerald. "Introduction to the Study of the Narratee." *Poetique* 14 (1973): 177–96. Reprinted in *Reader-Response Criticism,* ed. Jane Tompkins, 7–25. Baltimore: Johns Hopkins University Press, 1980.

Rosenblatt, Louise. *Literature as Exploration.* New York: Appleton-Century-Crofts, 1938.

Selzer, Jack. "More Meanings of *Audience.*" In *A Rhetoric of Doing,* ed. Roger Cherry, Neil Nakadate, and Stephen Witte. Carbondale: Southern Illinois University Press. In press.

Gay Gragson and Jack Selzer

Suleiman, Susan, and Inge Crosman. *The Reader in the Text.* Princeton: Princeton University Press, 1980.

Tompkins, Jane, ed. *Reader-Response Criticism: From Formalism to Post-Structuralism.* Baltimore: Johns Hopkins University Press, 1980.

Wilson, W. Daniel. "Readers in Texts." *Publications of the Modern Language Association* 96 (1981): 848–63.

11   A STUDY IN RHETORICAL READING

HOW EVOLUTIONISTS READ

"THE SPANDRELS OF SAN MARCO"

DAVIDA CHARNEY

In a review of a recently published collection of essays by Stephen Jay Gould, L. B. Slobodkin, a professor of ecology and evolution at the State University of New York at Stony Brook, criticized the style of Gould's writing as essentially unscientific. His basic objection has to do with the intrusion of Gould's personality and beliefs into the writing: "One of the most difficult aspects of scientific work is reporting what one sees without having it colored by one's own preconceptions. It may be ultimately impossible. Nevertheless, in most scientific prose the author strives for clarity in the dual sense of expository simplicity and in making oneself transparent so that the empirical world is visible through the text but the peculiarities of the author are invisible" (Slobodkin, 503). As readers of this volume are doubtless aware, Slobodkin's stance, as a scientist reading scientific discourse, has a distinguished pedigree; it is strongly reminiscent of the position of Francis Bacon and other scientists of the early Royal Society, who condemned ornate stylistic devices, metaphors, and other figures, schemes, and tropes for at best merely obscuring matters and at worst swaying a scientist against his reason, and presumably against the empirical evidence, to believe in some specious claim (see Halloran and Bradford). Since at least Bacon's time, scientists have taken as their ideal a form of scientific discourse that is straightforward, objective, and dispassionate, discourse that confines itself to describing independently confirmable observations and drawing dispassionately logical conclusions from them. As such, generations of scientists have conceived of their discourse as standing outside the realm of rhetoric, the classical art of persuasion.

Slobodkin, like many others, has cast the issue in terms of prose style, as a matter of whether or not the writing is clear and simple.

Underlying many injunctions for clarity and simplicity, however, is the far more basic issue of the purpose of scientific discourse and how scientists are expected to read it and act on it. An extreme empiricist-positivist position would have it that we live in a stable, uncontroversial, empirically knowable world and that science consists of theory-neutral facts and logical inferences that any competently trained—read, *invisible*—scientist could produce. Scientific discourse becomes controversial because, as fallible humans, scientists are apt to tinge, distort, and otherwise muddle their account of the world when they attempt to describe it in writing. Scientific language is meant to be a kind of windowpane, through which, if we keep it clean enough, we can see the world plainly (for similar accounts, see Bazerman, *Shaping Written Knowledge*; Miller, "A Humanistic Rationale"). In this view, the work of scientific readers is also quite simple: they need only comprehend a text, verify its technical accuracy, and integrate its results with their prior knowledge. The purpose of scientific discourse, then, can only be exposition, providing authoritative information—not argument. No rational person should require *persuasion* to accept an accurate description of the world. Rhetoric being therefore unnecessary, its use can only provoke suspicions of foul play.

Recent work by philosophers, rhetoricians, and sociologists of science, of course, has convincingly challenged the accuracy of this view, finding compelling evidence that science is in fact a richly rhetorical enterprise that reflects the complex, ambiguous, and probabilistic natural world that scientists and the rest of us actually inhabit (Bokeno). A wealth of rhetorical strategies have been found in constant use in scientific journal articles, proposals, and other standard forms of communication (Miller and Selzer; Fahnestock and Secor; Halloran and Bradford; Gragson and Selzer). Furthermore, ethnographers and sociologists observing scientists at work have found them quite adept (though perhaps unconsciously so) at weighing the appropriateness of specific rhetorical approaches for various audiences (Law and Williams; Myers; Latour and Woolgar).

As this research suggests, the aim of scientific discourse is profoundly argumentative and not merely expository; the goal is to persuade readers, to convince them of the validity and importance of the work, and to motivate them to acknowledge the force of the contribution by explicitly accepting and building upon it. The demands placed on the reader go well beyond simple comprehension and verification of neutral facts. As an equal participant in an argument, a reader is free to resist the writer's claims—even those claims that seem logically valid—by employing dissociation or devaluation or any of the other rhetorical techniques open to the writer (Perelman and Olbrechts-Tyteca). In this context, the traditional impersonal scientific stance becomes just one more tool in the writer's endeavor to stave off a reader's potential challenges to the writer's ethos: the very features that make scientific discourse seem most impersonal, dispassionate, and objective themselves amount to a deliberate rhetorical strategy to convince readers of the competence, thoroughness, and caution of the scientist. Paradoxically, the scientific discourse that is most successful as argumentation may be that which appears most disinterestedly expository.

The question that arises, of course, is what happens when scientists encounter texts that violate these rhetorical conventions of scientific dispassion? What happens, for example, when evolutionists read Gould and Lewontin's "Spandrels of San Marco"? "Spandrels" can in no way be considered a piece of dispassionate, objective, scientific prose. As readers of this book will be well aware, Gould and Lewontin's article is basically an attack on the mainstream approach to evolutionary biology, the so-called "adaptationist programme," which seeks an adaptive purpose for each feature of an organism's physiology or behavior, an explanation that optimizes the organism's chances for survival through natural selection. Gould and Lewontin argue that the adaptationist program is reductionist, incapable of falsification, subversive of Darwin's initial intent, and ultimately unscientific.

As several other chapters in this book argue, Gould and Lewontin try to shake up their readers, to loosen their paradigmatic preconceptions, by employing a wide array of unusual discourse strategies. The title of the article itself initiates one of the most significant moves, an extended literary ad hominem, branding the adaptationist program as the "Panglossian paradigm." This insulting characterization equates the adaptationists in their predilection for optimization with Dr. Pangloss, the ridiculous figure in Voltaire's *Candide* who sees every disaster as being ultimately for the best "in this the best of all possible worlds"; the phrase recurs—replete with quotes from *Candide*—throughout the first half of the article (149, 151, 153). Other atypical discourse strategies in the article include irony, invective, allusions to high Western culture, explicit references to its own rhetoric—all the elements, in short, of a vigorous intellectual debate. "Spandrels" is an atypical piece of scientific discourse, one that almost gleefully flouts the strictures for a plain style, one that is overtly and unrepentantly tendentious, while science normally represents itself as disinterested. Yet the "Spandrels" article is clearly intended as a serious contribution to evolutionary science. How can we understand Gould and Lewontin's choice of a rhetorical stance that apparently accommodates their scientific audience so little?

Davida Charney

In their reader-response analysis of "Spandrels," Gragson and Selzer argue that Gould and Lewontin's rhetorical strategies *invite* scientists to depart from their usual reading strategies, to read as more than " 'mere' biologists," to apply to the text the habits of mind of cultured intellectuals who are always ready for a lively debate. If scientists in fact adopt this role, the argument goes, then they might be more receptive to Gould and Lewontin's proposals. Halloran makes a similar argument in attributing some of the stunning effect of Watson and Crick's "A Structure for Deoxyribose Nucleic Acid" to its unusual rhetorical stance, in particular to the breezily confident ethos established by its authors. These text-based analyses describe hypothetical readers implied within the text and indicate ways in which real readers might approach it. But what of the real scientists who read "The Spandrels of San Marco"? Do they rise to Gould and Lewontin's rhetorical bait? Are they willing or able to abandon their "biological prejudices," to become the unconventional readers that the discourse requires? We cannot, of course, reconstruct with absolute certainty how scientists read "Spandrels" when it first appeared—though given the continuing controversy that the piece has inspired, we know that it was neither dismissed out of hand nor entirely successful in winning over its readers.[1] What we can study, as I investigate here, is how evolutionary biologists today read the article and, in particular, how they react to each of Gould and Lewontin's rhetorical strategies. Expressly because it departs from the conventions, "Spandrels" provides a particularly good opportunity to study how scientific readers cope with rhetorical strategies that generations of scientists have been trained to condemn as unscientific. For rhetoricians, this study also provides a valuable opportunity for illuminating how, and how well, the rhetoric of a specific scientific text works, in this case by means of tracing the responses of individual readers.

Method

I asked scientists from Gould and Lewontin's intended audience to read and react to the "Spandrels" article. The participants were seven evolutionists: five ecologists, one paleontologist, and one anthropologist. Readers from several disciplines were selected because "Spandrels" is not aimed exclusively at readers within one of the specialized subdisciplines of evolutionary studies, but rather addresses the theory and practice of evolutionary biology broadly conceived. The participants varied in standing within the academic scientific community: four were graduate students (two master's degree candidates and two Ph.D. candidates) and three were faculty members (one postdoctoral fellow, one associate professor, one full professor). My goal in selecting participants was to get a range of different levels of experience and different disciplines within evolutionary science, rather than to find a fully representative sample. While it is often desirable in reading research to limit the participants' previous familiarity with a text, all of these participants were well acquainted with at least some of Gould's work, and most had heard of "Spandrels" previously. Given Gould's reputation and the article's controversial nature, it would have been quite difficult to find completely "naive" readers, who would come to the text without preconceptions. And in fact, such an attempt would in some ways have defeated the purpose of this investigation into rhetorical reading. After all, argument rarely takes place between neutral parties who have no previous knowledge of their interlocutor or attitudes toward the issue in question; this is especially true when the parties in the debate are figures as well known and controversial as Gould and Lewontin. From the outset of this study, therefore, it was clear that the participants would read the article through the lens of history; they brought to it a wealth of knowledge and associations—both personal and professional—about the article itself, its subject matter, and its authors.

To enable observation of the participants' immediate reactions to specific rhetorical moves in "Spandrels," I asked the participants to read it using a method known as "thinking aloud." This methodology is widely used in cognitive psychology to study skill learning and problem solving; within English studies, the technique is used by many researchers to study both reading and writing processes.[2] To perform the think-aloud reading, participants were instructed to read the text aloud and to say aloud whatever thoughts went through their heads as they read. Although I was present during the think-aloud session, participants were told that they should ignore me as much as possible, that the purpose of the commenting was not to explain the text to me but to come to terms with it for themselves. They were told that many people mumble comments to themselves when they read; in a think-aloud session, they were simply to raise the volume of the mumbling to an audible level. After these brief instructions on the purpose and method of thinking aloud, participants were given an opportunity to practice it on a brief passage of a separate text. The participants were then asked to use the think-aloud method to read "The Spandrels of San Marco," treating the text as they would any other journal article that a colleague had recommended. While the participants varied greatly in the amount of their commenting, none had any apparent difficulty with the thinking-aloud method.

Davida Charney

The participants were not required to read the entire article, both because thinking aloud becomes fatiguing and because it provides such rich data that a sampling from the article was considered sufficient. After reading for thirty minutes, therefore, participants were asked to stop at the end of whatever section they had reached. Within this time frame, most managed to read about half the article. In an open-ended interview after the think-aloud reading, I asked participants about their backgrounds and their previous experience with the issues raised in the article, as well as about some of the comments they had made while reading. Both the think-aloud reading and the interview were tape recorded and later transcribed for analysis.

The analysis of the participants' comments was designed to uncover how scientists come to grips with the argument of a text, reading not simply to understand the information presented on the page but to decide whether or not to believe it, whether or not it matters, whether or not to act on it. A secondary goal was to compare the comments of faculty members and graduate students in order to learn more about how scientists learn to read like scientists. It may be that scientists have characteristic habits of mind that they bring to their discipline (or that help them survive within it); if so, we would expect graduate students and faculty members to comment in similar ways. Conversely, as evidence from other studies suggests (e.g., Herrington), students may require acculturation to the strategies of scientific literacy. This study begins to address these questions by comparing the kinds and quantities of comments from different participants in the act of reading "The Spandrels of San Marco."

To analyze the participants' comments, I segmented the transcripts or "protocols" of the think-aloud sessions into continuous episodes. The episodes were not defined as sentences or clauses but rather as "units of concentration in the [reader's] process," following Flower and Hayes. A new episode began whenever the reader shifted focus, changed a train of thought, or set up a new plan. The final segmentation of the protocols produced 664 commenting episodes, or an average of about ninety-five comments per participant. These episodes were then categorized according to a coding scheme (figure 11.1) that includes categories both for standard reading comprehension processes and for higher-level rhetorical processes. One set of categories (drawn from those used by Olson and his colleagues and by Bereiter and Bird) reflected the participants' efforts simply to comprehend the text (involving such processes as rereading, paraphrase, inference, and prediction). Another set of categories, developed inductively by me from the comments observed in the protocols, was established for participants' evaluations of the text (agreeing or disagreeing; commenting on importance, relevance, or interest; drawing on prior knowledge). A final set of categories, also developed inductively, was for comments about the structure of the text itself, in which participants called attention to the genre of the article ("it's a pedagogic piece") or to the nature of particular sections ("this isn't really a conclusion"). The comments were coded twice with this scheme, once by me and once by an independent rater. The interrater reliability was estimated with Cohen's Kappa and found to be acceptable: $K = .77$, $N = 664$.

Comprehension Processes

• Rereading	A verbatim repetition of text already read
• Paraphrase	A restatement or summary of essentially the same ideas as in the text ("so when we fail to explain something by natural selection, then we just ignore it")
• Inference	A conclusion or implication warranted by the text but not stated there explicitly ("in other words, we are missing the point"), or a problem-solving episode to figure out what the text means ("'neotenic' means it's retained in an early stage of growth and development . . .")
• Prediction	A reference to or neutral comment about earlier information in the text, a prediction about what the text will be about, or a confirmation or withdrawal of an earlier prediction ("that's what I thought he was going to conclude")
• Metacomment	A comment about the reader's own understanding or lack of understanding ("OK, got it," "unfamiliar term"); expletives or ambiguous comments ("OK," "Oh, Lord"); a comment about a plan for recovering from loss of comprehension ("better look at that again")

Evaluations of Content, Arguments, and Claims

• Validity judgment	An explicit assessment of the truth value or probability of an assertion ("OK, I can live with that," "yes, I believe that," "that's not true," "maybe," "bullshit")
• Value judgment	An explicit evaluation of an assertion's interest ("well, that's not surprising"), importance ("so what?" "that's a good point"), or relevance ("oh, this is a very orthodox view nowadays," "a rather poor example, at least not a familiar one")
• Reference to prior knowledge	New information introduced by the reader, including opinions, interpretations, examples, criteria, or associations with respect to the text, the authors, the subject matter, etc. ("this is a very deterministic argument," "I can probably think of some examples here," "I wish I knew a little more of Aztec cosmology," "that's overly simplistic," "yeah, but how do you test that?")

Davida Charney

- *Intratextual*   A comment explicitly identifying or referring to a structure *within* connected prose, such as the abstract, introduction, conclusion, summary, title, examples, wording, definitions ("this is really sort of a final discussion," "and here's the quote from David's article," "and this is still going on parenthetically?")

- Extratextual   A comment indicating a search for or reference to a structure *outside* the connected prose, such as works cited, tables or figures, the journal citation line ("what's the date on this?" "any more figures?")

- Holistic   A characterization or comment on the text as a whole, its genre, goal, aim, or approach ("this is an attack," "OK, it's a pedagogic piece")

Global Metacomments

- Metacomments   Comments on the reader's normal or habitual behavior ("I'd normally go look that up"), on reader's immediate behavior or circumstances ("this is a bad copy," "I should Xerox that actually," "I'm drifting from it"), or on current reading process or strategy ("OK . . . I'm just going to run through this paper")

Figure 11.1. Coding scheme for thinking-aloud comments

The data, consisting of the coded protocols and the interview transcripts, were analyzed both quantitatively and qualitatively. The analysis was designed to answer three questions particular to the rhetorical situation of reading Gould and Lewontin's article: 1. How did the participants go about reading the article? Is there any evidence that they accepted the textual invitation described by Gragson and Selzer to abandon their normal scientific reading strategies? 2. What aspects of the article provoked reactions from the participants? In particular, did participants react to the unusual rhetorical devices? 3. How did the participants react? Did they act as dispassionate logicians or inflamed partisans? On what basis did they accept or reject Gould and Lewontin's points?

Global Reading Strategies

Before describing the participants' reading behaviors, I would like to consider briefly how we might expect scientific readers to approach this article. Gragson and Selzer's analysis of "the reader in the text," because it is based on the sequential unfolding of the argument, seems to presuppose that readers would treat the article like an essay or piece of fiction, starting at the beginning and working through each successive point to the conclusion. The shape of the text itself certainly seems designed to encourage such an approach. The paper opens with a carefully plotted series of nonbiological examples that leads readers to question the assumptions of the adaptationist program. In the first of these examples, alluded to in the article's title, Gould and Lewontin draw an extended analogy between salient traits in biological organisms and the decoration of spandrels in Gothic cathedrals. While an adaptationist looking at the ingenious and intricate decoration of the spandrels might conclude that they were "the cause in some sense of the surrounding architecture," that the cathedral was designed to enable their creation, any reasonable reader, of course, would conclude that the artist of the spandrels simply made good use of a space that resulted conveniently and quite accidentally from the constraints of the architecture. Gould and Lewontin's point is that biological phenomena, such as the color of a land snail, may arise from similar circumstances—not from adaptations to the environment. The spandrels example is followed by another in the same vein concerning the development of human sacrifice in Aztec culture. At the end of the series, Gould and Lewontin comment openly on their strategy: "We deliberately chose non-biological examples in a sequence running from remote to more familiar: architecture to anthropology" because in these cases the relative triviality of local adaptations would not be "obscured by our biological prejudices" (147). The sequence of major sections in the article was surely planned with equal care, with the intention of gradually accumulating doubt in the reader about the logic of adaptationist reasoning (Introduction and section 2), about the scientific validity of its results (section 3), and about its Darwinist pedigree (section 4). Only after introducing a long list of such doubts about adaptationism do Gould and Lewontin turn to a systematic exploration of the logical alternatives to adaptationism (section 5), ending, full circle, with a nonadaptive approach to the very types of "architectural" phenomena presented in the introduction (section 6). The careful sequencing of these points suggests that the argument of the text works in a linear fashion and consequently may depend upon a linear reading to achieve its desired effect.

Linear readings, however, are not typical behavior for real scientists. From the little research we have on how scientists read, we know that they typically read selectively, starting from their decision about whether or not to read an article or report in the first place. For example, in a survey of more than eight hundred engineers and research scientists (mainly specializing in aeronautics) Pinelli, Cordle, and Vondran found that most respondents considering whether to read a technical report began by looking over its most general sections: the abstract, summary, introduction, and conclusion. In many cases, they stopped there—deciding that the report wasn't sufficiently relevant to their work. On those occasions when they went on to read the body of the report itself, the scientists again tended to read selectively; more than 90 percent reported reading the results and discussion, only 79 percent the research procedure, and a mere 67 percent the tables of data. The scientists also read parts out of order, reading the results before the experimental methods and the conclusions before either of those. Similar reading strategies were reported by the physicists whom Bazerman interviewed (*Shaping Written Knowledge,* chapter 8); they read relatively few articles in their entirety and almost none in order from start to finish. They also read selectively and purposively—some focusing on methods, some on theory, others on data. These global reading strategies reflect the small amount of time scientists have (or make) available for reading; like any other professionals, they rarely can devote attention to work that is irrelevant to their immediate purpose or their general research program.

Given that nonlinear reading strategies seem to be habitual for scientists, it would be surprising to find that the participants of this study read Gould and Lewontin's article in sequence. What I observed was a mixed bag. The least experienced participants—the four graduate students—all read in sequence. In contrast, two of the three faculty members—the two most senior ones—read nonlinearly. One faculty member began by previewing the organizational structure of the article: after reading the title, he leafed through the article, read the captions on the figures, noted the absence of a summary at the end, and then read the abstract and the introduction. When he reached the rhetorical question that ends the introduction, he again broke out of sequence, as shown in the excerpt below. (In this excerpt and other excerpts from thinking-aloud protocols, the text from "Spandrels" that the participant is reading at the time is represented in ordinary typeface. The reader's comments and observations are represented in italics. My observations on the reader's actions or transcription difficulties are presented in brackets. Excerpts from the transcripts of the interview, which took place after the thinking aloud was complete, are presented in a distinctive font. To preserve the anonymity of the participants, arbitrary two-letter codes rather than initials are used to identify the readers, while "DC" refers to me.)

> AN: But we trust that the message for biologists will not go unheeded: if these had been biological systems, would we not,

by force of habit, have regarded the epiphenomenal adaptation as primary and tried to build the whole structural system from it? *Nope. Well, I don't know. Straw man? What else have they got?* [Leafs through paper] *Part 2 is* The Adaptationist Programme . . . *Part 3 is* Telling Stories . . . *Part 4:* The Master's Voice Re-examined. *Right the* [unclear] *to Darwin* . . . *Part 5:* A Partial Typology of Alternatives to the Adaptationist Programme . . . 6: Another, and Unfairly Maligned Approach to Evolution. *That's the one they want to talk about. All right. Back to section 2.*

At that point, AN read through the remainder of the paper in order, except for occasionally turning to check citations in the list of references. In the interview afterward, it became clear that his previewing was a quite deliberate strategy to help him maintain his critical edge:

AN: OK, when I read a science paper, what I do . . . and what I'm always trying to get my graduate students to do . . . I read . . . I first find whether there's an abstract, a précis, or a summary and I read that first. Then I read all the legends to the tables and the figures. What I'm doing is trying to get keywords and concepts and things that will be there so that when I read the article it will not be the first time. . . .

DC: OK. So you said you didn't want to be victimized by [the text]. What did you mean by that?

AN: Well, I think if you go right into an article, and you read it word for word from the beginning, what happens is that you'll be pulled along by the author and you're not going to be critical . . . and you're going to be saying a lot of times: "Well, what is he saying? Oh yeah, I get that. And what's this word? And let me think about what's [unclear] modeling again." As opposed to, if you have an overview, then you can keep your critical facility alive. And that's what science is about. That's, it seems to me, that's what scholarship is about . . . is the ability to say: "No, no, that's not true. Or what about this? Or why are they saying that?" I mean, like that incredible . . . I mean these tactics they use, when he quotes Voltaire and then all of a sudden there's a cut in there about the adaptationist program. I mean, you know, it's really kind of transparent.

Throughout his reading of the article, AN remained skeptical but open-minded. His comment at the end of the introduction indicates that he had not been completely convinced of the illogic of adaptationist reasoning by the series of analogical cases. As I will show later, AN quite strongly rejected the characterization of the adaptationist program as "Panglossian" as he read the article. The other senior faculty member, PA, also began by previewing the article. After reading the

Davida Charney

title and byline, he leafed through the article *"to get an idea how long it is, because I usually have a class coming up."* Then he read the abstract and looked for a conclusion. Not finding one as such (*"This isn't really key. This is really sort of a final discussion"*), he read the final paragraph of the final section before turning back to the introduction. PA did not explicitly associate his previewing with the ability to mount a critique—but he had no trouble maintaining a critical position.

Previewing is known to be typical of strong, active readers, as many reading researchers will attest (e.g., Mayer). Previewing facilitates a critical stance but is probably not imperative for it—I doubt, for instance, that AN would have bought the Pangloss characterization even if he had read the article in sequence. In any case, it is not difficult to see how a preemptive previewing strategy can undercut the rhetorical force of an article. Through such reconnoitering, readers have opportunities to undermine a writer's argument that audiences in many other rhetorical situations do not enjoy. Unlike the auditors of a speech, an active reader need not wait for the orator to unfold the argument. And unlike a participant in a face-to-face dialogue, an active reader gives the writer no chance to adapt and respond to his or her specific objections. The writer, as represented in the text, can construct the most desirable sequence for promoting arguments, but the reader need not follow it.

In addition to previewing, PA disrupted the line of the argument in an even more drastic way—by skimming. He explained his strategy in the interview:

PA: You can see that I'm not reading every little thing in detail. It's possible to tell what's going on, by skimming through some of his examples.

DC: Is that what you would normally do with this kind of article?

PA: Yeah. Yeah, I read it as I would be reading it if I were. . . . Now if I were to be giving a seminar where I was going to specifically criticize the article in detail, I would sit down and read it much more closely. But I'm treating this as a reprint or as something I've seen in one of the journals I regularly keep up with. I'm trying to get enough of what he said so that I know what's going on in the field out there. So it's not essential, for instance, that I know any of the details about mountain bluebird aggression. You know. But I'm looking at his . . . at why he's introducing it and what he concludes out of it. . . . It's more important, for me anyway, in my own stage of development, to be able to look at the overall pattern of his argument and his thinking. And paying attention to the details, at least on this first skim through, would detract . . . would just bog me down.

The fine points of Gould and Lewontin's argument had no chance to sway this scientist because he never read them.

In contrast to the faculty members, the graduate students all read the main body of the article straight through. This behavior did not seem to derive from passive obedience to the instructions to read the article. Two of the four students (TO and OS) skipped the abstract, claiming always to read abstracts last. These students thus went out of their way to avoid overviewing the argument. One of them (TO) apparently adopted this strategy expressly because of the argumentative genre of the article. He said in the interview that he would normally "flip around" more in a "data paper" but that in a "thought paper like this . . . I would be more likely to read the whole paper, and then come back and read the abstract, and then do more thinking." So a linear reading does not preclude thoughtful critique. Several linear readers were quite capable of critiquing Gould and Lewontin's arguments—both positively and negatively. However, a linear strategy does indicate a greater willingness to follow the author's lead in this genre of scientific discourse. While the small size of the sample in this study precludes strong generalizations, the non-linear strategies of the more mature scientists may prepare them better to engage the writer's arguments.

Provocative Aspects of the "Spandrels" Article

As they read "The Spandrels of San Marco," the participants were actively engaged in comprehending as well as reacting to the text. Table 11.1 shows the distribution of comments among the coding categories, separating out summaries for graduate students and faculty members. Not surprisingly, a major proportion of the participants' attention was devoted simply to comprehending the text. Any scholarly text requires its readers, even disciplinary experts, to exert some energy to understand it; "Spandrels" was no exception. In fact, the comprehension processes of the faculty and graduate students looked very similar. On average, a faculty member tended to make about the same number of comprehension comments as a graduate student (fifty-seven versus forty-four comments) and engaged in the same kinds of comprehension processes in about the same proportions (see the middle portion of table 11.1; a statistical comparison of the distribution of the comprehension comments showed no significant differences between faculty and graduate students). All participants encountered some unfamiliar terminology, both scientific (e.g., "orthogenetic") and non-scientific (e.g., "portcullis"), and most had to work to unravel some

Davida Charney

Table 11.1. Distribution of protocol comments for graduate students and faculty participants

		Average number of comments per person		Proportion of total	
		Grads ($n = 4$)	Faculty ($n = 3$)	Grads (%)	Faculty (%)
All comment types					
Comprehension comments		44	57	66	42
Evaluations		19	59	29	45
Text structure/genre comments		3	12	4	9
Global metacomments		1	5	1	4
	Overall	67	133	100	100
Comprehension comments					
Rereading		19	22	44	39
Paraphrase		6	10	13	17
Inference		4	8	8	14
Prediction		3	4	6	7
Metacomment		12	13	28	23
	Overall	44	57	100	100
Evaluations					
Validity judgment		7	28	38	46
Value judgment		1	9	4	19
Prior knowledge		11	22	58	36
	Overall	19	59	100	100

tortuous syntax. In general, the participants had little difficulty understanding the text—their numerous comprehension comments simply indicate good active reading habits: monitoring their understanding, anticipating what will come next, and clearing up ambiguities.

While the graduate students largely confined their reading activities to comprehending the text (fully two-thirds of their comments fell in the comprehension category), the faculty were deeply involved in coming to grips with the text structurally and rhetorically. The faculty members' evaluations and genre comments were three or four times as frequent as those of the graduate students—the overall distribution of kinds and numbers of comments in the two groups was significantly different: $\chi^2(3) = 39.2$, $p < .01$. The graduate students' lack of evaluations does not seem due to their being overwhelmed by efforts to comprehend the text—as the similarity between their comprehension processes and those of the faculty attests. Rather, as will become apparent in the next section, the faculty's greater familiarity with the domain—and their greater stake in the debate—represented a stockpile of knowledge and attitudes against which they could weigh Gould and Lewontin's claims.

Not only the number but even the type of graduate students' evaluations of Gould and Lewontin's points differed from those of the faculty. As the bottom portion of table 11.1 indicates, while the graduate students primarily related what they were reading to what they already knew (prior knowledge), the faculty were significantly more often engaged than the graduates in assessing the validity and value of the text: $\chi^2(2) = 14.9$, $p < .01$. When graduate students did assess the truth value of the text, 84 percent of their comments indicated agreement with Gould and Lewontin. Faculty readers agreed 63 percent of the time. The faculty members often supported Gould and Lewontin's arguments with their evaluations (e.g., introducing more evidence or examples from their own domains of expertise), but they also frequently challenged the typicality of Gould and Lewontin's examples or their logic. In short, the activities of the two groups suggest that while both were capable of understanding the text, the faculty were much more willing or able to enter the rhetorical fray.

By sorting the evaluations according to location (i.e., the sentence being read just before a evaluation was uttered), I was able to identify a number of text segments that provoked the most numerous evaluations from the most participants. I will confine myself in this section to describing these "hot spots," leaving a more detailed description of the participants' evaluations to the next section.

In general, evaluations tended to cluster at structural junctures in the text, particularly at the beginnings and ends of paragraphs and sections. Topically, the most provocative segments were those in which Gould and Lewontin asserted some characterization of the adaptationist program. The single most provocative paragraph was the abstract, drawing a total of twenty-two evaluations, all from the three faculty participants. (For the article as a whole, the average number of evaluations per paragraph was eleven.) The evaluations of the abstract were concentrated largely around the second and third sentences, in which Gould and Lewontin give their definition of the adaptationist program ("It is based on faith in the power of natural selection as an optimizing agent. It proceeds by . . .") and the long sixth sentence, in which they present a list of charges against it ("We fault the adaptationist programme for its failure . . ."). The two next "hottest" locations were the opening paragraph of section 2 and the first full paragraph of section 3. The opening paragraph of section 2 essentially restates and elaborates on the abstract's definition of the adaptationist program. It drew a total of sixteen evaluations from six of the seven participants—from all but one graduate student. The opening paragraph of section 3 essentially characterizes the adaptationist program as incapable of empirical test or falsification, a charge tantamount to calling it unscientific. This paragraph drew a total of twenty-nine evaluations from all three faculty and two of the graduate students. Other characterizations of the adaptationist program also drew numerous evaluations. In particular, at the end of section 2, Gould and Lewontin present a numbered list of four "common styles of argument" used by adaptationists. Five sentences—the last sentence of each of the four items plus the last sentence of the section that immediately followed—together provoked a total of twenty-four evaluations from all participants.

Two other "hot" locations deserve mention here, in both of which Gould and Lewontin draw explicit attention to their own rhetoric. In the penultimate paragraph of the Introduction, Gould and Lewontin heap scorn on the adaptationist view of Aztec cannibalism and tie this example to the spandrels example. This paragraph drew twenty evaluations from five of the participants. In the middle of section 2, Gould and Lewontin anticipate that readers may accuse them of caricaturing the adaptationist program after explicitly labeling it Panglossian. This paragraph (beginning "At this point, some evolutionists will protest . . .") drew twenty-one evaluations from five participants. The sentence immediately preceding this paragraph, which caps the characterization of the adaptationist program as a Panglossian paradigm ("Each trait plays its part and must be as it is"), itself drew six evaluations from five participants.[3]

While each participant reacted to different aspects of "Spandrels," the concentration of evaluations at these particular locations suggests that the participants recognized that the crux of Gould and Lewontin's argument is the validity of their characterization of the adaptationist program. As such, the article (at least the first half of it, which all participants read) amounts to an argument from the rhetorical stases of definition and value (Fahnestock and Secor). If Gould and Lewontin can demonstrate that the adaptationist program is inherently unscientific and therefore unproductive, then they create the exigence for changing to a different and presumably more purely scientific alternative.

How Readers Engaged Gould and Lewontin's Arguments

In this section, I will describe in more detail the participants' evaluations of the article, focusing in particular on the "hot spots" identified above. The most complex commentary on the article was provided by AN (the faculty member who used the extensive previewing strategy), whose evaluations I will describe in some detail.

AN generally agreed with Gould and Lewontin's definition of the adaptationist program, as presented in the abstract. He accepted their assertion of the dominance of the adaptationist program in sentence 1 (*"I guess that's true"*), though in the next sentence he challenged their conflation of adaptationism with optimization (*"Well, I don't know if that's true. Because optimization theory had not been at it for forty years"*), and caught them a few sentences later in an important conceptual slide (*"Optimization is not perfection. Anyhow . . ."*). However, at the opening of section 2, when Gould and Lewontin reintroduce and elaborate the definition and explicitly apply the label "Panglossian paradigm," AN forcefully rejected it:

> AN: It is rooted in a notion popularized by A. R. Wallace and A. Weismann (but not, as we shall see, by Darwin) toward the end of the nineteenth century: the near omnipotence of natural selection in forging organic design and fashioning the best among possible worlds. *Oh shit. I don't know who the hell believes that. Who really believes that? If you asked them, put it to them that way? No one believes that, right? If you want to explain the appearance of the human knee or the human brain you go to natural selection, you don't go to genetic drift. We shall see.*

He was equally emphatic in rejecting the Panglossian label later in section 2:

> AN: The adaptationist programme is truly Panglossian. *That's incredible. I mean there's a quote from Voltaire and then the next line is: "the adaptationist programme is truly Panglossian." You know he just quoted Voltaire, Voltaire is making Pangloss Panglossian.* Our world may not be good in an abstract sense, but it is the very best we could have. Each trait plays its part and must be as it is. *No, that's bullshit. I don't think adaptationists believe in that. Yeah all right here we go, let's. . . .* At this point, some evolutionists will protest that we are caricaturing their view of adaptation. *Yeah, I think you're caricaturing their view.*

AN's objection here is not a sign that he rejected Gould and Lewontin's arguments out of hand. In fact, AN agreed with Gould and Lewontin throughout the introduction—he saw the point of the spandrels example (*"OK, so the point of that sentence is: here's something that was an architectural side effect, and when you see the whole structure you turn cause and effect around and start thinking that is what generates everything else. Right, makes sense, says it all"*). He agreed with their interpretation of Aztec cannibalism—based on his own familiarity with the researchers whom Gould and Lewontin cite: *"Ed Wilson would do that sort of thing because he doesn't know the details of anthropology or the issue"*; and *"Who's this?* Ortiz de Montellano. [Turns back to references] *Is that the chemist? Is that him?* Aztec cannibalism: an ecological . . . *Yeah, that's right in Science. A real good paper. It just shows it's a lot of bullshit."* AN even accepted Gould and Lewontin's use of nonbiological examples in the Introduction as a way of avoiding biological prejudices: *"In other words, we can see the argument that they're making really clearly. All right, fair enough. It's a pedagogic paper."*

So what is the basis of AN's disagreement with Gould and Lewontin? In certain cases, he contradicted their statement of the facts. In others, he agreed with their description of what adaptationists do, but disputed the valence or significance of these activities. Several of these responses emerge in his evaluations of the opening paragraph of section 3:

> AN: First, the rejection of one adaptive story usually leads to its replacement by another—*Yeah, because that's been a success in physiology and evolution for a century*—rather than to a suspicion that a different kind of explanation might be required. *Maybe.* Since the range of adaptive stories is as wide as our minds are fertile, new stories can always be postulated. *That's right. So what? They can be falsified.* And if a story is not immediately available, one can always plead temporary ignorance and trust that it will be forthcoming. *Well, maybe. I agree that might be what you would do.*

AN admitted that adaptationists explain traits with adaptive stories, but countered that that strategy is often successful. He admitted that adaptationists are reluctant to consider alternatives to natural selection, but denied that the cause is wrong-headedness:

> AN: We maintain that alternatives to selection for best overall design have generally been relegated to unimportance by this mode of argument. *No, they've been relegated to unimportance because selection is more powerful. That's why.*

Further, he saw no harm in a succession of stories, since he claimed that false stories are easily discovered. He denied outright Gould and Lewontin's claim that adaptationist stories are untestable, and in fact threw this charge back at their preferred alternatives:

> AN: We do wonder, though, whether the failure of one adaptive explanation should always simply inspire a search for another of the same general form, rather than a consideration of alternatives to the proposition that each part is 'for' some specific purpose. *The question is: how do you test the idea of nonfunction? And the answer is: you can't. That's why you move from one adaptive explanation to another. You can test those ideas. Oh well.*

AN's interpretation of Gould and Lewontin's claims was similar in many ways to that of PA, the other senior faculty member. Like AN, PA ended up agreeing to some extent with Gould and Lewontin's descriptions of what adaptationists do. He conceded that adaptationists try a succession of stories but countered that *"you try all the different possibilities and more stories is a reasonable approach."* Significantly, he flatly denied that adaptationists only consider adaptive explanations, that they always replace stories with more stories. He responded to two of the "four common styles of argument" attributed to adaptationists as follows:

PA: If one adaptive argument fails, assume that another must exist. *Well, no, we don't really do that. If an adaptive argument fails, try another—agreed, like the first choice. But then not necessarily assume another must exist. It may be that after you've tried several that . . . that you conclude that none of them work and therefore this probably isn't an adaptive feature. And that's what been recognized ever since Rudwig put out the paradigmatic approach to functional morphology in the late sixties.*

PA: In the absence of a good adaptive argument in the first place, attribute failure to imperfect understanding of where an organism lives and what it does. *Well, but that's true. We don't know what most organisms do in terms of life-style so that's a reasonable approach. But even so we don't automatically assume that there's going to be an adaptationist explanation.*

Thus like AN, PA admitted that Gould and Lewontin's alternatives are often ignored, but on reasonable grounds—not in principle:

PA: Under the adaptationist programme, the great historic themes of developmental morphology and *Bauplan* were largely abandoned. *Well, they were abandoned because they didn't pan out, in terms of evidence. While that doesn't mean they're totally wrong, it does mean that the emphasis on them in Europe may be misplaced. OK.*

In PA's alternative characterization of adaptationists, they are not nearly as inflexible and dogmatic as Gould and Lewontin paint them. While he conceded that some adaptationists have been unwilling to consider alternatives to adaptive stories (*"that's a legitimate criticism"*), PA reduced the significance of the problem by restricting its scope to second-rate researchers who unfortunately practice bad science:

PA: [I]f these had been biological systems, would we not, by force of habit, have regarded the epiphenomenal adaptation as primary and tried to build the whole structural system from it?

> *Perhaps an unsophisticated adaptationist would. But nobody who had some experience with real animals would have, if they had thought about it.*

PA: Second, the criteria for acceptance of a story are so loose that many pass without proper confirmation. *Well, that's true. But on the other hand, that's hardly a telling point against adaptation. It's a telling point against being sloppy and simplistic in your thinking. Which is in part what he's doing here.*

Finally, even after the concession of narrow-mindedness in this smaller group, PA denied the originality and importance of Gould and Lewontin's efforts to expose the problem by claiming that it had long since been recognized. To Gould and Lewontin's announcement in the abstract of their goal of reestablishing a focus on constraints that had long been popular in continental Europe, PA responded:

PA: *The sophisticated adaptationist in America has always thought this. This isn't something that has been ignored in America. But he's like a knight in shining armor trying to reintroduce, and against everybody else, with something that . . . to some extent he's not aware that everybody else has developed. [A bit later] In fact these things were already being talked about as long ago as early sixties. Nice that he's reemphasizing them, but he's not the only one who's thought about these things . . . in this way.*

It should be clear from these examples that these readers contested Gould and Lewontin's definition of the adaptationist program. The main points of outright disagreement concerned whether adaptive stories are in fact testable and whether adaptationists are willing in principle to consider nonadaptive explanations. Their own evidence, their experience that stories can be proven false, their acquaintance with adaptationists who are more broad-minded, is laid against Gould and Lewontin's rather absolute claims to the contrary. Even when these scientists conceded that Gould and Lewontin were accurate in their descriptions of some adaptationist practices, they ascribed quite different values to the practices—that's true, but it works; that's true, but only unsophisticated scientists do it; that's true, but it's not news; that's true, but it's no worse than your alternative. In essence, then, these scientists accepted as legitimate many of Gould and Lewontin's criticisms, but rejected the extreme form in which they were cast.[4]

Gould and Lewontin fail then to make a deductive case that adaptationists *in principle* are forced into unscientific methods. But this is not the only tack they take. They also try an inductive approach. In this context, Gould and Lewontin's use of nonbiological analogies in the

Introduction and their choice of examples of inappropriate adaptation-ist stories become quite important. It is patently absurd to take the spandrels as the starting principle of Gothic architecture. Their argu-ment in essence is that adaptationist stories are just as absurd. For the adaptationist program as a whole to be classed this way, as truly Pan-glossian, the examples of adaptive stories must be clearly ridiculous and must hold up as representative of the literature as a whole. Gould and Lewontin win the first half of this inductive argument—no one challenged the specifics of the examples themselves. But most partici-pants resisted the inductive leap to generalize from the examples to the adaptationist program as a whole. At the end of section 2, for example, Gould and Lewontin ridicule the adaptive explanation of the attenu-ated front legs of *Tyrannosaurus* and explicitly link this example to those presented in the Introduction. At the end of this paragraph, a graduate student commented:

> TO: One must not confuse the fact that a structure is used in some way (consider again the spandrels, ceiling spaces and Aztec bodies) with the primary evolutionary reason for its exis-tence and conformation. *I think that, while all of these examples are to the point they are, uh . . . not the way a lot of people do science. It would be very easy to find examples to the contrary.*

Another participant, the postdoctoral fellow, while sympathizing with Gould and Lewontin's general position, found the examples less compelling:

> KR: The examples are good, I mean they're very graphic. But they are caricatures, you know, in this case. Basically, the point of that is . . . "OK, here's a parallel that I want you to focus on here, its essentials, can you recognize that in biological cases?" Right, fine. The prob-lem I always have with something as broad as this is that I don't think there is a single sin in the practice of science that hasn't been committed at least once. So just the fact that they can find an in-stance of it doesn't mean that the whole literature is like that.

He did not reject outright Gould and Lewontin's examples, but de-ferred judgment until he could look at the literature himself. The examples spanned so many disciplines that no one was familiar with the issues in all of them. Lyne and Howe found that Gould ran into trouble in other contexts when he stepped outside paleontology to make claims about other specialities. That did not seem to be the case here. The partici-pants accepted his reasoning behind the rejection of each specific adap-tationist explanation. In fact, the two senior faculty members each could verify the examples within their own domains or supplement them with others: AN knew the research on Aztec cannibalism, and PA skimmed through the discussion of aggression in mountain bluebirds saying, "Yes, *consistency is not necessarily a confirmation. I think I can probably come up with some invertebrate examples that would be similar.*" But as we have seen, both denied that adaptationist stories *in general* were equally ill-founded. Interestingly, it was a doctoral student in ecology, OS, who most distrusted the importation of examples from other disciplines—particularly from a "soft science" like anthropology. In the interview, he said, "I never like the examples that he [Gould] uses," believing that Gould "used this example [Aztec cannibalism] that he knows that nobody likes and that nobody would argue with him . . . so people will get distracted looking at the quality of the example and he can say whatever he wants in the point that he wants to make."

The representativeness of the examples was weakened more significantly in other ways. Some participants challenged the use of the *Tyrannosaurus* example, which even Gould and Lewontin acknowledge comes from the popular arena rather than the professional literature. In the interview, TO commented that the examples were not from "classic papers," that the mountain bluebird example would be more convincing "if this were some big landmark study or something that people had built on, built a great deal on and then, these same mistakes are made in studies like that too." Several participants commented that the examples were typical of Gould, that he uses the same cases in many other articles. Their position seemed to be that if Gould and Lewontin's general characterization of adaptationism were true, then a wider range of examples in the professional literature should be easy to find and include in this and other articles.

In sum, Gould and Lewontin failed to win over most of these readers to the strong version of their position. Only one participant, LO, a master's student, agreed with them down the line. In her protocol comments, she characterized the adaptationist positions as *"deterministic"* and *"almost creationist."* On the other hand, it is not clear that "Spandrels" was what convinced her. She reported in the interview that she was already familiar with Gould's work and tended to "favor his views that the adaptationists are rather narrow, and they don't consider other explanations at all, and they fit their experiments into their preconceived ideas, and that's bad science to begin with." The other participants, regardless of their predispositions toward Gould (which ranged from sympathy to hostility), had a wider range of both positive and negative responses to the claims in the article. They resisted the generality of Gould and Lewontin's characterization of adaptationists, but most went further than PA to give them credit for increasing evolutionists' awareness of a wider range of possible explanations.

In closing, it is worth considering how the article's unusual prose style influenced these readers. All the participants were quite conscious—and sometimes wary—of Gould's facility with language. PA believed that Gould intentionally used style as a smoke-screen: *"by couching, by cloaking all his arguments in elaborate words, most people would be intimidated and go along with him."* But even those who were often sympathetic to Gould's position found his persuasive powers difficult to challenge:

> KR: Gould is one of these people who writes extremely well. And it makes him very slippery. I have found myself disagreeing with Gould on points in popular articles and having the darndest time figuring out why. Because everything he said was so reasonable. [Laughs]

While Gragson and Selzer suggested that the literary and cultural allusions in the introduction encouraged scientists to read as intellectual humanists rather than as " 'mere' biologists," there is little evidence that these readers accepted the invitation. Several participants commented that the sequence of examples in the introduction seemed far removed from the subject of evolutionary biology. KR commented at one point, *"hmm, should be getting to some biology here,"* and later, *"I'm getting tired of this analogical reasoning."* PA objected to the spandrels example even before completing the first sentence of the introduction:

> PA: The great central dome of St. Mark's Cathedral in Venice. . . . *What the hell has that got to do with evolution and biology? Except to promote Steve's erudition* [laughs].

But he also saw its point for evolution:

> PA: Such architectural constraints abound. In a sense, the design represents an 'adaptation', but the architectural constraint is clearly primary. *Well, OK, the obvious conclusion that he's going to come to, I think, is that the animal's anatomy represents an adaptation but what it's inherited from its ancestors is a constraint that's very important to understand. So, OK.*

None of the participants knew what spandrels were to begin with, and all had to turn to the illustrations to follow the discussion. Only one respondent, LO, verbalized an association with other cultural contexts. After reading the definition of spandrels, she said, *"I remember this from art history."* Voltaire was slightly more familiar to these readers. As described previously, AN understood and emphatically rejected the Panglossian designation. At the beginning of section 2, KR noted, *"They're, uh, smuggling in the phrasing from Voltaire here,"* and, after reading the quote on venereal disease from *Candide*, chuckled and said *"oh dear."* PA's response to the same quote was *"Steve and Dick are being a bit too philosophical here, rather than scientific. And I think they're blunting the sharpness of their attack by doing so."* And, of course, the allusions went right by a number of readers. After one session, in which a master's student voiced no reactions at all to the Introduction or to the allusions, I asked him if he knew who Dr. Pangloss was:

> MA: Well, I had an idea . . . in the context of the reading. First I was going to go in more detail and ask myself: "Well, what is he exactly talking about?" But then, you know, that would probably be missing the whole idea. So, I mean . . . it goes more to, you know, their points of view rather than . . . Like at the very beginning when he was talking about the architecture of these buildings . . . I was kind of surprised he was talking about something like that. But, you know, then you can get kind of trapped there and try to explain to yourself something that is not as important as the rest of the article. So I just considered that as part of the illustration.

While the literary allusions seem not to have lured these readers out of a scientific approach to the text, neither did they cause the readers to dismiss the text as a whole as nonscientific.

Conclusion

Gould and Lewontin, in order to persuade evolutionary biologists to change their theoretical assumptions, explicitly employ unusual examples to draw them out of their "biological prejudices." Arguably, the tightly organized structure of the article itself reflects an attempt to disrupt normal patterns of reading and thinking, patterns that presumably would tend to support the current adaptationist line. However, the scientists whom I observed seemed not to be drawn out of their normal reading strategies. Most of them were good, active, skeptical, scientific readers who took a quite serious approach to the arguments in the text—though there is evidence that these skills develop with experience in the discipline. The most senior participants read nonlinearly, as scientists typically do. The participants as a group seemed to refuse the rhetorical gambit of the extrascientific allusions. They continually brought to bear their knowledge of science, of scientific texts, and of the specific scientific debate at hand. On the other hand, they seemed remarkably tolerant of the unusual rhetorical moves in the piece: only one called them "unscientific," many attributed them to Gould's well-known stylistic proclivities or to the genre of "think pieces," and a few simply ignored them.

The major clash between writers and readers concerned the way in which Gould and Lewontin attempted to characterize adaptationists. Did Gould and Lewontin convince these readers that adaptationists *in general* are narrow-minded dogmatists who refuse to consider nonadaptive alternatives to their successions of far-fetched and unfalsifiable stories? Well, no. In their definition of the adaptationist program, Gould and Lewontin attempt to demonstrate that *in principle* it forces scientists into the absurd stance of a Pangloss. But the scientists rejected some key premises—that adaptationists cannot ever consider nonadaptive explanations and that adaptive stories are unfalsifiable. The argument devolved to an inductive one that rested on the strength of the examples of inappropriate adaptive stories. At this point, Gould and Lewontin came up against the classic difficulty of clinching inductive arguments. Yes, everyone admitted the validity of the examples; they acknowledged the analogical parallels and even supplied additional examples of the same kind from their own experience. But no one was willing to make the inductive leap to the generalization. They saw the examples as the product of second-rate scientists, as isolated "sins" of the kind that can be committed in any science, or as exceptions to the general success of the adaptationist program.

Does this study of readers' responses suggest that "The Spandrels of San Marco" was rhetorically ineffective? Certainly these readers found several ways to undermine Gould and Lewontin's rhetorical strategies. We don't often think much about the clash between rhetorically active readers and the rhetorical strategies of writers. Clearly, as a reader, the recipient of an argument enjoys distinct advantages that are unavailable to a listener in an audience or a participant in a dialogue. Readers have better opportunities to control the flow of the writer's argument, to scope out the writer's rhetorical strategies. Those who undermine the writer's rhetorical strategies may be uncooperative, but we cannot call them bad readers. (On the contrary; one is more apt to be critical of overly passive readers.) Nor does it mean that the text is rhetorically deficient. If we think of rhetoric in Aristotelian terms as the marshaling of the available means of persuasion, then it is obvious that *some* of the means of persuasion set out in this article did not work—at least for some readers. Gould and Lewontin's use of extreme terms and claims did not convince these readers of the general claim, but they did evoke emphatic response. According to this response and to the historical record (documented by Winsor in chapter 7 of this book), "Spandrels" succeeded according to the standards of scientific discourse. It was not ignored. So although the article is perhaps not optimally adapted to its audience, it has indeed survived.

It is difficult to draw general conclusions about scientific discourse from this study. "Spandrels" is an unusual piece, coauthored by scientific "stars" with well-known political and literary idiosyncrasies. The age and controversial nature of the article itself increase its atypicality. We cannot recreate the processes by which it was read in 1979, when it first appeared, in part because the article itself, Gould's subsequent work, and the work of his supporters and detractors have all influenced the nature of evolutionary science and thus any current readings by evolutionists. On the other hand, "Spandrels" is not at all atypical in its most essential qualities. Gould and Lewontin generalize about biological theory from empirical evidence in the standard scientific literature. This study suggests that scientists read such discourse rhetorically: they read as is convenient for their own purposes (they read parts selectively and out of order); they weigh the plausibility of claims and evidence; they struggle to understand unfamiliar technical terms; they cheer and get mad. In other words, they read their literature the way scholars in the humanities might read *PMLA*. This behavior seems so natural and so unremarkable that it is important to remember the special objective quality that some scientists have claimed for their discourse. If scientific discourse were really based on logical deduction alone, then readers should need to do no more than comprehend the text and integrate it with their prior knowledge, as the graduate students in this study tended to do. In fact, the more professionally advanced readers were *more* prone to treat the text rhetorically, as probabilistic argument about facts and values. They thought about who wrote the piece, where it was published, what kind of piece it is, the quality of the evidence, their own knowledge of the subject, and so on. This essay has explored the use of thinking-aloud protocols as a method of rhetorical analysis for scientific discourse. This method of analysis indeed illuminates the complex set of factors—prior knowledge, attitudes, motivation—that influence whether an argument carries for particular readers. The results of this exploration remind us that even the most well-conceived rhetorical moves in a text are gambits that have no guarantee of success. The normal reading strategies that scientists adopt, strategies that actively work to engage as well as to defuse the force of other strategies deployed by the writer, are themselves inescapably rhetorical.

NOTES

I am grateful to Carolyn Terry for her assistance in collecting and transcribing the data, to those who willingly participated and helped me find participants, and to Jack Selzer and Chris Neuwirth for commenting on earlier drafts. My thanks also to Robert Edwards, who drew my attention to the Slobodkin review that opens my essay. Earlier versions of this paper were presented at the Conference on College Composition and Communication and at the American Educational Research Association.

1. Reconstructing the response to any work of its "original readers" is the job of reception aesthetics, as pioneered by Jauss. An excellent example of reception theory, one that includes helpful references and background, is Mailloux's analysis of the reception of *Huckleberry Finn* (*Rhetorical Power,* chapters 3 and 4).

2. For a thorough discussion of the theory underlying the thinking-aloud method, its limitations, and the evidence for its reliability and validity, see Ericsson and Simon. For an introduction to its use in writing research, see Hayes and Flower. For other work in English studies on the responses of real readers to literary discourse—and other methods of turning up evidence of "readers' responses"—see Holland, Fish, and Bleich. For a good introductory overview of reader-response theory, see Mailloux's "The Turns of Reader-Response Criticism."

3. It is also interesting to consider these "hot spots" from the perspective of Latour and Woolgar's five categories for statement types in scientific discourse. Of the twenty claims by Gould and Lewontin that drew the most numerous reactions, half might be classified as type 4 statements (blanket assertions of the type usually found in textbooks), and half as type 3 statements (assertions qualified by explicit "modalities" or hedges). Without categorizing the claims of the entire paper, it is impossible to say whether the seemingly high proportion of type 4 statements here is due to the nature of the article itself (i.e., Gould and Lewontin make a large number of blanket assertions), or whether the participants picked out and reacted to the strongest assertions. For a fuller treatment of this issue, see Dorothy Winsor's essay in chapter 7 of this book; it analyzes the entirety of "Spandrels" in terms of Latour and Woolgar's statement types.

4. Interestingly enough, Lyne and Howe found a similar tendency toward extreme dichotomizing in Gould's controversial publications promoting his theory of "punctuated equilibrium," and a similar moderating in published responses.

WORKS CITED

Bazerman, Charles. *Shaping Written Knowledge: The Genre and Activity of the Experimental Article in Science.* Madison: University of Wisconsin Press, 1988.

Bereiter, Carl, and Marlene Bird. "Use of Thinking Aloud in Identification and Teaching of Reading Comprehension Strategies." *Cognition and Instruction* 2 (1985): 131–56.

230

Davida Charney

Bleich, David. *Subjective Criticism.* Baltimore: Johns Hopkins University Press, 1978.

Bokeno, R. Michael. "The Rhetorical Understanding of Science: An Explication and Critical Commentary." *Southern Speech Communication Journal* 52 (1987): 285–311.

Ericsson, Anders, and Herbert A. Simon. *Protocol Analysis: Verbal Reports as Data.* Cambridge: MIT Press, 1984.

Fahnestock, Jeanne, and Marie Secor. "The Stases in Scientific and Literary Argument." *Written Communication* 5 (1988): 427–43.

Fish, Stanley. *Surprised by Sin: The Reader in "Paradise Lost."* New York: St. Martin's Press, 1967.

Flower, Linda, and John R. Hayes. "The Pregnant Pause: An Inquiry into the Nature of Planning." *Research in the Teaching of English* 15 (1981): 229–243.

Gragson, Gay, and Jack Selzer. "Fictionalizing the Readers of Scholarly Articles in Biology." *Written Communication* 7 (1990): 25–58.

Halloran, S. Michael. "The Birth of Molecular Biology: An Essay in the Rhetorical Criticism of Scientific Discourse." *Rhetoric Review* 3 (1984): 70–83.

Halloran, S. Michael, and Annette N. Bradford. "Figures of Speech in the Rhetoric of Science and Technology." In *Classical Rhetoric and Modern Discourse,* ed. Robert Connors, Lisa Ede, and Andrea Lunsford, 179–92. Carbondale: Southern Illinois University Press, 1984.

Hayes, John R., and Linda Flower. "Uncovering Cognitive Processes in Writing: An Introduction to Protocol Analysis." In *Research on Writing: Principles and Methods,* ed. P. Mosenthal, L. Tamor, and S. Walmsley, 207–20. New York: Longman, 1983.

Herrington, Anne. "Composing One's Self in a Discipline: Students' and Teachers' Negotiations." In *Constructing Rhetorical Education,* ed. Marie Secor and Davida Charney, 91–115. Carbondale: Southern Illinois University Press, 1992.

Holland, Norman. *Five Readers Reading.* New Haven: Yale University Press, 1975.

Jauss, Hans R. *Toward an Aesthetic of Reception.* Trans. Timothy Bahti. Minneapolis: University of Minnesota Press, 1982.

Latour, Bruno, and Steven Woolgar. *Laboratory Life: The Construction of Scientific Facts.* Princeton: Princeton University Press, 1986.

Law, John, and R. J. Williams. "Putting Facts Together: A Study of Scientific Persuasion." *Social Studies of Science* 12 (1982): 535–58.

Lyne, John, and Henry F. Howe. " 'Punctuated Equilibria': Rhetorical Dynamics of a Scientific Controversy." *Quarterly Journal of Speech* 72 (1986): 132–47.

Mailloux, Steven. *Rhetorical Power.* Ithaca: Cornell University Press, 1989.

Mailloux, Steven. "The Turns of Reader-Response Criticism." In *Conversations: Contemporary Critical Theory and the Teaching of Literature,* ed. Charles Moran and Elizabeth Penfield, 38–54. Urbana, Ill.: National Council of Teachers of English, 1990.

Mayer, Richard. "The Sequencing of Instruction and the Concept of Assimilation-To-Schema." *Instructional Science* 6 (1977): 369–88.

Miller, Carolyn. "A Humanistic Rationale for Technical Writing." *College English* 40 (1979): 610–17.

Miller, Carolyn, and Jack Selzer. "Special Topics of Argument in Engineering Reports." In *Writing in Nonacademic Settings*, ed. Lee Odell and Dixie Goswami, 309–341. New York: Guilford, 1985.

Myers, Greg. "The Social Construction of Two Biologists' Proposals." *Written Communication* 2 (1985): 219–45.

Olson, Gary, Susan Duffy, and Robert Mack. "Thinking-Out-Loud as a Method for Studying Real-Time Comprehension Processes." In *New Methods in Reading Comprehension Research*, ed. David Kieras and Marcel Just, 253–86. Hillsdale, N.J.: Lawrence Erlbaum Associates, 1984.

Olson, Gary, Robert Mack, and Susan Duffy. "Cognitive Aspects of Genre." *Poetics* 10 (1981): 283–315.

Perelman, Chaim, and Lucie Olbrechts-Tyteca. *The New Rhetoric*. Notre Dame, Ind.: Notre Dame University Press, 1959.

Pinelli, Thomas, Virginia Cordle, and Raymond Vondran. "The Function of Report Components in the Screening and Reading of Technical Reports." *Journal of Technical Writing and Communication* 14 (1984): 87–94.

Slobodkin, L. B. "Foxes and Hedgehogs: A Look at Four Books by Celebrated Scientists." *American Scientist* 76 (September–October 1988): 503–4.

Winsor, Dorothy. "The Construction of Knowledge in Organizations: Asking the Right Questions about the Challenger." *Journal of Business and Technical Communication* 4 (1990): 7–20.

12 DECONSTRUCTING

"THE SPANDRELS OF SAN MARCO"

DEBRA JOURNET

The topic announced by the title of this chapter suggests the applicability of a method of analysis called deconstruction to a particular scientific article. But there is another meaning playing in the chapter's title that depends on puns: deconstruction not only as a method of critical analysis, but as a word somehow about building, or de-building, that is, construction or de-construction; and "spandrels" not only as part of the title of an article about an "adaptationist programme," but also as a word about part of a building, something that has been constructed and can also be de-constructed. The pun is, of course, accidental, clearly intended neither by the authors of "The Spandrels of San Marco," who were probably not thinking about deconstruction when they wrote the article in 1978, nor by the author of this chapter, who "inherited" both the term "deconstruction" and the title of the article. Rather than being something intended by the writers, the architectural puns generated by this fortuitous linking of terms are produced by what a deconstructionist would call the "play of signifying references that constitute language" (Derrida, *Of Grammatology,* 7)—that is, by the contingent associations and relations inherent in language itself.

Deconstruction, Jacques Derrida tells us, is not a method or a technique or a system of thought; rather, it is an activity, a "protocol of reading" (*Margins of Philosophy,* 246). But it is an activity that depends on a cluster of theoretical assumptions about reading and writing that are complex and difficult to summarize.[1] Though there are many different forms of deconstruction, probably the most influential is that of Derrida himself who, in his own writing, "deconstructs" or "de-sediments" or "unravels" a set of themes that he sees as defining Western philosophy from Plato to the present. These themes, which Derrida gathers under the label the "metaphysics of presence," all posit a foundation for truth or reality that is independent of language.[2] Although competing philosophies may differ in how they define such foundations, they are alike in their "desire for a firm and ultimate ground, a terrain to build on" (*Margins of Philosophy,* 224). The source of that foundational grounding can vary widely: for example, Platonic idealism, seventeenth-century empiricism, or Cartesian subjectivism; or, in more modern terms, dialectical materialism or the psychoanalytic self. Indeed, it is the assertion of such " 'fundamental,' 'structuring,' 'original' philosophical oppositions" as "thought and language" that, in Derrida's terms, constitutes "philosophy's *unique thesis*" (*Margins of Philosophy,* 229). But all such foundations are said to exist prior to and separate from any linguistic representation.

The key word here is "representation," because the presence of a foundational truth means that language is simply a tool to represent or re-present something else that already exists. Such a position leads to mimetic or "transparent" theories of language based on notions of identity—that is, meaning is present outside of language and is re-presented through language, most directly through speech, secondarily through writing.[3] Writing becomes a copy of a copy of an original reality. The thrust of deconstruction is to destabilize that hierarchy and to complicate the metaphysics of presence by suggesting a self-reflexive or "opaque" view of writing that depends not on presence but absence, not on identity but difference.[4]

This deconstructive view of writing extends linguist Ferdinand de Saussure's definition of the sign as *arbitrarily* connected signifier (sound or graphic image) and signified (concept evoked), and his understanding of language as a set of differential relations among signs. In Derrida's narrative, the goal of Western philosophy is to find a "transcendental signified"—a foundational truth or meaning that exists independently of any signifier. A prime example, though Derrida does not say much about it, would be the "truth" provided by the scientific method, a process that aims at discovering the voice of nature itself. Deconstruction, however, argues that no meaning exists prior to or independent of a chain of signifiers. Thus, a deconstructive version of science would describe science's activities as the inscription of texts (or signifiers) rather than the investigation of an extratextual, nonlinguistic reality (or transcendental signified).[5] Instead of positing an originary meaning outside language, Derrida claims that reality—or what we constitute as reality (self, culture, nature)—is always already language; indeed only language makes such concepts as self, culture, and nature possible. Thus, writing—rather than being an imitation of an imitation of a preexisting meaning—is the very constitution of meaning.

Writing in this view is based not on presence, identity, and substitution, but on absence, difference, and supplement. That is, a sign is understood not in terms of natural identity but in terms of its differences from other signs, a potentially endless process of interpretation. Derrida calls this phenomenon "différance"—a French pun and neologism combining ideas of difference and deferring. Différance, which is defined primarily in terms of what it is not, evokes all the ways in which language resists or defers identification with any transcendental signified or extratextual referent but instead constantly implicates itself in other acts of signification or other texts. Meaning slips, as it were, ceaselessly from signifier to signifier, from text to text.

Deconstruction, thus, describes language as "indeterminate," in that no reader (including a text's author) can ever locate a complete meaning that is present in or identical with the author's intention (or with anything else). Writing or reading, rather than being a finite act of decoding, is a limitless process of interpretation: "The absence of the transcendental signified extends the domain and the play of signification infinitely" (*Writing and Difference,* 280). Deconstructive readings thus resist the goal of finding a full or final meaning for any text, either in authorial intention or in a referent such as scientific "truth":

> [A deconstructive reading] cannot legitimately transgress the text
> toward something other than it, toward a referent (a reality that
> is metaphysical, historical, psychobiographical, etc.) or toward a
> signified outside the text whose content could take place, could
> have taken place outside of language, that is to say, in the sense
> that we give here to that word, outside of writing in general.
> (Derrida, *Of Grammatology,* 158)

Instead of looking toward a referent or toward a transcendental signified, deconstructive readings concentrate on the signifying system of the text, and on the inevitable tension between what Barbara Johnson calls "the rhetoric of an assertion" and its "explicit meaning" (cited in Crowley, 7). And because such tensions or differences are seen as significant, deconstructive readings look away from the central argument toward the text's margins, to buried metaphors or seemingly contingent details—those places where the text deconstructs itself. Deconstructive readings, though, as Sharon Crowley explains, are not simply "destructive" or aimed at pointing out flaws or weaknesses of an author (7). Instead, deconstruction looks at the way *all* texts escape the intentions of their authors and instead disseminate meaning endlessly through the inevitable différance of language.

Deconstruction offers a metatheory about *all* writing, not just a theory about literary or philosophical writing. "Theoretically," deconstruction should allow us to see tensions in any kind of text—fictional or nonfictional—and should be relevant to any genre and any discipline. Thus, deconstruction should be applicable, at least to some extent, to a scientific article such as "The Spandrels of San Marco." But deconstruction has typically dealt with literary and philosophical texts that are themselves concerned with issues of representation and interpretation, not with texts primarily concerned with making substantive statements about empirical data.[6] Science, in some sense, has remained a privileged discourse, even for deconstruction. However, since "The Spandrels of San Marco" is closely involved with issues of representation and interpretation (in ways that are unusual in scientific discourse), it is particularly open to a deconstructive reading; indeed, I will suggest in this chapter that Gould and Lewontin themselves often function as deconstructive readers of the adaptationist texts they challenge. I will in particular locate three textual concerns where deconstruction seems especially useful in illuminating some of the rhetorical issues and tensions of "The Spandrels of San Marco": 1. the figurative language through which much of the argument is presented; 2. the hierarchical oppositions by which the argument is constructed and deconstructed; and 3. the foregrounding of issues of analysis and interpretation that often complicates and defers the scientific argument. But I will also suggest that to the extent to which "The Spandrels of San Marco" concerns itself with what Robert Scholes calls "realities, whether physical or social" (*Textual Power,* 96), it may resist the insights of a deconstructive analysis, and in the process may reveal something about the relation of scientific writing to other cultural artifacts.

Deconstruction and Figurative Language

In its concern with the tension and interplay between a text's logic and rhetoric, or between its thought and language, deconstruction pays particular attention to the figurative language through which apparently literal meaning is articulated. But because it denies any absolute distinction between the literal and the metaphoric, deconstruction offers a view of metaphor that makes it not merely decorative, transparent, or contingent, but an essential part of all language.

In his consideration in "White Mythology" of the role of metaphor in philosophy, Derrida points to both its inescapability and its paradoxical nature. On the one hand, he explains, metaphor as a trope of *mimesis* or imitation is "based on the *theoretical* perception of resemblance or similarity. . . . The condition for metaphor (for good and true metaphor) is the condition for truth" (*Margins of Philosophy,* 237). But, on the other hand, "the energy of metaphor" supposes that "the resemblance is not an identity" (239). That is, though metaphor promises a relation built on truth and identity, it functions, as metaphor, only by deferring or evading that relation and creating an "energetic absence" (239). As a result, metaphor "can always miss the true," can always miss access to a referential or nonmetaphorical meaning: "By virtue of its power of metaphoric displacement, signification will be in a kind of state of availability" (241).

Though "White Mythology" is primarily a consideration of metaphor in philosophical language, Derrida understands all language as metaphorical: "Before being a rhetorical procedure within language, metaphor would be the emergence of language itself" (*Writing and Difference*, 112). Language, by its nature, is metaphorical in that it can never be completely referential, mimetic, or identical with extralinguistic meaning. Through this recognition, deconstruction wants to undermine the hierarchical opposition between literal and metaphorical and call into question any attempt to distinguish them. Deconstructive readings, then, look at the figurative quality of the texts they investigate, but they typically do not focus on the overt metaphors. Rather, deconstructive readings attempt to uncover a text's concealed metaphors, or those places where a text denies its own figural nature.

The question of metaphor is particularly interesting in regard to science, which has always been uneasy about the role of figurative writing and which has a long history of attempting to excise metaphor and achieve instead a "plain style" or, in more modern terms, an "observation language." But as the positivistic models that lie behind such attempts have become superseded by constructivist models of scientific activity, metaphor has become increasingly understood as an inescapable constituent of scientific language (Halloran and Bradford; Hoffmann). Indeed, Derrida's claim in "White Mythology" that the very notion of concept, theory, or foundation is itself metaphoric (*Margins of Philosophy,* 224) is in some ways parallel to the quite different argument of philosophers such as Mary Hesse and Max Black, who also see scientific theory as metaphoric.

Nevertheless, "The Spandrels of San Marco" is unusual for scientific writing in the way it foregrounds figurative language. In fact, Gould and Lewontin's first move is to present two ways of interpreting evolutionary change through analogies, though the metaphoric status of those analogies is not acknowledged. Gould and Lewontin twice call these opening moves "nonbiological examples," but what do they exemplify? Clearly neither is an example of genetic adaptation or selection in the Darwinian sense of evolution: the spandrels and dome are human artifacts, and though the story of Aztec rituals might be an example of cultural evolution, Gould and Lewontin explicitly distinguish this kind of "cultural adaptation" from "Darwinian adaptation based on genetic variation," as in their account of the confused thinking in human sociobiology (159). Thus, though the term "example" suggests identity (example as typical part or subset of the whole), both spandrels and rituals function in some sense through difference, since neither is a subset of genetic evolution. Both the dome and the Aztec rites are offered as analogies or metaphors: in analyzing the causes of these architectural or ritual systems, readers employ processes of interpretation that are similar (but obviously not identical) to the processes used to interpret evolutionary change. A deconstructive reading would highlight the metaphorical cast of this explanation by analogy and would examine the implications of these metaphors as a way of unraveling the text's rhetoric.[7]

In the rhetoric of the text, the dome and spandrels of San Marco become associated with alternativist arguments[8] (which the authors want to validate), just as the Aztec rites of human sacrifice become linked with the adaptationist program (which they reject or denigrate).[9] Though Gould and Lewontin present, as a straw man, an (obviously architecturally unacceptable) "adaptationist" explanation for the dome and spandrels, it is quickly demolished so that they may describe in detail a (more architecturally reasonable) "alternativist" explanation based on architectural constraint; these architectural terms are echoed in the later image of the *Bauplan*. Conversely, though Gould and Lewontin offer alternativist explanations of Aztec rites of human sacrifice, they devote much more time to critiquing and undermining a more persuasive (and possibly anthropologically acceptable) "adaptationist" explanation.

Rhetorically, the two opening images are particularly interesting because of the further associations they create. The "great central dome of St. Mark's Cathedral," which comes to stand in for the alternativist model, offers a "detailed iconography expressing the mainstays of Christian faith" that is "elaborate, harmonious and purposeful" (147–48). Through this example of angels, disciples, and evangelists and the following example of King's College Chapel in Cambridge, with its "bosses alternately embellished with the Tudor rose and portcullis" (148–49), Gould and Lewontin illustrate an alternativist explanation for evolutionary change that focuses not on adaptation but on "constraint." By way of contrast, Gould and Lewontin illustrate an adaptationist explanation through the example of Aztec human sacrifice which, adaptationists hypothesize, "arose as a solution to the chronic shortage of meat." Rather than seeing Aztec rites as "an elaborate social system and a complex set of explicit justifications involving myth, symbol and tradition," as an alternativist would, adaptationists are shown as reducing "an entire culture" to "an unusual way to beef up the meat supply" or to a "mighty poor way to run a butchery" (149–50).

These two extended figures or analogies of alternative and adaptationist explanations are clearly not rhetorically neutral. On the one hand, the alternativist position, as imaged in San Marco, here called St. Mark's,[10] is associated with the divinity and order of Christian faith as it is expressed in architectural harmony and design and the high culture of Renaissance art, both Italian and English. The adaptationist position, on the other hand, as imaged in the Aztec rituals, is associated with the barbarism of "human sacrifice" and "butchery" as practiced in the third world.[11] Interestingly, alternativist explanations of Aztec human sacrifice would save it from butchery and would see it, like the dome of San Marco, in the religious terms of myth, symbol, and tradition. That is, the opposition between alternative and adaptationist explanations is metaphorically extended so that they are also contrasted in terms of culture versus butchery, natural harmony versus the disorder of cannibalism, Christ versus Aztec barbarism, good versus evil. These opposing connotations continue throughout the essay with Gould and Lewontin's generation of positive associations for the alternativist model, such as the "resurrection" of Darwin's "sainthood (if not divinity)" (155), and negative associations for the adaptationist program, such as venereal disease (151) or cuckoldry (155).

Interestingly, Gould and Lewontin's critique of the adaptationist program is an attack on a method of explanation that is, they claim, well established in evolutionary biology. Their own alternativist model is put forth as one less established, and thus competing for the recognition now given the adaptationist model by the community of scientists that make up the article's readers. Nevertheless, Gould and Lewontin constantly describe their own alternativist position through metaphors of behavior sanctioned by established social structures, particularly religion, while they consistently connect the adaptationist model with metaphors that imply socially unacceptable behavior, such as cannibalism, or the consequences of unacceptable behavior, such as venereal disease. Thus, Gould and Lewontin's figurative language amounts to an effective rhetorical ploy that allows them to realign metaphorically what is revolutionary and what is established for the social community that constitutes their audience.

But though the images that serve to differentiate the two positions are quite explicit, there are other more hidden or implicit ways in which the images echo one another and undercut the apparent oppositions. Just as San Marco is an artifact of Christian religion, human sacrifice is an artifact of Aztec religious rites; similarly, the cannibalism of the Aztecs is paralleled by the symbolic cannibalism of the Eucharist. Furthermore, both the adaptationist and alternativist positions are explicitly figured through images of eating. From the first, Aztec human sacrifice appears as either a story of "an unusual way to beef up the meat supply" or "a mighty poor way to run a butchery." Similarly, there is the analogy between the adaptationist's position and Dr. Pangloss's story of venereal disease, which "poisons the source of generation," but allows us to have chocolate and cochineal. These examples of disruption through consumption of life's most basic processes are echoed in a later example used to buttress the alternativist position: the particularly grisly story of the arthropod "females that reproduce as larvae and grow the next generation within their bodies. Offspring eat their mother from inside and emerge from her hollow shell, only to be devoured a few days later by their own progeny" (157). And just as the opening adaptationist image of cannibalistic Aztec sacrifice connects with a later alternativist example of cannibalistic arthropods, there is a sense in which the "harmonious and purposeful beauty" of San Marco's dome, used as an alternativist analogy, will parallel the harmonious and purposeful nature of the adaptationist stories. The effect of this metaphoric blurring is to undermine the easy identification of opposing positions, and suggest instead an ongoing process of metaphoric signification.

Deconstruction and Hierarchical Oppositions

Examining patterns of imagery in "The Spandrels of San Marco" suggests a tension between that text's rhetoric and its explicit argument, or between what Jonathan Culler calls "structures of languages or texts and structures of thought" (146). But there are other deconstructive reasons for distrusting an argument defined around binary opposites in which one term is so steadily privileged. The construction of such hierarchical oppositions is, in Derrida's narrative, the strategy of Western philosophy and its foundational metaphysics of presence. To establish possibilities for presence, truth, and meaning, foundationalist theories have to suppress the potential of absence, language, and expression. While the first set of terms can be seen as analogous to science's attempts to find and represent the truth of the natural world, the second may suggest ways in which that project will always be undercut by science's inability to circumvent language and get to "something beyond representation" (Myers, *Writing Biology,* 26).

"The Spandrels of San Marco" is structured around a set of binary opposites that in many ways resembles the kinds of hierarchy that deconstruction typically undermines. On the one hand, there is the adaptationist program, in which "natural selection is so powerful and the constraints upon it so few that direct production of adaptation through its operation becomes the primary cause of nearly all organic form, function, and behavior" (150–51). This almost universally applicable cause-and-effect argument offers a monistic and homogeneous explanation—a kind of master narrative—for evolutionary change. Moreover, it provides a foundational system or metaphysics by which virtually all organic phenomena can be given meaning through a teleological vision of the evolutionary process as goal-directed and progressive. Such teleology is parodied in the parallel with Voltaire's Dr. Pangloss but is presented more directly in the examples of adaptationist arguments. Commisures of clams, suites of external structures (horns, antlers, tusks), and the Eskimo face are all explained by the adaptationist program in terms of their fit to problems (152). This theory provides the natural world with not only a scientific "meaning" but also a moral (and hence ideological) one: there is a tacit sense in which these stories justify nature, as in the Panglossian attempt to see everything as expressing or creating the best of all possible worlds.

On the other hand, the alternativist positions, initially at least, do not promise a universal theory of natural order but offer instead a pluralistic range of possibilities. Furthermore, those possibilities describe a world much more random and less ordered than the adaptationist world. Rather than being based on an a priori sense of evolution as progress, alternativist positions are based on pragmatic needs to recognize different explanations that work in different cases: a hierarchy of alternatives to immediate adaptation, from "no adaptation and selection at all" (156) to adaptation as "secondary utilization of parts present for reasons of architecture, development or history" (159). Thus, instead of offering a teleological narrative of progress based on an inherently ordered world, alternativists explain phenomena through stochastic processes of change and historical contingency.

A deconstructive reading would first note the hierarchical relation of these positions by showing how one half of the opposition is always seen as privileged because of the way it suggests an extratextual meaning (in this case the telos of evolutionary perfectibility) that can be directly perceived and represented through language. In a sense, this is exactly what Gould and Lewontin themselves do; for much of the article, they are themselves deconstructive readers of adaptationist texts. As deconstructive readers, Gould and Lewontin first point out the adaptationists' foundational gesture of positing a particular structure in the natural world (one based on progress toward perfectibility) and then deconstruct that position by labeling it "story-telling." Gould and Lewontin criticize this position because such stories can never lead to "rejection for want of evidence." Instead, "rejection of one adaptive story usually leads to its replacement by another, rather than to a suspicion that a different kind of explanation might be required" (153). This is one of Gould and Lewontin's most cogent rhetorical arguments against the adaptationist program: they show how explanations are stories or narratives constructed by biologists and written onto natural phenomena, a point further made in the way they frequently describe the adaptationist in terms of another constructed narrative, Voltaire's *Candide*. Such an argument helps Gould and Lewontin deconstruct the adaptationist's foundational premise that evolutionary change is progressive and goal-directed, and suggests that the structure and meaning implied by these narratives may also be a construction of the writer.

A deconstructive reading, however, does not simply replace one hierarchy with another. Instead, such readings try to destabilize hierarchies by seeing how binary opposites collapse into one another. In this case, it is important to note that though Gould and Lewontin may want to deconstruct the adaptationist program, both that program and the alternative positions they offer in its place are their own constructions. That is, as Fahnestock (chapter 9) and Winsor (chapter 7) also emphasize in this book, the texts of the adaptationist program appear in the text of "The Spandrels of San Marco" only through the interpretations offered by Gould and Lewontin. This is a point they themselves remind us of several times, as in their acknowledgement that "some evolutionists will protest that we are caricaturing their view of adaptation. After all, do they not admit genetic drift, allometry, and a variety of reasons for non-adaptive evolution?" (151). But it is also interesting to note that Gould and Lewontin refute the charge of caricature by immediately—and more overtly—caricaturing once again. Adaptationists, like all scientists, they explain, try to extend the range and power of their own theory, in part by minimizing alternatives. This explanation, though, is couched in an oddly parodic way:

In natural history, all possible things happen sometimes; you generally do not support your favoured phenomenon by declar-

> ing rivals impossible in theory. Rather, you acknowledge the rival,
> but circumscribe its domain of action so narrowly that it cannot
> have any importance in the affairs of nature. Then, you often
> congratulate yourself for being such an undogmatic and
> ecumenical chap. (151)

The shift in tone achieved by the unusual turn to the second person, as well as the obviously sarcastic concluding phrase of "undogmatic and ecumenical chap" is an oddly caricaturing way to refute the charge of caricature.

In attacking the adaptationist program as story-telling, Gould and Lewontin implicitly set up another binary opposition: mere story-telling versus real science. But Gould and Lewontin's deconstructive reading of the adaptationist program may itself deconstruct through certain textual suggestions that their own alternativist positions are also a form of story-telling. By "story," Gould and Lewontin seem to mean an explanation that is plausible but that has not yet been tested. They criticize the adaptationist program for offering as established truth stories that have only been shown to be plausible. And in response they argue that the biologist must go further than mere plausibility: "The key to historical research lies in devising criteria to identify proper explanations among the substantial set of plausible pathways to any modern result" (154). Though Gould and Lewontin do not define "proper" here, one can infer from later examples that they mean experimentally testable and potentially disprovable: "But what good is a theory that cannot fail in careful study" (155). It is now the alternativist position that claims privileged access to knowledge of the natural world through the experimental methods of science.

By "story-telling," then, Gould and Lewontin seem to mean offering explanations that are unproven; but that sense quickly glides into connotations of explanations that are false or deceptive, as opposed to the "proper explanations" that they themselves advocate. In this sense, a story is an explanation that is not only plausible (because it is consistent) and unproven, but also probably wrong: all the adaptationist stories are offered in order to be rejected.

But the term "story-telling" carries other connotations. Most important, it suggests the function of narrative, an important mode of explanation and understanding in both fictional and nonfictional discourse, particularly—as Gould was himself to argue ten years later in *Wonderful Life*—in historically oriented sciences such as evolutionary biology and geology. It is outside the scope of this chapter to detail the studies suggesting that nonfictional narratives (from both the human and physical sciences) are constructions of patterns or interpretations of data, rather than revelations of preexisting patterns inherent in data;[12] and in chapter 3 of this book Susan Wells attends particularly to the narrative lines of "The Spandrels of San Marco." But it is important to note here that part of a deconstructive agenda is to show how we write narratives into reality rather than experience their presence directly.

Gould and Lewontin are also story-tellers, both in the way they characterize adaptationists and in the larger way they, like all other scientists, construct narratives. Occasionally, like adaptationists, they too rely on plausible but unproven examples. In one case, their example—their story—is completely "hypothetical": "A mutation which doubles the fecundity of individuals will sweep through a population rapidly," a scenario for which they offer possible endings (158). It is particularly interesting to note that this hypothetical scenario is the only example they offer for the position of nonadaptation, or selection without adaptation—that position diametrically opposed to the adaptationist program. This raises the question of whether it was not possible, especially here in this crucial stage of the argument, to find some supporting evidence other than the kind of "hypothetical example," or plausible narrative, that they accuse adaptationists of depending on.

Gould and Lewontin are also story-tellers in the larger sense because they are involved in the narrative construction of historical explanations. Gould and Lewontin emphasize historicity in the explanations that are their alternatives to the adaptationist program: from examples that exhibit "purely random factors" (finite, isolated populations that change by *genetic drift*), through examples of phenomena that are "primarily a by-product of selection" for something else (the arthropods that reproduce as larvae and grow the next generation within their bodies), to examples that clearly show "historical contingency" (the various adaptations of the land snail *Cerion*) (156–59). In fact, in their emphasis on history, Gould and Lewontin may be more fundamentally connected to narrative or "story" than adaptationists. Moreover, like adaptationists, Gould and Lewontin consistently buttress their arguments with examples that are almost always narrative in form. In this sense, their alternative explanation to Barash's story about aggression in birds (154) is also a narrative. (That it can be subject to other kinds of criteria does not diminish its power as a narrative explanation.)

Finally, Gould and Lewontin are story-tellers in the construction of their own teleological master narrative: the *Bauplan*. Though the image of the *Bauplan* is rooted in architecture, not stories, it evokes just as strongly the notion of something made: if adaptationists construct stories, alternativists construct buildings. And while the *Bauplan* is a more spatial than temporal image, it too is open to narrative interpretation, both because it results in something that is built over time and because that resulting building can itself be read as a narrative, as Gould and Lewontin read the dome and spandrels of San Marco. Just as the opening image of San Marco in its "harmonious and purposeful beauty" recalls the purposeful teleology of the adaptationist narrative, the *Bauplan* narrative offers another teleological explanation of evolutionary change, but one built on constraint rather than adaptation: "[The argument of the *Bauplan*] holds that constraints restrict possible paths and modes of change so strongly that the constraints themselves become much the most interesting aspect of evolution" (160). In this narrative, evolutionary processes are controlled by " '*bautechnischer*', or *architectural*, constraints (Seilacher 1970). These arise not from former adaptations retained in a new ecological setting (phyletic constraints as usually understood), but as architectural restrictions that never were adaptations, but rather the consequences of materials and designs selected to build basic *Baupläne*" (161). Like the adaptationist narrative, this explanation is illustrated primarily by examples or stories—the evolution of the divaricate form in molluscs and brachiopods (161–62)—interpreted through field observation and general principles.

Gould and Lewontin's own reliance on narrative may suggest that all biologists are, in some sense, story-tellers, not because they deceive but because they order experience in a way that is essential to the construction of scientific knowledge—just as it is important to the construction of stories and architecture. Thus, though Gould and Lewontin may not tell Panglossian stories of the utter perfection of the natural world (though it is not clear that adaptationists do this either), they do tell stories that construct and deconstruct order and pattern.

Deconstruction and Interpretation

Gould and Lewontin's emphasis on story-telling calls attention to a third way in which "The Spandrels of San Marco" is particularly open to deconstructive analysis. Much deconstructive criticism tends to be self-reflexive in the curious sense that many of the texts that deconstruction unravels are themselves texts about processes of reading and interpretation and the problematic status of language: one thinks of Derrida on Plato, Saussure, Freud, or Austin; DeMan on Locke, Rousseau, or Romanticism. To analyze texts that themselves seem to be about analysis and interpretation is a common deconstructive move. Barbara Johnson, for example, in the course of deconstructing Derrida's reading of Lacan's reading of Poe's "The Purloined Letter," calls attention to "the *act of analysis* which seems to occupy the center of the discursive stage, and the *act of analysis of the act of analysis* which in some way disrupts that centrality" (110; emphasis in original). Many deconstructive readings are about such acts of analysis.

John Schilb claims that the "oppositions most often cited by deconstructionists include referentiality and textuality, the literal and the figural, cause and effect, and inside and outside" (266), all of which are concerned in one way or another with the status of language and interpretation. The opposition between referentiality and textuality is perhaps most significant to "The Spandrels of San Marco," because of what might appear to be a conflict between the privileged relation to a referential reality often accorded to science and the textual focus usually identified with deconstruction. But science (in most of its current, nonpositivist versions) and deconstruction acknowledge both halves of the referentiality/textuality opposition.

Though science privileges the referential, it also recognizes the inevitability of textuality. If referentiality suggests the matching up of a literal language with the "external" world and textuality suggests the range of rhetorical structures through which language will always show its own "internal" workings, then science should be for many the most referential of discourses, offering a version of reality closest to the "truth."[13] To the extent to which scientific language can offer a "transparent" window to reality, and to the extent to which it can avoid its implication in its own linguistic structures—in its status as a textual object—scientific writing might be said to be inimical to deconstructive analysis. But much recent work in philosophy of science, as well as the rhetoric of scientific texts (including all the essays in this book), demonstrates the degree to which scientific language is itself a textual and rhetorical construct.[14] Furthermore, though deconstruction privileges the figural status of language, it also recognizes language's referential function. One of Derrida's most famous statements, *"il n'y a pas de hors-texte"* ("there is nothing outside the text" [*Of Grammatology*, 158]), is sometimes taken as a denial of physical reality. But rather than denying language any referential function, deconstruction wants to critique what Derrida calls "naive" attempts to make language correspond in any straightforward way to an external referent: whatever there is outside the text, it can only be known as text, that is, as a system of signs.

"The Spandrels of San Marco" in many ways foregrounds the referentiality/textuality opposition by highlighting issues of interpretation. It is, like many deconstructive texts, concerned with the act of analysis, and the act of analysis of the act of analysis, as much as—or more than—it is concerned with what is being analyzed. The opening images of St. Mark's as a "detailed iconography" and the Aztec rites as a "complex set of explicit justifications involving myth, symbol and tradition" immediately announce this concern with interpretation. The reader is asked not just to explicate this iconography or these myths, but to understand the architectural and religious systems that generated them. This task is complicated, however, by the way cause and effect are quickly blurred when the reader is told to abandon the natural "temptation" to view the dome's mosaic design as "the starting point of the analysis" (148) and to regard such an analysis as an "inversion of explanation" (149). What was first offered as cause is now seen as effect.

The opening section of the essay further problematizes the reader's task by introducing a number of conflicting interpretations to be negotiated. In the case of the Aztec rituals, the reader is given Harner's initial interpretation, Wilson's elaboration of this interpretation, Sahlins's qualification of this interpretation, Gould and Lewontin's counterinterpretation (presented as a strong suspicion), Sahlins's further qualification of Harner's interpretation, and Ortiz de Montellano's argument against the original premise on which Harner's interpretation was built. The initial effect of this interpretive nesting is similar to Johnson's description of Poe, Lacan, and Derrida as nested texts: "the subversion of any possibility of a position of analytical mastery" (110).

After this initial deferral of the subject of analysis through various versions of the act of analysis, Gould and Lewontin explain that their intention in this essay is "to question a deeply engrained habit of thinking among students of evolution," a habit of thinking that they label "the adaptationist programme or the Panglossian paradigm" (150). In a sense, as I have suggested, they deconstruct the adaptationist program by showing its status as a constructed narrative, even while they reveal themselves as constructors. But though the binary opposition between adaptationists as story-tellers and alternativists as truth-tellers does not survive, differences remain in the kinds of story they offer.

"The Spandrels of San Marco" is not only concerned with two ways of analyzing natural systems or interpreting sets of physical data. In the article, the natural world functions as a kind of text that is open to interpretation, and adaptationists and alternativists become different kinds of readers with different interpretive strategies.[15] Adaptationists, as presented by Gould and Lewontin, look for monistic and homogeneous explanations; in a sense, they read the text of the natural world in search of a single determinate meaning. In example after example, Gould and Lewontin show the adaptationist interpreting the natural world as a narrative of almost purposeful cause and effect in which every evolutionary change is the result of adaptation through the operation of natural selection. In contrast, though nonadaptationists are also readers of the natural world, they interpret natural processes in rather more indeterminate ways. Instead of searching for a single, universal explanation, alternativists accept a plurality of interpretations based on random processes, contingency, and history.

"The Spandrels of San Marco" further calls attention to the textuality of the supposedly referential discourse of science by foregounding what a deconstructionist would call its "intertextuality." Intertextuality, as Charles Bazerman explains in chapter 2, suggests that no text is closed, autonomous, or full, but that each is inevitably interwoven with the traces and echoes of other texts. Indeed, the ability to understand a text—its diction, syntax, organization, genre, or topic—depends on the ability to differentiate it from other texts. But because such a process of differentiation is endless, any final meaning will be constantly deferred through a kind of "genetic indetermination" (Derrida, *Writing and Difference*, 292). The act of interpretation is thus inexhaustible.

Through its intertextuality, "The Spandrels of San Marco" continually pushes away any possibility of analytic certainty. As several other chapters in this volume indicate, Gould and Lewontin spend a good deal of time referring to, discussing, and arguing against the work of other scientists. In scientific discourse, the writer naturally depends heavily on citation to establish the relationship of his or her text to existing knowledge as well as to carve out a space for new knowledge. Some of Gould and Lewontin's citations function this way, though because their paper offers not new research but a new orientation to established research, an unusual number of their citations and the discussion of those citations are agonistic. But citations are also used in a more curious way in "The Spandrels of San Marco." Often, as in the nested explanations of Aztec rites, the reader is given several levels of interpretation of a single text, as when Gould and Lewontin interpret Romanes interpreting Sir Wyville Thomson interpreting Darwin (155–56). The effect of this analysis of the act of analysis is to problematize the possibility of a final meaning and defer any interpretive certainty.

The nested citations of "The Spandrels of San Marco" create a kind of intertextuality that is unusual in scientific writing.[16] Even more unusual, though, for scientific writing is the intertextuality created by the invocation of other discourses, such as archeology and art history, anthropology and literature. The most obvious of these are the references to Voltaire's *Candide:* citations or paraphrases of Voltaire in the title, as well as other key places, such as the first section and the epigraph to third section, mark this as one of the text's most prominent narratives.

These intertextual references to Voltaire call further attention to Gould and Lewontin's paper as a textual object and further establish them as readers and writers of texts—both biological and fictional.

But it is not just an ordinary text that is being evoked, for Voltaire's *Candide* is itself a work that raises issues of referentiality and textuality and in the process reinforces a sense of the inexhaustibility of interpretation. First, in citing Voltaire, Gould and Lewontin are not performing an interpretively neutral act. Through their selection of parts of *Candide* to quote and paraphrase, Gould and Lewontin have interpreted Voltaire's text in terms of what they call the "Panglossian paradigm," itself an interpretive statement. (Other interpretations of *Candide* might center on other characters or read the character of Dr. Pangloss in a different way.) Furthermore, those parts of Voltaire's text that are included in Gould and Lewontin's article tell of Dr. Pangloss's attempt to interpret stories of the Lisbon earthquake and Columbus's introduction of venereal disease into the Americas, interpretations that will, in turn, allow him to offer an interpretation of "the best of all possible worlds." (These interpretations, and the philosophical systems behind them, are themselves interpreted through explicit debate and satirical presentation in *Candide*, as they are in "The Spandrels of San Marco.")

Voltaire's *Candide*, in fact, raises most of the issues of concern to deconstruction. Pangloss's name ("all-tongue," or explainer away of everything) privileges speech, just as his originary myth of Leibnitzian "preestablished harmony" privileges a metaphysics of presence. By yoking such a metaphysic with that of the adaptationist program, Gould and Lewontin reinforce the teleological foundationalism of the adaptationist narrative. But Pangloss's philosophy is an object of parody in Voltaire's novel, as it is in Gould and Lewontin's article: Voltaire uses Pangloss to parody Leibnitz as Gould and Lewontin use him to parody adaptationist arguments. This parodic doubling and redoubling of textual attempts to construct and deconstruct foundational narratives marks "The Spandrels of San Marco" as itself a deconstructive and deconstructing text.

Deconstructing "Deconstructing 'The Spandrels of San Marco'"

By way of a conclusion, one might suggest that a deconstructive reading of "The Spandrels of San Marco" draws attention to the way scientific writing, like other discourses, resists being pinned down to one single and determinate meaning. The rhetorical dimensions of the article's inevitably metaphoric language, deconstructing hierarchies, and interpretive processes all call attention to its status as a textual object. But deconstruction resists both origins and ends. So instead of offering a conclusion, I want to end by attempting to turn this reading around. If "The Spandrels of San Marco" offers an intertextual nesting of interpretations, then this deconstructive reading obviously offers only one more layer; its effect, like that of all the other nested interpretations, is not to close off but to defer meaning. Deconstruction, in its rejection of totalizing systems of interpretation, cannot offer itself as a replacement.

In place of a conclusion, then, I wish to ask what this deconstructive reading suppresses. And my reply would be "the basic business of science," which I would define as the attempt to make substantive statements about the physical world in ways that are persuasive to other scientists. This provisional definition of science emphasizes two constraints that seem to me of little importance in deconstruction's play of signification: language as influenced by empirical experience (whatever it is that lies outside of language—even though it may only be known through language), and as affected by social forces (whatever it is that makes a discipline's texts recognizably distinctive and gains them an audience).[17]

This deemphasis of the physical and social implications of language has sometimes been explained as a consequence of deconstruction's privileging of Saussure's concept of *langue* (language as a closed system of signs) over *parole* (individual utterances and texts). Wendell Harris, for example, describes the way deconstruction extends Saussure's notion that

> language is a self-regulating system of signs in which the values or meanings of signs are generated by the relations between them, not by the correspondence between the signs and an independent or pre-existing reality. . . . [Deconstruction] pursues the implications of regarding *langue* as a system that cannot be corrected by comparison with reality because the concepts by which we grasp reality are partly its products. (116–17)

Thus, while deconstruction does not deny that reality exists, it may diminish the power of language to negotiate or affect that reality, a consequence that becomes increasingly problematic in a discourse like science, which depends so heavily on accountability to shared versions of reality. Similarly, Robert Scholes finds that "Saussure's definition of the sign as a sound-image and a concept is inadequate to describe those elements in language that depend upon the relationship between concepts and objects or referents" (*Textual Power,* 96). Scholes thus describes the limitations of deconstruction in dealing with any discourse that attempts to ground signifieds upon phenomena:

> From the perspective of deconstruction there is nothing upon which we can ground an argument for evolutionary biology as opposed to fundamentalist creationism, since both are discourses, with their blindnesses and insights, and neither one can be said to be more or less accurate than the other, there being no pathway open from the text to the world. (*Textual Power,* 99)

The implications for science of this inadequacy are most fully argued by Charles Bazerman, who in *Shaping Written Knowledge* sees science as a social process formed by the dialectical relationship between disciplines and texts, and as a project defined by its social desire to be accountable to empirical experience. "The problem of reference haunts all studies of scientific language" (187), Bazerman claims:

> Users of the scientific linguistic system seemed to believe their language was useful in gaining some control over the natural world, and many of their behaviors as writers and readers seemed constructed out of that belief. . . . Specifically, the individual is placed within a communicative context that constantly encourages and demands that the individual at many junctures considers how empirical results either can advance the claim-making procedure or call for reconsideration of the claims and representation of phenomena. (188)

For Bazerman, the high value that scientific communities place on holding representation accountable to experience necessitates going beyond inadequate "current concepts of language" (295). Rejecting theories of language and interpretation based on Saussure's separation of linguistic code from contextual forces of use and meaning, Bazerman welcomes the move of linguistics toward the pragmatic study of "how people use language in real life to do things" (298).[18] Turning to Vygotsky, Bazerman finds a model "of scientific use of language [that] will suggest how the work of science can be accomplished through the unfolding social and empirical activity of individuals coordinated (cognitively and behaviorly) within groups" (302).

Focusing on the way science values claims that seem to be in accord with empirical experience would allow one to ask questions about the relative value of the claims reported in "The Spandrels of San Marco." Adaptationists and alternativists offer interpretations of experience that can be adjudicated through the shared norms of the scientific community—in this case, how claims can be shown to solve problems by explaining past empirical experiences and predicting future ones. That is, while adaptationists and alternativists may both tell stories, the scientific community can value one story over another because of the story-teller's ability to persuade the reader—through descriptions and interpretations of data, methodology, experimental design, relevant literature, etc.—of the story's referential connection with reality; these are concerns that have not been of much interest to deconstruction.

Similarly, focusing on the social groups that constitute scientific communities would allow one to ask questions about the relationships of those groups both to one another and to groups larger than the scientific community. That is, alternativists and adaptationists may offer explanations laden with a whole range of other competing social values that will affect any interpretation of their texts. Elsewhere, for example, Richard Lewontin (this time with coauthor Richard Levins) has argued that the doctrine of adaptation is part of

> an ideology of equilibrium and stability that characterizes modern
> evolutionary theory as much as it does bourgeois economics and
> political theory. . . . There is a striking similarity between this
> view of evolution and the claim that modern market society is the
> most rational organization possible, that although individuals
> may rise or fall in the social hierarchy on their individual merits,
> there is a dynamic equilibrium of social classes, and that tech-
> nological and social changes occur only insofar as they are needed
> to keep up with a decaying environment. (84)

While such an explanation is itself ideological and open to debate—and would probably be rejected as inadequate or inaccurate by someone subscribing to adaptationist arguments—deconstruction may not be helpful in understanding or evaluating such a debate.[19]

These questions suggest that the insights derived from deconstruction alone may be inadequate in understanding Gould and Lewontin's paper, because those insights diminish both the social context out of which science arises and the way that context is created around science's accountability to empirical experience. But deconstruction is nevertheless useful in analysis of "The Spandrels of San Marco" because it offers another way to move beyond the positivist notion of scientific language as literal description of a preexisting reality and because it calls into question the degree to which authorial intention can be translated into discourse. It thus contributes to our on-going understanding of science as rhetoric.

Debra Journet

NOTES

My thanks to Tony O'Keeffe, Tom Byers, and Brenda Brueggemann, as well as to the members of the 1991 seminar in rhetorical approaches to literature at the University of Louisville, for their very helpful readings of this essay.

1. And deconstruction would emphasize that summary is not a neutral act; any reading of a text is a revision of that text. Extensive interpretations of deconstruction are offered by Culler and Norris. For two other versions of deconstruction of particular interest to those in rhetoric, see Crowley and Neel.

2. Deconstruction has created a somewhat specialized terminology but resists attempts to define that terminology, because such definitions would imply notions of paraphraseable identity. It is difficult, though, to avoid such notions, because assumptions of translatability, substitution, identity, and definition are built into language. (The attempt to undermine these assumptions accounts, in part at least, for the infamous difficulty of Derrida's prose style.)

3. Derrida argues throughout *Of Grammatology* that the privileging of speech over writing is an essential consequence of a metaphysics of presence. Speech is seen to stand in closer relationship to thought or to be more closely identified with it. Writing thus becomes an inferior version of speech, more distant from an originary meaning. It is this privileging of speech and its foundational implications that Derrida wants to deconstruct by arguing that speech as presence is an illusion. For Derrida, speech carries the same secondariness or absence as writing.

4. For Derrida, the term "writing" is used to describe not just graphic inscription, but any system of signs: "And thus we say 'writing' for all that gives rise to an inscription in general, whether it is literal or not and even if what it distributes in space is alien to the order of the voice: cinematography, choreography, of course, but also pictorial, musical, sculptural 'writing'" (*Of Grammatology*, 9).

5. See Myers, *Writing Biology*, for a discussion of the deconstructive focus of much recent work in sociology of science.

6. Some interesting examples of deconstructive readings of other kinds of texts include Brooke and Crowley (deconstructing composition pedagogy); Brueggemann (deconstructing philosophy of science), Schilb (deconstructing literary nonfiction), and Shapiro (deconstructing public policy writing).

7. "Analogy," Derrida explains, "is metaphor par excellence" (*Margins of Philosophy*, 242). A metaphysics of presence privileges analogy because it is so clearly based on similarity and substitution.

8. Gould and Lewontin do not provide a name for the plurality of positions they offer as alternatives to the adaptationist program. For convenience, I have grouped these alternative positions under the term "alternativist." But it is important to note that Gould and Lewontin are not putting them forward as a unified school, program, or paradigm.

9. In the *logic* of the text, of course, Aztec cannibalism, like spandrels, is a

"secondary epiphenomenon" (150) that can also be used to refute adaptationist arguments. But in the *rhetoric* of the text, Aztec rites are more closely associated with the inadequate "adaptationist programme," just as spandrels more closely reflect alternativist explanations.

10. That the examples and terminology used to create the analogy of the alternativist position become increasingly English is another way to make them more familiar and acceptable.

11. I am obviously not arguing that Gould and Lewontin label Aztecs as barbaric or third world or even that they intend, consciously or unconsciously, these associations. But that such connotations can be engendered through their language is part of deconstructionist theory.

12. Reviews of some of this literature are offered in Journet and in Myers.

13. It is not altogether clear whether science has this cachet for deconstructionists. Terry Eagleton has claimed that the model of science frequently derided by poststructuralism "is usually a positivist one—some version of the nineteenth-century rationalistic claim to a transcendental, value-free knowledge of 'the facts.' This model is actually a straw target" (144). But science seems to remain a kind of privileged discourse in Derrida's works. Scholes has further argued that for Derrida, the concept of "rigor" becomes a "talisman against relativism and historicism. . . . It can be so used, of course, only if it is itself outside the dialectic, beyond deconstruction—only, in short, if it is indeed an Absolute, functioning in much the same way as Reason has functioned in traditional philosophy" (*Protocols*, 86).

14. See R. Allen Harris for a review of much of this literature.

15. Derrida describes two such kinds of interpretation: "There are thus two interpretations of interpretation, of structure, of sign, of play. The one seeks to decipher, dreams of deciphering a truth or an origin which escapes play and the order of the sign, and which lives the necessity of interpretation as an exile. The other, which is no longer turned toward the origin, affirms play and tries to pass beyond man and humanism, the name of man being the name of that being who, throughout the history of metaphysics or of ontotheology—in other words, throughout his entire history—has dreamed of full presence, the reassuring foundation, the origin and the end of play" (*Writing and Difference*, 292).

16. But see Myers, *Writing Biology*, for an account of a similar kind of intertextuality in papers written about scientific controversies.

17. This is essentially a pragmatist argument. Richard Rorty, for example, agrees with Derrida in his deconstruction of foundational epistemology, but disagrees with his thorough-going textualism. Rorty, like Kuhn, continues to see knowledge as a symbolic construct, but one that is a product of consensus, to be accepted or rejected by social groups because of its ability to solve problems, or to keep the conversation going.

18. One of Bazerman's examples of such a pragmatic project is Searle's theory of speech acts—a notion of language use that was the center of the celebrated argument between Derrida and Searle (see Derrida, "Signature";

Searle, "Reiterating"; Derrida, "Limited")—showing how opposed such a view is to deconstruction.

19. The relation of deconstruction, in all its different guises, to political thought and action is a large question. Terry Eagleton, for example, argues that while American deconstruction is the "latest form of liberal skepticism" (147), the later work of Derrida "is clearly out to do more than develop new techniques of reading: deconstruction is for him an ultimately *political* practice, an attempt to dismantle the logic by which a particular system of thought, and behind that a whole system of political structures and social institutions, maintains its force" (148). Michael Shapiro, on the other hand, argues that there is not much room for political analysis in deconstruction. Shapiro claims that "although the deconstructive critiques have gone a long way toward producing a pedagogy for unraveling texts, if a literary analysis is to yield significant results for social and political thinking, one needs more of a political frame of understanding than has come out of such studies" (369).

WORKS CITED

Bazerman, Charles. *Shaping Written Knowledge: The Genre and Activity of the Experimental Article in Science.* Madison: University of Wisconsin Press, 1988.

Black, Max. *Models and Metaphors: Studies in Language and Philosophy.* Ithaca: Cornell University Press, 1962.

Brooke, Robert. "Control in Writing: Flower, Derrida, and the Images of the Writer." *College English* 51 (1989): 405–17.

Brueggemann, Brenda. "The Collapsing Structure of Thomas Kuhn's (R)Evolutionary Text." Paper presented at the 1990 Twentieth Century Literature Conference, Louisville, Ky.

Crowley, Sharon. *A Teacher's Introduction to Deconstruction.* Urbana, Ill.: National Council of Teachers of English, 1989.

Culler, Jonathan. *On Deconstruction: Theory and Criticism after Poststructuralism.* Ithaca: Cornell University Press, 1982.

Derrida, Jacques. "Limited Inc ABC." *Glyph* 2 (1977): 162–254.

Derrida, Jacques. *Margins of Philosophy.* Trans. Alan Bass. Chicago: University of Chicago Press, 1982.

Derrida, Jacques. *Of Grammatology.* Trans. Gayatri C. Spivak. Baltimore: Johns Hopkins University Press, 1976.

Derrida, Jacques. "Signature Event Context." *Glyph* 1 (1977): 172–97.

Derrida, Jacques. *Writing and Difference.* Trans. Alan Bass. Chicago: University of Chicago Press, 1978.

Eagleton, Terry. *Literary Theory: An Introduction.* Minneapolis: University of Minnesota Press, 1983.

Gould, Stephen Jay. *Wonderful Life: The Burgess Shale and the Nature of History.* New York: Norton, 1989.

Halloran, S. Michael, and Annette Norris Bradford. "Figures of Speech in the Rhetoric of Science and Technology." In *Classical Rhetoric and Modern Discourse,* ed. Robert J. Connors, Lisa S. Ede, and Andrea Lunsford, 179–92, 284–86. Carbondale: Southern Illinois University Press, 1984.

Harris, R. Allen. "Rhetoric of Science." *College English* 53 (1991): 282–307.

Harris, Wendell. "Toward an Ecological Criticism: Contextual Versus Unconditioned Literary Theory." *College English* 48 (1986): 116–31.

Hesse, Mary. "The Explanatory Function of Metaphor." In *Revolutions and Reconstructions in the Philosophy of Science,* 111–24. Bloomington: Indiana University Press, 1980.

Hoffman, Robert. "Metaphor in Science." In *Cognition and Figurative Language,* ed. R. P. Honeck and R. R. Hoffman, 393–23. Hillsdale, N.J.: Erlbaum, 1980.

Johnson, Barbara. *The Critical Difference: Essays in the Contemporary Rhetoric of Reading.* Baltimore: Johns Hopkins University Press, 1980.

Journet, Debra. "Ecological Theories as Cultural Narratives: F. E. Clements's and H. A. Gleason's 'Stories' of Community Succession." *Written Communication* 8 (1991): 446–71.

Journet, Debra. "Forms of Discourse and the Sciences of the Mind: Luria, Sacks, and the Role of Narrative in Neurological Case Histories." *Written Communication* 7 (1990): 171–99.

Levins, Richard, and Richard C. Lewontin. *The Dialectical Biologist.* Cambridge: Harvard University Press, 1985.

Myers, Greg. *Writing Biology: Texts in the Social Construction of Scientific Knowledge.* Madison: University of Wisconsin Press, 1990.

Neel, Jasper. *Plato, Derrida, and Writing.* Carbondale: Southern Illinois University Press, 1988.

Norris, Christopher. *Deconstruction: Theory and Practice.* London: Methuen, 1982.

Norris, Christopher. *Derrida.* Cambridge: Harvard University Press, 1987.

Rorty, Richard. "Philosophy as a Kind of Writing: An Essay on Derrida." In *Consequences of Pragmatism,* 90–109. Minneapolis: University of Minnesota Press, 1982.

Saussure, Ferdinand de. *Course in General Linguistics.* New York: Philosophical Library, 1959.

Schilb, John. "Deconstructing Didion: Poststructuralist Rhetorical Theory in the Composition Class." In *Literary Nonfiction: Theory, Criticism, Pedagogy,* ed. Chris Anderson, 262–86. Carbondale: Southern Illinois University Press, 1989.

Scholes, Robert. *Protocols of Reading.* New Haven: Yale University Press, 1989.

Scholes, Robert. *Textual Power: Literary Theory and the Teaching of English.* New Haven: Yale University Press, 1985.

Searle, John. "Reiterating the Differences." *Glyph* 1 (1977): 198–208.

Shapiro, Michael J. "The Rhetoric of Social Science: The Political Responsibilities of the Scholar." In *Rhetoric of the Human Sciences: Language and Argument in Scholarship and Public Affairs,* ed. John S. Nelson, Allan Megill, and Donald N. McCloskey, 363–80. Madison: University of Wisconsin Press, 1987.

13 MAKING ENEMIES

HOW GOULD AND

LEWONTIN CRITICIZE

GREG MYERS

Fortunately my contribution to this book does not have to sum up the wide range of applied linguistic work on scientific texts, because this task has already been done by John Swales, whose readable book *Genre Analysis* includes a daunting bibliography of studies that will either lure readers into the field or terrify them thoroughly. What these studies have in common is the assumption that a detailed understanding of the systems of language in general can help us to understand, and especially to teach, the particular kinds of language that we find in academic writing. Apart from that, they differ in the kinds of linguistics they apply, the applications to which they put it, the level on which they analyze language, and the motivations (and financial support) behind the studies.

My main interest when I encounter a scientific text is to correlate linguistic features with specific social acts. So I start with some attempt to characterize the social acts involved in a text (such as making a claim, making a criticism, referring to the work of others for support, naming, popularizing), always trying to find an act that is recognizable and that has a name for the participants. Of course these are purely informal characterizations—I would not want to argue that there is some exhaustive set of categories for the acts scientists perform. These characterizations are useful to me, though, in choosing texts to compare and focusing on possible differences between them. With some social act in mind, I give the text a highly selective (and quite unnatural) reading, looking for linguistic features I can relate to that act, such as pronouns, references to people, hedges on the strength of a verb, verb tense sequences, cohesion, noun phrase construction, format of titles, overall organization, illustrations. To do so I often assume some stylistic expectations on the part of the reader—say, that a research article title won't include the names of people, or that a claim is likely to be hedged—as a background against which some irregularity is foregrounded.[1] But I look for some atypical text in which the linguistic feature that interests me stands out, not some typical text or some general description of scientific English—which is fortunate, because "The Spandrels of San Marco" is hardly typical.

The social act I would like to focus on in Gould and Lewontin's article is suggested by the word *critique* in their title. Of course criticism is common in scientific articles; in almost every research report the authors criticize, at least by implication, earlier research (including their own earlier work) and competing interpretations. But it is rare for such criticism to be presented as the main purpose of a scientific article, and rarer still for this act to be foregrounded in the title as it is in "Spandrels." (In contrast, this title, with its allusive first part and critical second part, would be quite usual in a journal of sociology or literary theory.) I will investigate the linguistic correlates of this act of criticism by drawing on what is called, somewhat misleadingly, politeness theory.[2] The term is misleading because it suggests to nonlinguists a narrow range of entirely conventional behavior such as saying "please" and keeping one's elbows off the table. There is very little room for politeness, thus narrowly conceived, in scientific texts (except perhaps in marginal parts of the text, such as the acknowledgements). In linguistics, though, the term has become a handy name for a much broader range of studies of all the ways speakers define their relations to hearers in the performance of actions in a given social and linguistic system. Since Brown and Levinson published a vast catalogue of such features in three languages in 1978, there have been thousands of studies of politeness in cross-cultural communication, literature, and academic texts.[3] Brown and Levinson present a model in which conventions of politeness are rational strategies for the redress of Face Threatening Acts (FTAs) that are an essential but potentially disruptive part of any human interaction, acts such as requests, apologies, and criticisms.

This admittedly deductive (and reductive) characterization of interaction is relevant to scientific writing if we see scientists as competing in their papers for credit before an observing audience of their peers. So scientific articles dealing with current controversies in specialized journals are addressed both to the handful of people doing very similar research, who may have reason to feel annoyance or solidarity, and to a broader audience that will judge the participants by whether they have behaved in an appropriately scientific way. This helps us explain why direct criticism is so rare in biology articles, while it is very common in

Greg Myers

linguistics and is almost essential in literary criticism. Criticism must seem impersonal if its effect is not to rebound on the critic for breaching the accepted convention of distance between participants.

There is a good example of this turn in the article immediately following Gould's and Lewontin's in the *Proceedings of the Royal Society,* the introduction to general discussion by the British zoologist A. J. Cain, who has devoted his career to the study of generations of land molluscs in the field. Cain comments on an earlier review by Lewontin:

> Lewontin (1972) characterized the sort of determination of actual selection coefficients which is the only way of accumulating evidence on the importance of natural selection as a British upper middle class activity. Presumably when prejudice is strong, facts can be dispensed with as well: my own background and upbringing can only be distinguished by the extreme purist from working class. Lewontin's statement is irrelevant; the only point scientifically is how far this activity stands up to scientific, not sociological, explanation. If not, then Lewontin's own conclusions require sociological investigation before we can accept them.[4]

Cain's response to a criticism of his work as class-biased is that the criticism reflects on Lewontin, not on him; quite apart from the sociological accuracy of the criticism, the resort to sociological territory is a retreat from grounds Cain considers properly scientific. The mere fact that Lewontin would say such a thing is offered to the reader as reason for dismissal of Lewontin's view. Of course the exchanges do not usually get this far, at least in public, even though the private rhetoric may become bitter indeed.

Public exchanges between scientists rarely get this personal. But it would be a great mistake to assume that the subject matter of science makes the texts impersonal; the creation of this sense of impersonality is part of the strategies of assertion and criticism. Often scientists identify their work with the consensus view of the research community, shown by citations, so that rival interpretations and researchers, whether named or not, are left outside. That is, scientists make friends to define enemies. Gould and Lewontin take what seems to be an opposite and very unusual strategy of identifying error with the established disciplinary consensus, inviting the imaginary reader to join them outside it. They make enemies to define their friends. It is a high-risk strategy. But when successful (as in, say, the early rhetoric of generative grammar), the gamble has a high payoff; the new center of a field forms around the position that was previously defined as outside it.

Politeness in "The Spandrels of San Marco"

I am interested in the linguistic devices Gould and Lewontin use to define *us* and *them*, scientific allies and opponents. I will start with a linguistic feature already noted by Gragson and Selzer (in chapter 10): the ambiguity of the use of *we*, split between the exclusive use, meaning only the authors Gould and Lewontin, and the inclusive use, designating the reader also.[5] While I entirely accept Gragson and Selzer's reading, I would also complicate matters slightly by proposing that the inclusive use might cover a range of persons:

> we$_1$—Gould and Lewontin, or only one of them ("We criticize this approach. . . ." [147]; "We have, for example [Gould 1978], criticized Barash's [1976] work. . . ." [154]).
> we$_2$—all "pluralists" who agree with Gould and Lewontin.
> we$_3$—the whole relevant community of biologists ("Have we not all heard the catechism about genetic drift . . .?" [151]).
> we$_4$—any reader with common sense ("It is rooted in a notion popularized by A. R. Wallace and A. Weismann [but not, as we shall see, by Darwin] . . ." [150]).

Now we (that is, a reader with common sense) might expect that the textual problem for Gould and Lewontin is to move readers from we$_3$ to we$_2$, from an uncommitted position to identification with their position. But in fact the we$_2$ I have imagined does not occur at all. In fact Gould and Lewontin hardly even give this position a name; it is "a pluralistic view" only in the last sentence. So they are not presenting a manifesto for a party that already has a name and a membership list, or at least that is not what they say they are doing; they are creating such a party or a position in the course of the text, and deliberately leaving its platform open.[6]

The construction of a point of view is not just a matter of pronouns and noun phrases.[7] We can see it more subtly in the adverbs and adverbials that suggest the attitude of the authors (and presumably the readers) to a statement:

> But they are usually dismissed as unimportant or else, and more *frustratingly,* simply acknowledged and then not taken to heart and invoked. (151)

> *Unfortunately,* a common procedure among evolutionists does not allow such definable rejection. . . . (153)

> Why not *at least* perform the obvious test? (154)

Greg Myers

To interpret these adverbials (emphases added), the reader needs to enter into the authors' point of view. More subtly, the reader has to be following their point of view to link and evaluate correctly sentences that, out of context, seem to be flat statements:

> Often, evolutionists use *consistency* with natural selection as the sole criterion and consider their work done when they concoct a plausible story. But plausible stories can always be told. The key to historical research lies in devising criteria to identify proper explanations among the substantial set of plausible pathways to any modern result. (154; emphasis in the original)

The first sentence is to be taken as a negative evaluation, the second as a criticism, the third as a positive statement of *our* method. But these links can be made only by entering into the authors' viewpoint; there is no other indication of how each sentence is to be related to the one before it.

The construction of *us* takes on meaning only in relation to the construction of *them*. The other side does have a name here ("An adaptationist programme"), but it is a name that, as we will see, may be rejected by its proponents (and the use of "an" suggests that it may be new to readers). The other side is referred to in general terms, so that no one can be specifically singled out (emphases added):

> *Anyone* who tried to argue that the structure exists because the alternation of rose and portcullis makes so much sense in a Tudor chapel . . . (149)

> *Some evolutionists* will protest that we are caricaturing their view of adaptation. (151)

> Rather, *you* acknowledge the rival, but circumscribe its domain of action. (151)

> [*T*]*hey* clearly mean that the action of natural selection applied to particular cases . . . (155)

Or the agent of "the adaptationist programme" is deleted entirely, in a passive or imperative formulation:

> An organism is atomized . . . [and] interaction is acknowledged. (151)

> If one adaptive argument fails, try another. (152)

> If one adaptive argument fails, assume another must exist. (152)

[A]ttribute failure to an imperfect understanding. (152)

Emphasize immediate utility and exclude other attributes of form. (153)

To be sure, Gould and Lewontin are not shy about naming such opponents as Harner, Wilson, Barash, and others. But in this article (in contrast to Gould's and Lewontin's specific critiques of Wilson), these researchers are to be criticized only as representatives of a more general tendency. The effect of the attack on a whole approach to research would be lost if the reader saw this as an attack on any specific researcher.

This elaborate strategy of taking sides runs through the first half of the article. Then with section 5, Gould and Lewontin turn to a different and more typically scientific way of presenting a position, with almost no personal subjects. One paragraph can serve as an example—it is a series of statements of fact, almost all with forms of *to be*, or stative verbs:

> In a distinct minority of cases, the divaricate pattern becomes functional in each of the four categories (figure 3). Divaricate ribs may act as scoops and anchors in burrowing (Stanley 1970), but they are not properly arranged for such function in many clams. The colour chevrons are mimetic in one species *(Pteria zebra)* that lives in hydrozoan branches; here the variability is strongly reduced. The mineralization chevrons are probably adaptive in only one remarkable creature, the peculiar bivalve *Corculum cardissa* (in other species, they either appear in odd specimens or only as post-mortem products of shell erosion). This clam is uniquely flattened in an anterior-posterior direction. It lies on the substrate, posterior up. Distributed over its rear end are divaricate triangles of mineralization. They are translucent, while the rest of the shell is opaque. Under these windows dwell endosymbiotic algae! (162).

Out of context, it might be taken from a textbook. There would seem to be no issue of taking sides here. The only clue that we have taken on a particular point of view in following this information is the way the paragraph ends—with an exclamation point. Apparently the reader who has followed the argument and found that most of the divaricate patterns are not adaptive is then in a position to be surprised by the one very peculiar exception, where the divaricate pattern in fact seems to be a matter of design.

After this textbook-like section, Gould and Lewontin finish with a

Greg Myers

strongly adversarial conclusion. I needn't quote the entire passage: the beginnings of the sentences will make it clear enough:

> We feel . . .
> We do not offer . . .
> We welcome . . .
> Under the adaptationist programme, the great historic themes . . .
> Too often, the adaptationist programme . . .
> It assumed . . .
> A pluralistic view could put . . .

We (I mean we$_4$ this time) have moved from the division of sides, to a characterization of the opposition, to an acceptance of the assumptions of the *we* side, and finally to the division of sides again.

Politeness and Positions in the Royal Society Symposium

The explanation of linguistic features in terms of politeness may be criticized in the same way Gould and Lewontin criticize the explanation of biological features in terms of adaptation: whatever the data, one can always find a plausible story for it. I cannot say whether the criticism is justified as applied to evolutionary biology, but I am willing to grant that it holds for politeness. One can always attribute linguistic features to politeness, and thus support the strategic model of interaction on which the theory is based. If a writer uses personal subjects and no hedging, that may be one form of politeness; but if the writer uses impersonal forms and hedging, that's another form of politeness. The assertion that these features are due to politeness is not a testable claim about language; the theory provides at best an unusually detailed checklist to guide interpretation. But that does not mean it allows unconstrained interpretations, and in textual analysis, as in biology, the constraint is in comparisons. Anyone who claims that a linguistic feature functions in a politeness strategy should be able to show similar features in similar situations created by various authors, and different features in different situations created by the same authors. That is what I will try to do here.

I am not sure exactly how to classify the discourse of the published proceedings of a Royal Society Symposium; the papers are not review articles, nor research reports, nor popularizations. They seem to address both a broad audience of scientists and a fairly specialized audience, and some seem to address each other while others are less closely related. Unfortunately we do not have the text of the discussion, even in the brief and formal form in which Royal Society discussions sometimes appear. So the simplest way to find a comparable situation to that in Gould and Lewontin's text and a sense of how it worked in interaction is to look elsewhere in the same symposium. We can see how some authors respond to "The Spandrels of San Marco" (several seem to have revised their papers after the symposium to include such a response) or to Gould's and Lewontin's earlier papers. And we can see a range of politeness strategies in criticism, from cautious impersonality, to various ways of including Gould and/or Lewontin in the consensus, to more direct criticism.

I would hesitate to say whether any strategy for redressing criticism can be considered typical of science, but there is citation analysis evidence to suggest that the most common strategy is to avoid making critical references at all.[8] In this Royal Society collection, for instance, D. Charlesworth and B. Charlesworth seem (to the naive reader at least) to avoid negative comments in their essay, "The Evolutionary Genetics of Sexual Systems in Flowering Plants" (79–96). John Maynard Smith is not quite so reserved. But as a reader of Gragson and Selzer (in this volume) might expect, Maynard Smith is very carefully impersonal in his references to himself in "Game Theory and the Evolution of Behaviour" (41–54), and he stresses consensus in his references to others:

> The theory of games was applied to the evolution of fighting behaviour (Maynard Smith and Price 1973) in the hope of explaining the conventional aspects of this behaviour in terms of natural selection. (41)

> In fact, the disagreement between Geist (1974) and Maynard Smith and Price (1973) was more apparent than real. Part of the difficulty arose because of our failure to note that Geist had earlier put forward part of our argument. (51)

It should be stressed that this style is not available only to someone as distinguished as Maynard Smith; other contributors use similar tactics.

At the other extreme would be the hostile response of A. J. Cain, from which I have already quoted, or the less personal but still directly critical response of B. C. Clarke in "The Evolution of Genetic Diversity" (19–40). Clarke starts with a paragraph of general (though not universally accepted) methodological principles; he then contrasts these principles with those of Gould and Lewontin, reinterpreting Occam's

Greg Myers

razor ("prefer the simple to the complex") to make it a criticism of what Gould and Lewontin call *pluralism*.

> Nature is often complicated, but we come to understand it by making and testing simple hypotheses. . . . Operationally, we should always prefer the simple to the complex, and the singular to the plural.
> Gould and Lewontin (this symposium) argue for pluralism in the explanation of evolutionary change. . . . It is my purpose here, in discussing contemporary genetic variation, not only to reject this argument, but also to attempt a Darwinian synthesis of recent observations and experiments. (19)

Here there is direct criticism, but the "not only . . . but also" subordinates it to the positive project of a synthesis. Except in this one rather roundabout sentence, the source of the criticism is the impersonal precepts of methodology. The rest of the article avoids such direct criticism. When the criticism is taken up again in the last sentence, it is directed at Gould and Lewontin only by implication. Here, typically, it is the evidence rather than the author that criticizes the other side: "The accumulation of evidence for the theory of frequency dependence, and its general consistency with the facts of variation, give no encouragement to those who argue that we should abandon the Darwinian canon, even to the extent of a 'pluralistic' compromise" (44). Of course, Gould and Lewontin specifically argue that they are *not* abandoning Darwin, but his characterization of them as "those who argue that we should abandon the Darwinian canon" can still be used to attack their position.

Between the two extremes of ignoring opponents and attacking them directly, there is the strategy of refusing to accept the division of *us* and *them* that those who would call themselves opponents have drawn. For instance, R. Dawkins and J. R. Krebs in "Arms Races between and within Species" (55–77) praise the work of Gould, who would present himself as their opponent, in a way that seems to avoid all controversy or offense: "The arms race interpretation favoured by Jerison and Gould seems very plausible to us" (66). There is a long quotation from Gould's popular writing to support this point, but it is still possible that Gould would be unhappy with the anthropomorphic metaphor of the "arms race," the sort of thing he has criticized in Dawkins's writing. In a more complex example, the authors agree with Gould and Lewontin's paper, but in doing so yoke it to an example of the very "adaptationist" position Gould and Lewontin criticize: "The arms race idea provides one positive reason (there are, of course, many more; Maynard Smith 1978b; Gould and Lewontin, this symposium) for avoiding naive perfectionism when we look at individual adaptation" (73). These words imply that there was no basic disagreement for Gould and Lewontin to build on; they were attacking a position held only by the naive. This is more direct than Maynard Smith's assertions of a deeper consensus underlying apparent controversies, and I assume it gives some pleasure at its wit. It not only redresses an apparent criticism but also reduces any claim Gould and Lewontin have to novelty and significance in their critique.[9]

Perhaps the clearest contrast to Gould's and Lewontin's strategies of *us* and *them* in the symposium is the paper "Comparison and Adaptation" (113–31) by T. H. Clutton–Brock and P. H. Harvey, which presents criticisms only in an impersonal form and makes personal references only with apparent praise. Brock and Harvey present themselves in the abstract as responding to "critics of the 'adaptationist programme' " (113), thus identifying Gould, Lewontin, and others only indirectly and keeping the name *adaptationist* only in scare quotes. Later, after a reference to Gould, they say, "The problem that allometry raises is not insuperable" (121); it is the theory, not its proponents, that must be addressed. Their first sentence is a general methodological observation, like that of Clarke, but it is not designed to reflect badly on the other side: "Most adaptive explanations for traits or trait complexes are founded initially on comparison" (113). The other strategy used by Clutton–Brock and Harvey is to refer frequently to both Gould and Lewontin, as in this example: "adaptive differences can arise through a variety of processes other than natural selection (see Tinbergen 1972; Gould 1978)" (116). Here Gould is linked to their own hero figure, Tinbergen. When Gould is cited for a view that they do not accept, he also finds himself in good company: "In ungulates it is commonly claimed that . . . (see, for example, Beninde 1937; Geist 1971; Gould 1974; Maynard Smith, this symposium)" (118). Criticism that Gould might have intended to be directed at them, as sociobiologists, can be cited and treated as encouragement to further refinement of the program, not as a call for its abandonment: "As Gould (1978) has clearly pointed out, it is important that evolutionary biologists should seek to solve these problems if adaptive arguments are to retain credibility" (127). In a similar response, made earlier to a comment by Lewontin, it is noted as "unfortunate" that not enough is known about the function of the trait he discussed (rhinoceros horns) to make his point persuasive. Criticisms of adaptation models are listed in a way that does not make explicit which view the authors hold. The most unequivocal criticism in the article is reserved for a "straw biologist" holding a position on baboon behavior (115); this tactic both avoids direct criticism and jokingly acknowledges

Greg Myers

the danger of being seen to be constructing opposition. Gould and Lewontin deflect criticism in a similar way with their own remark about not using glass beasts as straw men (153).

In summary then, the other contributors to the Royal Society Symposium do engage in criticism, even if it is not foregrounded in the way that Gould and Lewontin's is. But they tend to more impersonal strategies for marking the sides, and they tend to respond to criticisms by asserting an underlying consensus. Some British biologists I have interviewed have pointed to differences in national scientific style, and indeed the only other American in the symposium, G. C. Williams, did engage in direct criticism in his "The Question of Adaptive Sex Ratio in Outcrossed Vertebrates" (133–46). But I am wary of taking such participants' explanations at face value, for they are themselves rhetorical devices, ways of separating friends and enemies on a level of national traditions in science, as Gould and Lewontin show: "In continental Europe, evolutionists have never been much attracted to the Anglo-American penchant for atomizing organisms into parts and trying to explain each as a direct adaptation" (159). Other explanations for differences in politeness strategies might turn on the participants' ages, specialties, or personalities. But we must remember that the sociology of biology offered by biologists is part of the battle, not an explanation of the battle.

Gould in Paleontology and in Literary Theory

Gould and Lewontin are, of course, well-known figures with distinctive personalities. To show that the linguistic features here tell us about the acts they are performing, not just about them, we need to show that they can use different strategies—different personalities—when the situation calls for them. One could presumably do this with Lewontin, who has published widely, but it is particularly easy to show with Gould because Gould has written articles for quite different audiences and cites two of them in "The Spandrels of San Marco": "Allometry in Primates, with Emphasis on Scaling and the Evolution of the Brain" in *Approaches to Primate Paleobiology,* for a specialized paleontological audience; and "D'Arcy Thompson and the Science of Form" in *New Literary History,* for an audience of literary theorists.[10] In the first he is addressing fellow researchers on technical issues rather than on the broad programmatic issues of the Royal Society Symposium. In the second he is speaking for biology as a whole in a collection with contributions from various disciplines. Neither divides *us* and *them* in the way he and Lewontin do in "The Spandrels of San Marco."

Gould's complex use of *we* and implied criticism of positions can be seen in a passage from "Allometry in Primates" that lays out explanations of a ratio in intraspecific scaling:

> Several proposals work for some cases, but cannot provide a general explanation.
> (1) For intrapopulational curves, *we* can argue that. . . . [But] [t]his argument will not work for interpopulation allometries. . . .
> (2) *Jerison* [1963] argued that . . . [but] it cannot work for reptiles and. . . .
> (3) *Geist* [1973] and *Jerison* [1973] have argued that . . . [but] it will not explain the intrapopulational exponents.
> (4) I would argue that. . . . (557–58; emphasis added)

Here the *we* introduces a hypothetical suggestion, offered without the author's backing; *I* is reserved for the author's favored position, which as usual seems to be the last on the list of alternatives. But there is no clear line between *us* and *them;* there appears to be no criticism of the cited authors. While this may seem to be a version of the pluralism of "The Spandrels of San Marco" on the level of narrower technical arguments, it seems to differ fundamentally from the broad methodological polemics of that essay. For one thing, not as much is at stake.

In addition to the *we*s of Gould and Lewontin, there are other *we*s in Gould's "Allometry" that have a broader, joking referent in some cases: "Among the more persistent effects of anthropocentrism, *we* must count our desire to measure levels of cephalization, and thus to affirm in rigourous terms *our* obvious superiority" (263; emphasis added). The *we* might include the readers (we$_4$), and the first *our* probably extends as far as the whole primatological community; but the second *our* can only be read to refer to members of *homo sapiens* as a species. The joke is carried further when the author identifies himself with an even broader biological group: "Like any good primate, I will go out on a limb and predict . . ." (280). This kind of joke presumably identifies him not only with primates but also with primatologists, who are used to thinking of humans in relation to other mammals.

As the *us* in the specialized article is more complex and less definite than in "The Spandrels of San Marco," so is the *them*. The only direct criticism is aimed at someone long dead: "This dubious claim remained impervious to future test because Dubois, thenceforward, merely passed lines of 5/9 slope through single points" ("Allometry," 248). I assume that puns on researchers' names are normally taboo (and for good reason) but that the taboo does not apply to errors of the past. When criticizing errors of the present, Gould, like other scientists, sometimes includes himself in the criticism: "Many authors, myself included, have advocated the use of 'regression' formulae that consider error in both variates. . . . This can lead to problems of the following type. . . ." (253). Potentially controversial or divisive statements are almost always hedged, as in these examples (emphasis added) from the first paragraphs:

> When Julian Huxley established the quantitative study of allometry in the 1920s, he pursued a strategy that proved excellent in the short run but *rather* restrictive in longer perspective. (244)

> Allometry is *perhaps* the major principle relating basic differences in form among related animals. (245)

> Moreover, when applied, it is *often* applied badly. (245)

But it is apparently unnecessary to hedge criticisms based on now-accepted methodology:

> We might defend this [approximation of curves with straight lines] as a descriptive simplification, but its perpetrators generally imbued the points at which their straight lines met with special significance as "critical stages" (Alcobe and Prevosti, 1951, p. 17) marking sudden transitions from one type of growth to another. *They are artifacts.* (252; emphasis added)

General methodological statements, like those opening Clarke's symposium article, can remain unhedged; they are on a level of generality where it can be assumed that opponents could not disagree: "An appropriate multivariate technique depends upon the questions asked" (272). Gould also hedges his own statements strongly, often more strongly than he might hedge those of another researcher: "*I am intrigued with* the idea that intraspecific scaling may have played an important role in the evolution of man" (280); "The *most outrageous* extension of this hypothesis would attempt to apply it to the evolution of man" (284) (emphases added). Criticisms of the approach Gould shares here are put in their strongest form (but note that they are phrased as questions, as Holloway might put them out of politeness):

> Can the gross size of the brain (or merely of the endocast infossils) mean anything at all when we deal with such an immensely complicated computer of such subtle and variable structure? Is it not even a vulgarization to make such an attempt?

> Holloway (1969, 1970, 1972) has been particularly forceful in
> asserting these criticisms. (253)

To put the criticisms this way strongly grants their apparent reason-
ableness before Gould responds to them. The opponents in "The Spandrels of San Marco" are not given this much credit.

As in "The Spandrels of San Marco," the adverbs in the "Allometry"
article give more subtle indications of the attitudes the author and reader
are to share (emphases added):

> Radinsky (1967) suggested that the area of the foramen magnum
> be used as a surrogate for body size, but *(unfortunately)* the pri-
> mary correlate of this area seems to be with brain size itself. (261)

> Weidenrich (1941) argued, *overenthusiastically,* to be sure, (Starck,
> 1953), that the relative size of the brain plays a dominant role in
> the morphology of the skull. (262).

> In practice, however *(and thankfully),* the basal insectivore curve
> seems to serve its designated function remarkably well. (268)

While adverbs are often used in scientific writing to show sympathy for
the point of view of the criticized writers (so "unfortunately" can mean
"unfortunately for them"), they here suggest a desire, shared by all
readers, for some common and practical methodological guidelines. It
is the reader who has to decide what point of view is indicated by each
adverb.

Indeed, the problem of attributing attitudes is a major difference
between the article for the Royal Society Symposium and that for *Approaches to Primate Paleobiology.* The paleontology article ends with a
long quotation from Count (1947) that gives no indication whether the
uninformed reader should take it as a prescient forerunner of Gould's
opinion or as a dated belief now superseded. The paleontological reader
of the more specialized article is assumed to have the knowledge necessary to see which side a statement is supporting, and thus to know
whether Gould's attitude toward the statement is favorable or unfavor-
able. The reader of the article for a broader audience of evolutionists has
to be given much stronger signals, especially at the opening and in the
conclusion.

All the articles I have considered so far have been addressed by biologists to other biologists, members of wider or narrower groups within
the discipline. We would expect to see different politeness strategies
when an author is writing as a representative of biology as a whole, for
an audience of nonbiologists. Fortunately (and I mean that adverb to

refer to both my good fortune and that of my readers as textual analysts), Gould wrote another article on the subject of adaptationism for *New Literary History,* one of the two or three most influential journals for articles on the theory of literary criticism. Of course Gould is one of the acknowledged masters of the popular science essay, but the task in a literary journal is different from popularization; he has to take on the conventions of another discipline. For instance, he gives his article a literary title of the characteristic type, "Name of Author and Name of Concept": "D'Arcy Thompson and the Science of Form." The word *science* here distinguishes his article from other contributions to the journal's special interdisciplinary issue on "Form"; Gould is to be the token scientist, just as someone else is to be the art historian or linguist. He also gives the article a characteristically "lit crit" opening, with an epigraph, some learned observations, and a very long wait until the reader finds out (on the third page) what the article is about.[11] As in most articles in literary criticism, Gould grounds his reflections in a classic and insufficiently understood text; he will defend this text against its critics and establish it in its true place in the canon.

With the issue defined in this way, disputes within paleontology, and even the ongoing battle for resources between molecular biology and the rest of biology, are worth only passing mentions. All the attention is on D'Arcy Thompson; Thompson's critics remain completely nameless:

> Many critics have failed to grasp this view of action of physical forces. (241)

> For his critics have missed a central point: that D'Arcy Thompson was interested in the deformed net, not primarily in the animal that it generated. (245)

> The critics of *Growth and Form* are numerous. Their major objections can be condensed to three categories. (247)

But Gould must give some explanation of the methods and views of those critics, to provide a background against which Thompson can be seen as remarkable: "All biologists must deal with form, but it does not follow that they treat it adequately. In my own field of evolutionary biology, I detect three approaches that seem especially insufficient when compared with the insights of D'Arcy Thompson: . . ." (231). The complete lack of references to critics could be frustrating for any readers who want to leave the arts library and pursue these issues at the biology library on the other side of the campus. But it is appropriate to the role Gould takes on here as representative biologist. There is no point in naming the opposition to an audience that won't recognize the names.

On the other hand, there is no point in hedging one's criticisms before an audience that doesn't include those criticized.

When Gould criticizes Thompson, it is with the carefully hedged respect of a scholar toward a canonical author: "When he applies it [Thompson's theory of form] to large and complex forms, I begin to doubt its propriety while continuing to admire its sheer ingenuity" (251). The criticism that the method is improperly applied is hedged by the use of *I*, by the hedge *begin to*, by the coordinated praise *continuing to admire*, and by the impersonal evidence that follows the sentence. The key split on which the article is built is not between one view of evolution and another, or one mathematical formula for allometry and another, but between the broadly educated inquirer (Thompson, Gould, and now the reader) and the narrow pursuer of scientific fashion.

Conclusion

What is the point of such a comparison? First, it is important for nonscientists like me to improve our skills as critical readers of science. Gould and Lewontin (and Maynard Smith and many others) have pointed out the need to look at the complexity of scientific opinion on issues of public interest, and not to take it as one voice. Other contributions to this collection explore the nature of such a critical reading.

But since applied linguistics is typically applied to education, I will suggest why analyses of politeness are of interest to teachers. Students new to academic discourse, first venturing beyond textbooks to journal articles, for instance, often miss the subtle cues that show the taking of sides. In their essays I sometimes find attribution of views, taken out of context, that I know to be exactly the opposite of those with which the cited author is usually identified. To an experienced reader of the discipline's literature, it seems that the reader has carelessly missed the critical note in a reference. But both assertions and criticisms are made within constraints of linguistic politeness, with impersonal constructions and very complex hedges and cohesive relations that rely on a detailed knowledge of disciplinary discourse. It is not much help to give the students general descriptions of the scientific register—that it is characterized by passives or hedges—because the problem lies in recognizing when and why an author might use the passive voice or hedge a statement. It is not even much help to give students general illustrations of politeness strategies, since as we have seen they may vary with the kind of text and since the same author may use different strategies in different situations. The purely linguistic analysis will be of no use unless we also have a model of scientific discourse as an agonistic and cooperative field in which researchers make facts by making alliances and sometimes making enemies.[12]

NOTES

My thanks to the Stylistics Research Group, University of Lancaster, for comments on a version of the first part of this paper, to students in the linguistics M.A. course "Understanding Academic Writing" for comments on the last part, to Geoff Thompson and a seminar at the English Language Unit at the University of Liverpool for their comments, and to Robert Cameron for help with the background in the evolutionary biology community.

1. There has long been a tendency to define scientific style by what it is not—for instance, in making points about literary style. Studies of scientific style as a single register of co-occurring features include C. L. Barber, "Some Measurable Characteristics of Modern Scientific Prose"; R. D. Huddleston, *The Sentence in Written English;* Myrna Gopnik, *Linguistic Structures in Scientific Texts;* and M. A. K. Halliday, "On the Language of Physical Science." There is an influential critique of such studies in Henry Widdowson, *Explorations in Applied Linguistics,* but Widdowson, though calling for rhetorical analysis, sees the rhetorical function of scientific texts as simply conveying information, a view that limits his analysis.

2. The classic study is Penelope Brown and Stephen Levinson, *Politeness.* Other influential studies include Robin Lakoff, "The Logic of Politeness, or, Minding Your P's and Q's"; and Geoffrey Leech, *Principles of Pragmatics.*

3. For politeness in cross-cultural communication, see a special issue of the *Journal of Pragmatics* 14 (April 1990), ed. Shoshana Blum-Kulka and Gabriele Kasper. For studies of politeness in literary texts, see Paul Simpson, "Politeness Phenomena in Ionesco's *The Lesson*" and its references. For politeness in letters to a university on a tenure decision, see Roger Cherry, "Politeness in Written Persuasion." See also my "Pragmatics of Politeness in Scientific Writing" and "Politeness and Certainty: The Language of Collaboration in an AI Group."

4. A. J. Cain, "Introduction to General Discussion," *Proceedings of the Royal Society of London, B: Biological Sciences* 205 (1979): 165–69. Further references to this and other papers in the *Proceedings* (including the one by Gould and Lewontin) are identified by title and by page number in the text.

5. On pronouns and politeness, see the classic study by R. Brown and A. Gilman, "Pronouns of Distance and Solidarity"; and Peter Muhlhausler and Rom Harré, *Pronouns and People.*

6. Compare, in the same volume of *Proceedings of the Royal Society,* John Maynard Smith's introduction to "Game Theory and the Evolution of Behav-

iour." His approach does have a name and leaders recognizable by any biologist—and it presents a clear target at which to shoot.

7. A good analysis of linguistic correlates of point of view is Roger Fowler, *Linguistic Criticism*, chapters 9 and 10. Other textbooks on the stylistics of prose are Geoffrey Leech and Michael Short, *Style in Fiction*; and Ronald Carter and Walter Nash, *Seeing Through Language.*

8. Winsor discusses the citation history of "The Spandrels of San Marco" in chapter 7 of this book. For a review of citation context analysis from the perspective of applied linguistics, see John Swales, "Citation Analysis and Discourse Analysis." For a critical dialogue between citation analysis and the sociological (rather than linguistic) strand of discourse analysis, see Diana Hicks and Jonathan Potter, "Sociology of Scientific Knowledge: A Reflexive Citation Analysis *or* Scientific Disciplines and Disciplining Science."

9. We cannot tell from the *Proceedings of the Royal Society* alone whether in fact a consensus was forming in 1979 around a view like Gould and Lewontin's. The problem is that the claim of consensus—whether for or against one's own views—is a powerful rhetorical tool. Any two views on an issue may be portrayed as like or different, depending on the context in which they are considered. We cannot even tell whether there was consensus by reconstructing the history of the late 1970s, say, through interviews with participants. Some now say that by 1979 many biologists had criticized adaptive explanations and that Gould and Lewontin were setting up a straw man. Gould and Lewontin might say that this consensus became apparent only after their efforts. For varying retrospective accounts of consensus formation in another controversy, see Nigel Gilbert and Michael Mulkay, *Opening Pandora's Box.*

10. The pair of essays by the same author makes a good text for classroom discussion of the differences between disciplines; one can only hope that Gould someday takes time to publish on allometry in *Econometrica* or *Sociology.* (Indeed he has published on social science in *The Mismeasure of Man.*) Gould's recent book, *Wonderful Life*, makes use of some of the same devices of constructing the opposing sides.

11. On article introductions in several disciplines, see Charles Bazerman, *Shaping Written Knowledge*; John Swales, *Genre Analysis*; Graham Crookes, "Towards a Validated Analysis of Scientific Text"; and Tony Dudley-Evans and Willie Henderson, "The Organisation of Article Introductions."

12. I have assumed throughout an analysis of science based on the sociology of knowledge, as developed in some other essays in this collection. For a range of approaches to this field, see Michael Mulkay, *Science and the Sociology of Knowledge*; or Bruno Latour, *Science in Action: How to Follow Scientists through Society*; or Harry Collins, *Changing Order*; or Steve Woolgar, *Science: The Very Idea.*

Greg Myers

WORKS CITED

Barber, C. L. "Some Measurable Characteristics of Modern Scientific Prose." In *Contributions to English Syntax and Phonology.* Stockholm: Almquist and Wiksell, 1962. Reprinted, with commentary, in *Episodes in ESP,* ed. John Swales, 1–16. Oxford: Pergamon, 1985.

Bazerman, Charles. *Shaping Written Knowledge: The Genre and Activity of the Experimental Article in Science.* Madison: University of Wisconsin Press, 1988.

Brown, Penelope, and Stephen Levinson. "Politeness." In *Questions and Politeness,* ed. Esther Goody. Cambridge: Cambridge University Press, 1978. Reprinted, with updated bibliography, as *Politeness.* Cambridge: Cambridge University Press, 1987.

Brown, Roger, and Albert Gilman. "Pronouns of Distance and Solidarity." In *Style in Language,* ed. T. A. Sebeok, 253–76. Cambridge: MIT Press, 1960. Reprinted in *Language and Social Context,* ed. P. Giglioli, 252–82. Harmondsworth, England: Penguin, 1972.

Carter, Ronald, and Walter Nash. *Seeing Through Language: A Guide to Styles of English Writing.* Oxford: Basil Blackwell, 1990.

Cherry, Roger. "Politeness in Written Persuasion." *Journal of Pragmatics* 12 (1988): 63–81.

Collins, Harry. *Changing Order: Replication and Induction in Scientific Practice.* Beverly Hills: Sage, 1985.

Crookes, Graham. "Towards a Validated Analysis of Scientific Text." *Applied Linguistics* 7 (1986): 57–70.

Dudley-Evans, Tony, and Willie Henderson. "The Organisation of Article Introductions: Evidence of Change in Academic Writing." In *The Language of Economics: The Analysis of Economics Discourse,* ed. Tony Dudley-Evans and Willie Henderson, 67–78. London: Modern English Publications and the British Council, 1990.

Fowler, Roger. *Linguistic Criticism.* Oxford: Oxford University Press, 1986.

Gilbert, G. Nigel, and Michael Mulkay. *Opening Pandora's Box: A Sociological Analysis of Scientists' Discourse.* Cambridge: Cambridge University Press, 1984.

Gopnik, Myrna. *Linguistic Structures in Scientific Texts.* The Hague: Mouton, 1972.

Gould, Stephen Jay. "Allometry in Primates, with Emphasis on Scaling and the Evolution of the Brain." In *Approaches to Primate Paleobiology,* ed. F. S. Szalay, 5244–92. Basel: Karger, 1975.

Gould, Stephen Jay. "D'Arcy Thompson and the Science of Form," *New Literary History* 2 (1971): 229–58.

Halliday, Michael A. K. "On the Language of Physical Science." In *Registers of Written English,* ed. M. Ghadessy, 162–78. London: Frances Pinter, 1988.

Hicks, Diana, and Jonathan Potter. "Sociology of Scientific Knowledge: A Reflexive Citation Analysis *or* Scientific Disciplines and Disciplining Science." *Social Studies of Science* 21 (1991): 459–501.

Huddleston, R. D. *The Sentence in Written English*. Cambridge: Cambridge University Press, 1971.

Lakoff, Robin. "The Logic of Politeness, or, Minding Your P's and Q's." *Papers from the Ninth Regional Meeting of the Chicago Linguistics Society* 9 (1974): 292–305.

Latour, Bruno. *Science in Action: How to Follow Scientists and Engineers through Society*. Cambridge: Harvard University Press, 1987.

Leech, Geoffrey. *Principles of Pragmatics*. Harlow, England: Longman, 1983.

Leech, Geoffrey, and Michael Short. *Style in Fiction*. Harlow England: Longman, 1983.

Muhlhausler, Peter, and Rom Harré. *Pronouns and People: The Linguistic Construction of Social and Personal Identity*. Oxford: Blackwell, 1990.

Mulkay, Michael. *Science and the Sociology of Knowledge*. London: George Allen and Unwin, 1979.

Myers, Greg. "Politeness and Certainty: The Language of Collaboration in an AI Group," *Social Studies of Science* 21 (1991): 37–73.

Myers, Greg. "The Pragmatics of Politeness in Scientific Writing." *Applied Linguistics* 10 (1989): 1–35.

Simpson, Paul. "Politeness Phenomena in Ionesco's *The Lesson*." In *Language, Discourse, and Literature: An Introductory Reader in Discourse Stylistics*, ed. Ronald Carter and Paul Simpson, 171–94. London: George Allen and Unwin, 1989.

Swales, John. "Citation Analysis and Discourse Analysis." *Applied Linguistics* 7 (1986): 39–56.

Swales, John. *Genre Analysis: Analysing Academic and Research Texts*. Cambridge: Cambridge University Press, 1990.

Widdowson, Henry. *Explorations in Applied Linguistics*. Oxford: Oxford University Press, 1979.

Woolgar, Steve. *Science: The Very Idea*. London: Tavistock, 1988.

14 PROVOCATIVE ARCHITECTURE

A STRUCTURAL ANALYSIS

OF GOULD AND LEWONTIN'S

"THE SPANDRELS OF SAN MARCO"

BARBARA COUTURE

The authors of "The Spandrels of San Marco" play upon structures: architectural, biological, and textual. They build a text that provokes readers to disassemble a foundational claim of evolutionary theory: the belief that natural selection produces biological "structures optimally designed . . . for their function" to the exclusion of other constraints. The architectural metaphor developed in "Spandrels" suggests that meaning in this essay is revealed in its structures—a suggestion that can be demonstrated through a structuralist analysis of the text. My purpose in this chapter is to apply some techniques of structural textual analysis to "Spandrels" to show how it builds systems of meaning that provoke the reader to disassemble foundational claims in evolutionary theory. I begin with a brief review of structuralism as an interpretive technique, focusing in detail on systemic functional analysis, a variety of structural analysis particularly suited to interpreting meanings that are socially construed. My discussion continues with an analysis of particular structures—lexical and grammatical features, participant exchange structures, and macrostructuring devices—in a few key passages of "Spandrels," an analysis that includes a comparison of "Spandrels" with another article appearing in the same issue of the Royal Society's *Proceedings*. Through these textual analyses I hope to show how structuralism can articulate the ways scientists (like Gould and Lewontin) choose to build, expand, and even "retrofit" scientific knowledge about evolution by manipulating the structures of written language.

Background: Structuralism, Systemics, and Interpretation

Structuralist analysis, as a technique used in textual interpretation, has its roots in the work of linguist Ferdinand de Saussure. In a revolutionary theory of meaning in language, Saussure asserted that meaning is a function not only of what interpretation is given to a linguistic sign, but also of the value of that sign relative to others within the same language system. This amounts to asserting that one interprets the meaning of a word or a grammatical marker through a comparison with other words or grammatical markers in the same language with which it contrasts or shares a similar meaning. Previous to Saussure, units of language were regarded as independent entities, "each of which somehow has a separate 'meaning' attached to it" (Hawkes, 19). Saussure's unique contribution to linguistics was his assertion that language has meaning because of the relationships we perceive between its various parts. These contrasts and similarities contribute to its "current adequacy" as a tool to represent our experience (Hawkes, 20).

Saussure's theory of meaning in language is forwarded in his *Course in General Linguistics*, which explains a principle that has guided literary interpretation since the abandonment of the New Criticism: "The linguistic sign is arbitrary" (67). As Saussure explains, the meaning of linguistic signs depends upon their relationship both to the ideas assigned to them by the language community and to other signs in a language system. These principles are basic to structuralist theories of verbal text: linguistic meaning is developed socially and linguistic signs are assigned a value relative to other socially developed meanings in a language system. Saussure notes that "values always involve" two variables: "(1) something *dissimilar* which can be *exchanged* for the item whose value is under consideration, and (2) *similar* things which can be *compared* with the items whose value is under consideration" (113). The condition of *value* defined by Saussure implies that meaning itself is a property of recognized similarities and differences; hence, meaning and value are interdependent.

The influence of value upon meaning can be seen within both lexical and grammatical systems in a language. As Saussure explains, the "difference in value between [the lexical items] *sheep* and *mouton* hinges on the fact that in English there is also another word *mutton* for the meat, whereas *mouton* in French covers both" (114). Likewise, the meaning of the plural marker in one language may have a different grammatical sense than in another; the different values of *plural* depend upon how many grammatical categories exist in a language for a number more than one. In Sanskrit, for instance, a third category covers the expression "my eyes," which is considered to be "neither in the singular nor in the plural but in the dual" (114).

Saussure's notions of the arbitrariness of the linguistic sign and of the value of meaning within a linguistic system have been applied to various forms of communication, including special forms of language such as literary writing. Jakobson and other linguists of the Prague School applied structuralist principles to the analysis of literature and concluded that the function of literary art, in effect, is to expose the arbitrary (that is, culturally assigned) nature of the linguistic sign (Hawkes, 69–71). Such exposure is achieved by manipulating the function of a sign within the system which assigns it value.

Formalist critics, as Jakobson and others were called, developed "mathematical" formulas for determining the "poetical" value of a literary text. Jakobson, for instance, claimed: "The poetic function [of language] projects the principle of equivalence from the axis of selection into the axis of combination" (358). In essence, this means that poetic language superimposes the method by which meanings and values are assigned onto syntactic placement in a sentence. This manipulation creates a new meaning dependent on values established in each linguistic system. Jakobson's principle is illustrated by Hawkes with the following example: When the word *beetle* is used in a sentence, it generally plays the grammatical role of a subject, object, or predicate nominative, as in the sentence: My car is a beetle. This expression represents an "ordinary" use of *beetle* in English. By contrast, a "poetic" use of *beetle* might project the relationship of equivalence suggested between the words *beetle* and *car* in the sentence above onto the syntax and grammatical structure of a sentence to suggest how the action of the car is like a beetle, as in the following: My car beetles along. This "poetic" use of the word *beetle* manipulates relationships among the assigned name of an insect in English, the conventional juxtaposition of subject and verb in English, and the assigned grammatical meaning of a verb in English to suggest the equivalence of *car* and *beetle* in a novel way (see Hawkes, 79; Jakobson, 370).

Critics of the formalist interpretation of *poetic* meaning as illustrated above argued that such analysis assumes that the meaning of words and grammatical systems is fixed, a notion denied by the poststructuralist, deconstructionist, and reader-response criticism that followed structuralism. But recent analysts of the structuralist position, of which formalism was a variant, have taken a kinder view. As Attridge claims in a retrospective review, Jakobson's formula did not specify that meaning was determined or fixed in the literary text. Rather his theory of the poetic function showed how meaning systems in language could be manipulated to *invite* "the reader to treat as equivalent items in the sequence which would have simply been contiguous in nonpoetic language" (21). In short, linguistic signs conventionally associated with certain kinds of meaning can be manipulated to invite the reader to interpret these meanings in another way. This perspective supports both the readers' autonomy in interpreting text and the possibility of socially constructed meaning in language.

In recent years, structuralist theory, particularly in its more social guise, has had its most complete explanation in the work of M. A. K. Halliday. Though American audiences are quite familiar with Halliday and Hasan's *Cohesion in English,* not enough are aware that Halliday's theory of cohesion is but one part of a more comprehensive theory of how language represents social meaning: systemic functional linguistics. Systemic functional linguistics aims to explain language in terms of choices that have "potential" for meaning within a structured system. The systemic approach to grammar elaborates Saussure's distinction between *langue* (linguistic competence) and *parole* (linguistic performance) by maintaining a link between social reality and linguistic structure. Saussure held that language both reflects the functions of social behavior within the community and changes over time to accommodate those functions. To account for the integral relationships between the structure of language and the structure of social relations, systemicists accordingly interpret language or language competence as "*linguistic behaviour potential* [or] the range of options from which a person's language and the culture to which he belongs allow him to select the range of possible things that he 'can do' linguistically" (Berry, 24). This view of language allows systemic structuralists to describe language genres, both oral and written, as subsets within a language that limit the speakers' choices in certain socially prescribed ways. Hence the scientific experimental article or, as in the case of "Spandrels," theoretical critique would be viewed by systemicists as a variety of language that restricts the range of linguistic possibilities that a speaker may employ to express meaning. Genres thus express a potential for meaning in themselves because the restricted range of linguistic features that they employ differs from the full range of possibilities within a language. Hence this difference in *value* within a system, as Saussure would put it, creates a scene for *meaning.*

Because language is viewed by systemicists as social behavior, the lexicon and grammar of languages are analyzed by systemicists as correlates of the social properties of a communication event in the culture where the language is spoken. The speaker's choice of one lexical or grammatical feature over another limits the meaning potential of the resulting expression in specific ways that are understood by speakers of the same language. Systemic grammarians have developed fairly elaborate networks that describe these choices. Those who follow Halliday's model generally assume that all languages express the potential for three kinds of meaning: ideational meaning, interpersonal meaning, and textual meaning.

Halliday's systemic grammar classifies *ideational* meaning as that which structures experience and logical relations.[1] The transitivity function of verbs and the naming function of the lexicon create the potential for ideational meaning, or meaning about our experience of the world ("what is going on") (Halliday, "Functions," 26). Halliday asserts that certain components of the sentence "can be thought of as representing the real world as it is apprehended in our experience" ("Functions," 19). For instance, words or grammatical markers can be classified as expressing actions that are performed or received or that represent mental or physical processes. Likewise, linguistic features naming persons can be said to identify doers, sayers, recipients of action, or some other participant role in human experience; such naming features can also identify attributes associated with these participants. Further, logical relationships between experiences so expressed can be identified through linguistic features expressing coordination, subordination, equivalence, and other kinds of connections. Halliday's gloss of the verse "Or leave a kiss within the cup, and I'll not ask for wine" serves as a typical example of how systemicists account for ideational meaning in language. As he explains, in this text the first verb, *leave*, expresses a physical process, while the second verb, *ask*, expresses a mental process. Surrounding these mental-process and physical-process verbs are elements that identify doers (*you* and *I*), goals *(kiss)*, and locatives *(in cup)*. The first clause is logically connected to the second by the relationship *if . . . then* ("Functions," 18–21).

Within Halliday's scheme, *interpersonal* meanings are defined as those expressed through linguistic features that "establish and maintain social relations." Through the interpersonal function "social groups are delimited, and the individual is identified and reinforced" ("Language Structure," 143). The grammatical features of mood, modality, and person express potential for interpersonal meaning, or meaning about the tenor of discourse ("who are taking part") ("Functions," 26). Hence, in the example in verse, the verb *leave* might be classified as expressing a physical process ideationally and as expressing a demand made by the speaker of his audience interpersonally. The latter interpretation is based on the elliptical presence of the second person pronoun *you*, signalling the imperative mood. Further, the verb *ask* may be interpreted as a mental process ideationally and as a voluntary activity on the speaker's part interpersonally (as reflected in the use of the first person pronoun *I* and the modal *will*).

Finally, *textual* meaning for Halliday is expressed through those features of language by which it makes "links with itself and with features of the situation in which it is used" ("Language Structure," 143). Textual theme and cohesion are among the grammatical features that express potential for meaning about the "mode of discourse" or the "role assigned to language" in a particular communication event ("Functions," 26). Again returning to the well-known verse by Ben Jonson, the speech function of "request followed by an offer" is highlighted textually through a repeated pattern: "Drink to me only with thine eyes [request] / And I will pledge with mine [offer] / Or leave a kiss within the cup [request] / And I'll not ask for wine [offer]." A cohesive link between these elements is also formed through repetition of the pronoun *I*.

Systemic structural analysis of text has attracted a number of scholars looking for ways to describe written language in terms of its function. Not all systemic analyses emphasize a meticulous correspondence between linguistic features of a text and Halliday's grammatical scheme for interpreting meaning potential. In fact, scholars have interpreted the relationship between Halliday's three semantic functions and the lexical and grammatical features of a text in very diverse ways. Though the varieties of analytic schemes that claim to have origins in systemic theory are quite eclectic, most incorporate these general principles:

1. Language expresses three semantic functions, which correspond to the situational context: the ideational function (which expresses meaning about the way we experience the world in terms of actions and participants and the logical relations between events), the interpersonal function (which expresses meaning about the social relationship between a speaker and an audience), and the textual function (which expresses meaning about how parts of a text relate to each other, to the context, and to other texts).

2. These three semantic functions are communicated through a wide range of grammatical/lexical features, the chief among them being these: a. verbal transitivity, naming, logical connectors (express ideational meaning); b. mood, modals, person (express interpersonal meaning); c. informational theme, cohesion (express textual meaning).

3. Texts can be described in terms of how they represent language

genres, or ways of speaking within a language. This involves
identifying a range of choices within the grammar and lexicon of
a language that correspond to ideational, interpersonal, and textual functions that are invoked in certain repeated situations.

Analytic schemes revealing these three properties have been devised
by systemicists to demonstrate many social functions of both oral and
written English. Some applications to nonliterary writing include analysis of the ideational, interpersonal, and textual functions that characterize reporting, narrating, and evaluating genres (see Martin and
Rothery); that characterize valued academic writing (see Peters); and
that characterize scientific research reporting as opposed to popular
writing (see Francis and Kramer-Dahl). In most cases, the researcher
completing the analysis has attempted to identify the social function
achieved by a variety of writing (e.g., social recognition as a variety of
writing, social recognition as an exemplary form of writing, social recognition as a form of writing expressing professional ethos). Many but
not all systemic analyses are limited to lexical and grammatical descriptions at the sentence level. Some apply to conversational exchange structures (i.e., the patterns of talking between participants in a conversation), to macrostructuring devices (i.e., the patterns for organizing
sequences of discourse larger than a sentence), and to clause relations
(i.e., the logical relations established across clauses to create textual
cohesion and coherence). Systemic analyses of discourse also can range
from highly organized computer-generated networks of meaning systems that are invoked by a single sentence to fairly discursive interpretations of how certain textual features convey ideational, interpersonal, or textual meanings in a given essay or poem.

The applications of systemic analysis to "Spandrels" that I will be
presenting in this essay illustrate some of the interpretive range described above; they also demonstrate the systemicist premise that linguistic features have "generic" potential to convey meanings about real
world experiences and relationships between them (ideational function); about relationships between authors and readers (interpersonal
function); and about relationships among parts of a text, a text and a
context, and a text and other texts (textual function). In particular, the
"Spandrels" analysis shows how its linguistic features identify it as a
genre of scholarly writing—the theoretical critique. Through its specific
ideational, interpersonal, and textual functions, this genre highlights
"ways of knowing" that are characteristic of scholarship in the humanities and that oppose the scientific method, the epistemology most commonly forwarded in the genre of scientific reporting.

The Written Genres of Scientific Reporting and Critiquing

Gross distinctions between the social values of science and other scholarly disciplines have been articulated popularly by C. P. Snow in his famous essay "The Two Cultures." In the expanded version of his argument, published in 1963, Snow makes several distinctions between humanistic inquiry (characterized by literary intellectuals at one extreme) and scientific inquiry (represented by physical scientists at the other). According to Snow, humanistic inquiry is skeptical, critical, elitist, and antagonistic to technology. Humanists, that is, have little faith (perhaps too little) in the power of social forces to improve the condition of man, are apt to make distinctions between abilities based on individual judgment, and have little respect for the power of technology to advance the human condition. Scientists, in contrast, characteristically have an optimism (perhaps naive) about the ability of society to work together to improve the human condition; they judge arguments on the basis of precise definitions of what is objective or subjective, and they are said to understand much of the world in terms of tools that can be used to increase knowledge. Books, which for humanists explore the vast divergence of perspectives on "psychological or moral or social life," for scientists are yet more "tools," instruments used along with others to assist one in meeting some predefined end (Snow 18–19). Though Snow's argument is somewhat facile and often contentious, it consistently follows the structuralist principle of identifying an element of contrast that allows us to assign meaning to the two concepts: scientific and humanistic inquiry.

Refinements of this difference have continued to be developed, as well as arguments for its existence. Scholars of the sociology of science have claimed that the scientific community's acceptance of the epistemology of the scientific method is based on four commonly held "moral imperatives" (Merton, quoted in Stehr, 174): *universalism,* or the acceptance or reception of scientific claims according to "impersonal cognitive criteria"; *communism,* or the communal agreement among scientists not to withhold information; *disinterestedness,* or the conflict over individual motivation; and *organized skepticism,* or the social agreement to subject knowledge to common scrutiny against "technical norms" (Stehr, 174). The lexical and grammatical features of text that have been traditionally associated with these values include:

1. Use of the passive (disguises agency; promotes "objective stance");

2. Avoidance of evaluative language, that is, language that expresses judgment *not* subject to "technical norms";
3. Use of declarative statements (declaratives oppose imperatives and interrogatives, which assert interpersonal criteria for acceptance rather than "impersonal cognitive criteria");
4. Reliance on third-person pronouns (avoids reference to self or audience as individuals invested in an outcome, thus expressing "disinterestedness"); and
5. Use of problem-solution, general to particular, and enumerative organization patterns (these strategies all reinforce the epistemology of the scientific method—the bastion of "organized skepticism").

By contrast, humanistic inquiry is characterized by subjective criteria, contentious disagreement, eclectic skepticism, and individual autonomy. Rorty offers a wry look at this difference, claiming that humanists, unable to muster a common analytic method that gives their disciplines " 'cognitive status' without the necessity of discovering facts" (38), are forced to "either describe themselves as concerned with 'value' as opposed to facts, or as developing and inculcating habits of 'critical reflection' " (38–39). Rorty notes that humanists are placed in this untenable position because rationality is associated with the scientific method; hence, "to be rational means to be able to lay down criteria in advance" of making a judgment (40). This stipulation, of course, is antithetical to most humanistic inquiry, which invites critics, artists, and poets to "make up new standards of achievement as they go along" (39).

Rorty suggests that we adopt "another meaning" for rationality, one that associates it with "a set of moral virtues: tolerance, respect for the opinions of those around one, willingness to listen, reliance on persuasion rather than force" (40). In applying this " 'weaker' conception of rationality" (41), there need not be a polar distinction between humanistic inquiry and the sciences: "On this construction, to be rational is simply to discuss any topic—religious, literary, or scientific—in a way which eschews dogmatism, defensiveness, and righteous indignation" (40). Despite Rorty's argument, I would argue that scholarship in the humanities is more tolerant of critique that authoritatively asserts an opinion, is putatively defensive, and expresses acerbic indignation—all positions not far from the "morally" incorrect positions Rorty abhors. These techniques are socially accepted and even admired as evidence of the quick wit and "smart sense" of an author who believes polite disagreement is an appropriate response to faulty or wrong-headed thinking. The humanist critique, in short, far from preserving the scientist's ideal of dispassionate proof, is more likely to honor the intellectual prowess of the individual who dares to challenge the status quo. Writers can signal their intention to uphold these humanist values and develop a critique by choosing linguistic features that are associated with the values of individual identity and evaluation. Among them:

1. Use of active voice, with subjects of actions clearly identified (features agency—that is, who is doing something, saying something, or claiming to do or say something);
2. Use of evaluative language that expresses personal judgment (emphasizes the value of intuitive thinking and individual point of view);
3. Use of rhetorical questions, imperatives, and subjunctives (involves the author and reader in an exchange of points of view, emphasizes the persuasive power of discourse);
4. Use of first- and second-person pronouns (identifies the author's perspective and signals an intention to involve the audience directly); and
5. Use of organizational patterns that move from particular to general or develop a point discursively (emphasizes reflective development of an idea as opposed to proof of a preestablished hypothesis).

The oppositional epistemologies of scientific and humanistic inquiry can be invoked by writers in either sphere to call into question assumptions about how reality is interpreted. Gould and Lewontin employ this technique of "arguing from another place" to challenge a claim that has resisted refutation through strict application of the scientific method. In doing so, they call attention to the fact that they are not simply contributing knowledge to the profession; rather, they are shaking the foundation upon which that knowledge has been built. Their criticism of an agreed-upon truth must test the social meaning of "tolerant disagreement" as it contrasts with "unforced agreement" in the scholarly pursuit of truth (Rorty, 48). In fact, Gould and Lewontin must reject the stance of "tolerant disagreement" altogether if their goal is to persuade the scientific community to regard its former agreement as an error.

To do its job, a critique like "Spandrels" must accomplish two aims: to challenge the grounds upon which social agreement has been attained (in the case of science, this would be to challenge claims of validity founded in scientific reasoning) and express intolerant disagreement, which by virtue of its intolerance calls for a new basis of agreement. Thus critiquing—in contrast to theorizing or reporting em-

Barbara Couture

pirical research in the arts or the sciences—must undermine assumptions and provoke response. The analysis that follows is designed to show how this is accomplished.

A thorough examination of text by means of linguistic analysis can threaten to produce a report of encyclopedic proportions. To keep my systemic structural analysis of "Spandrels" manageable, I have chosen to examine three portions of continuous text: a passage introducing the aim of the paper, a passage summarizing its subject of critique, and a passage concluding its argument (figure 14.1). I also have analyzed the title, subtitle, and topic sentences following subtitles (figure 14.2). To demonstrate how "Spandrels" constitutes a variety of scientific writing that differs from others in the same genre, I shall compare it in some instances with "Game Theory and the Evolution of Behaviour" by J. Maynard Smith, printed in the same issue of the *Proceedings of the Royal Society*. This text is also a critique, but it is a less extreme example of the form's function of challenging existing social values. The passages selected for analysis from "Game Theory" have the same organizational functions as those selected from the "Spandrels" text (figures 14.3 and 14.4). My analyses examine three kinds of structures: lexical and grammatical features, participant-exchange structures, and macrostructuring devices. Each of these varieties of structural devices is examined with reference to the ideational, interpersonal, and textual functions as defined in Halliday's systemic theory.

A. Passage introducing purpose of paper

We wish to question a deeply engrained habit of thinking among students of evolution. We call it the adaptationist programme, or the Panglossian paradigm. It is rooted in a notion popularized by A. R. Wallace and A. Weismann (but not, as we shall see, by Darwin) towards the end of the nineteenth century: the near omnipotence of natural selection in forging organic design and fashioning the best among possible worlds. This programme regards natural selection as so powerful and the constraints upon it so few that direct production of adaptation through its operation becomes the primary cause of nearly all organic form, function, and behaviour. Constraints upon the pervasive power of natural selection are recognized of course (phyletic inertia primarily among them, although immediate architectural constraints, as discussed in the last section, are rarely acknowledged). But they are usually dismissed as unimportant or else, and more frustratingly, simply acknowledged and then not taken to heart and invoked. [154 words]

B. Passage summarizing subject of critique

Studies under the adaptationist programme generally proceed in two steps:

(1) An organism is atomized into 'traits' and these traits are explained as structures optimally designed by natural selection for their functions. For lack of space, we must omit an extended discussion of the vital issue: 'what is a trait?' Some evolutionists may regard this as a trivial, or merely a semantic problem. It is not. Organisms are integrated

entities, not collections of discrete objects. Evolutionists have often been led astray by inappropriate atomization, as D'Arcy Thompson (1942) loved to point out. Our favourite example involves the human chin (Gould 1977, pp. 381–382; Lewontin 1978). If we regard the chin as a 'thing', rather than as a product of interaction between two growth fields (alveolar and mandibular), then we are led to an interpretation of its origin (recapitulatory) exactly opposite to the one now generally favoured (neotenic). [146 words]

C. Passage concluding the argument

We feel that the potential rewards of abandoning exclusive focus on the adaptationist programme are very great indeed. We do not offer a council of despair, as adaptationists have charged; for non-adaptive does not mean non-intelligible. We welcome the richness that a pluralistic approach, so akin to Darwin's spirit, can provide. Under the adaptationist programme, the great historic themes of developmental morphology and *Bauplan* were largely abandoned; for if selection can break any correlation and optimize parts separately, then an organism's integration counts for little. Too often, the adaptationist programme gave us an evolutionary biology of parts and genes, but not of organisms. It assumed that all transitions could occur step by step and underrated the importance of integrated developmental blocks and pervasive constraints of history and architecture. A pluralistic view could put organisms, with all their recalcitrant, yet intelligible, complexity, back into evolutionary theory. [145 words]

Figure 14.1. Selected passages from "The Spandrels of San Marco and the Panglossian Paradigm"

The Spandrels of San Marco and the Panglossian Paradigm
1. Introduction

The great central dome of St Mark's Cathedral in Venice presents in its mosaic design a detailed iconography expressing the mainstays of Christian faith.

2. The adaptationist programme

We wish to question a deeply engrained habit of thinking among students of evolution.

3. Telling stories

'All this is a manifestation of the rightness of things, since if there is a volcano at Lisbon it could not be anywhere else.'

4. The master's voice re-examined

Since Darwin has attained sainthood (if not divinity) among evolutionary biologists, and since all sides invoke God's allegiance, Darwin has often been depicted as a radical selectionist at heart who invoked other mechanisms only in retreat, and only as a result of his age's own lamented ignorance about the mechanisms of heredity.

5. A partial typology of alternatives to the adaptationist programme

In Darwin's pluralistic spirit, we present an incomplete hierarchy of alternatives to immediate adaptation for the explanation of form, function, and behaviour.

6. Another, and unfairly maligned, approach to evolution

In continental Europe, evolutionists have never been much attracted to the Anglo-American penchant for atomizing organisms into parts and trying to explain each as a direct adaptation.

Figure 14.2. Title and subtitles, with sentences following, from "The Spandrels of San Marco"

Barbara Couture

A. Passage introducing purpose of paper

The aim of this paper is to discuss how far the application of game theory, and in particular the concept of an 'evolutionarily stable strategy' is useful in interpreting animal contests. By 'interpreting' I mean giving an account of the selective forces responsible for the evolution of fighting behaviour: the short-term developmental and physiological mechanisms call for a different type of explanation. That is, I am seeking a 'functional' rather than a 'causal' explanation. I am not concerned to show that any particular model—for example the Hawk-Dove game—is an adequate account of any particular case, because clearly it is not and was not intended to be. Rather, I ask whether certain general predictions of the game theory approach are borne out, and whether there are categories of observation that cannot at present be understood. [135 words]

B. Passage summarizing subject of critique

The following three general predictions follow from game theory models of evolution:
  (i) In symmetric contests, we expect to find 'mixed' strategies.
  (ii) In asymmetric contests, we expect the asymmetry to be used as a cue to settle the contest.
  (iii) We do not expect contestants to exchange information about their intentions, although they certainly exchange information about their size and fighting ability.

These predictions are now derived and explained in terms of a simple model. Suppose a contest takes place between two individuals over some resource. Individuals can adopt two 'strategies': H (Hawk), i.e., 'fight until victory or defeat', and D (Dove), i.e. 'display, but retreat before being hurt if one's opponent escalates'. The changes in fitness after a contest can be expressed in a 'payoff matrix', which might be as in table 1. [136 words]

C. Passage concluding the argument

Evolutionary game theory is, I believe, the appropriate framework for a functional interpretation of contest behaviour. There is no shortage of problems, but progress has been rapid. There are, however, some general criticisms of the approach that should be discussed. . . . [40 words] In addition to attempting a general formulation for the analysis of contests, we proposed that a specific strategy, namely 'retaliation', was likely to be a component of evolutionarily stable behaviour. I think that this conclusion is correct (despite some mathematical difficulties; see Gale & Eaves 1975), but it had been stated quite explicitly by Geist (1966). In a later analysis of contest behaviour in ungulates, V. Geist has given the concept of retaliation a more central position and has extended his earlier argument in an attempt to relate the types of weapons and behaviour to the ecology of different species. Had there been space, the role of retaliation would have been treated here as a fourth prediction of game theory; it was omitted only because it is observationally less controversial. [130 words]

Figure 14.3. Selected passages from "Game Theory and the Evolution of Behaviour"

Game Theory and the Evolution of Behaviour
Introduction

The theory of games was applied to the evolution of fighting behaviour (Maynard Smith & Price 1973) in the hope of explaining the conventional aspects of this behaviour (Lorenz 1966 and earlier) in terms of individual selection.

The predictions of game theory

The following three general predictions follow from game theory models of evolution: [three points follow]

Are there mixed e.s.s.'s?

Contest behaviour is often very variable, but this does not prove that we are looking at a mixed e.s.s.

Are asymmetries taken as clues?

The most contra-intuitive prediction of game theory is that 'ownership' may be taken as a cue to settle contests, even if it does not affect the likelihood of winning an escalated fight or the payoff for winning.

Is information transferred during contests?

This question, and the related ones of bluff and lying, are the main problem areas for a game theoretical analysis of contest behaviour.

Discussion

Evolutionary game theory is, I believe, the appropriate framework for a functional interpretation of contest behaviour.

Figure 14.4. Title and subtitles, with sentences following, from "Game Theory"

The Lexical and Grammatical Features of "Spandrels"

An analysis of the lexical and grammatical features of "Spandrels" shows that it exploits the critique genre by provoking a response from the reader in two of the three areas of meaning identified by Halliday: the ideational and the interpersonal. In representing experience (the ideational function), "Spandrels" relies on lexical features that express judgments about persons and events, and it contains a high proportion of mental-process verbs that emphasize the evaluative stance of the authors. In expressing a relationship between the authors and readers of the text (interpersonal function), "Spandrels" highlights the dubious claims of those who support the theory the authors oppose, baiting the audience into discontent with the opposition's view. Further, features achieving these functions are more pronounced in the "Spandrels" text selections than in comparable selections from the "Game Theory" critique by Maynard Smith.

Table 14.1 names the lexical and grammatical features that communicate ideational and interpersonal functions (column 1). Those specific features that signal adherence to the social values of the scientific method are shown in column 2, and those that oppose these values and provoke critique are shown in column 3. Table 14.2 lists all the lexical and grammatical features identified in table 14.1 that support a

Barbara Couture

Table 14.1. Lexico-grammatical features expressing ideational and interpersonal functions

Lexico-grammatical categories	Specific features supporting scientific text	Specific features provoking critique of scientific text
Ideational function Naming	Lexical items that express evaluation on the basis of technical norms	Lexical items that express arguable judgments in themselves or in specific contexts
	Definitions that are standard	Idiosyncratic definitions
	Lexical items that correspond to standard definitions	Idiosyncratic definitions
Transitivity (mental-process verbs)	Passive voice—hidden agency ("doer" of the action)	Active voice—agency ("doer") specifically identified
Interpersonal function Person	3rd Person Definite references, establishing impartial reference to persons, places, and things	1st Person Definite references, identifying author; 1st Person Limited references, identifying author and exclusive group of others; 1st Person Indefinite references, identifying author and inclusive group of others; 2nd Person Definite, Limited, and Indefinite references, appealing directly to reader
Phatic moves (signal beginning, ending, and continuing of the discourse for the reader)	Introductory problem statement, concluding statement, identification of and/or enumeration of discussion	Asides to audience
Mood	Declarative	Imperative, interrogative, or exclamatory
Modals	Modals employed to identify tense only	Modals used to imply social obligation

Table 14.2. Lexico-grammatical items supporting critique genre in "Spandrels" and "Game Theory"

Lexico-grammatical feature	"Spandrels"				"Game Theory"			
	Passages		Titles		Passages		Titles	
	No.	%	No.	%	No.	%	No.	%
Ideational function								
Lexical items expressing arguable judgments								
In context	11	2	8	4	4	1	1	1
Standing alone	55	12	12	6	22	5	10	6
Idiosyncratic terms	7	2	6	3	4	1	1	1
Idiosyncratic definitions	1	0	0	0	1	0	0	0
Active mental process	18	4	10	5	17	4	5	3
Subtotal	92	21	36	18	48	11	17	10
Interpersonal function								
Personal references								
1st Person Definite	7	2	2	1	6	1	1	1
1st Person Limited	1	0	0	0	0	0	0	0
1st Person Indefinite	2	0	0	0	3	1	0	0
2nd Person Definite	0	0	0	0	0	0	0	0
2nd Person Limited	0	0	0	0	0	0	0	0
2nd Person Indefinite	0	0	0	0	0	0	0	0
3rd Person Definite	7	2	7	3	5	1	3	2
3rd Person Limited	3	1	2	1	10	2	0	0
3rd Person Limited	0	0	1	0	0	0	0	0
Subtotal for personal references	20	4	12	6	24	5	4	2
Modals (including modals that imply or create social obligation)	9	2	0	0	6	1	3	2
Mood and other devices								
Imperatives	0	0	0	0	1	0	0	0
Rhetorical questions	1	0	0	0	0	0	3	2
Exclamations	0	0	0	0	0	0	0	0
Indirect questions	0	0	0	0	0	0	0	0
Quotations	0	0	1	0	0	0	0	0
Indirect quotations	2	0	0	0	0	0	0	0
Asides	2	0	1	0	1	0	0	0
Total provocative features	126	28	50	25	80	18	27	16
Total words	445	100	203	100	441	100	171	100

Note: All counts show number of instances; features in more than one category are counted only once. Percentages shown in columns may not add up because of rounding.

critique, providing a quantitative analysis of how these features are realized in equal and comparable portions of the "Spandrels" and "Game Theory" texts.

The "Spandrels" text passages (totaling 445 words), when contrasted with the "Game Theory" text passages (totaling 441 words), contain far more provocative features—that is, linguistic signals that invite the reader to see this text as opposing the social values of scientific reporting and supporting the critique genre. As table 14.2 shows, in expression of ideational content, 21 percent of the "Spandrels" text passages (figure 14.1) contain provocative features that name the speaker's evaluative judgments, as opposed to 11 percent of the "Game Theory" text passages (figure 14.3). Similarly, in the titles, subtitles, and first sentences of "Spandrels" (figure 14.2), 18 percent of the features are provocative, as opposed to 10 percent of the "Game Theory" titles, subtitles, and first sentences (figure 14.4). Further, the "Spandrels" selections contain more idiosyncratic terms (such as "adaptationist programme," "Panglossian paradigm," and "atomization")—terms that indicate the authors' individualist stance—than do the "Game Theory" passages and titles. Overall, a gross analysis of portions of both texts reveals that 28 percent of the "Spandrels" text passages and 25 percent of its titles contain provocative features, as opposed to 18 percent of the "Game Theory" passages and 16 percent of its titles (table 14.2). This difference is notable, but not entirely startling. Both texts contain a healthy proportion of provocative features, supporting their allegiance to the genre of theoretical critique. What may be of greater interest, however, is the difference between features that upon gross analysis look the same, but with detailed study appear different. For instance, the gross analysis shows little difference between the two works in the proportion of personal references to total text and in the use of modals that imply or create social obligation. Though hardly large enough to prove significant, the "Spandrels" passages and titles contain two indirect quotations and three asides, appealing to the audience, as opposed to one aside in the "Game Theory" selections. The difference in the impact of these features on the interpersonal functioning of the two texts becomes clear if we examine them in the context of surrounding text.

Look, for instance, at the category "3rd Person" personal references. I included this category, which does not ordinarily support provocative critique, because differences in the contexts in which these features are placed in "Spandrels" and "Game Theory" demonstrate how "Spandrels" pushes the edge of the critique genre. Of the seven "3rd Person Definite" personal references in "Spandrels," all are cited within the context of opposing points of view: 1. "It is rooted in a notion popularized by A. R. Wallace [1] and A. Wiesmann [2] (but not as we shall see by Darwin) [3]"; 2. "Evolutionists have often been led astray by inappropriate atomization as D'Arcy Thompson [4] (1942) loved to point out. Our favorite example involves the human chin (Gould 1977, pp. 381–382 [5]; Lewontin 1978 [6])"; 3. "We welcome the richness that a pluralistic approach, so akin to Darwin's [7] spirit, can provide." Two of the references are to the authors themselves in parenthetical citation, documenting previous publications opposing atomistic analysis; and the two references to Darwin invoke the spirit of the father of evolution himself to oppose the adaptationist program.

In the "Game Theory" sample, by contrast, the five "3rd Person Definite" personal references are all made in support of the author's argument, with none citing the author himself:

> I think that this conclusion is correct (despite some mathematical difficulties; see Gale [1] & Eaves [2] 1975), but it had been stated quite explicitly by Geist [3] (1966). In a later analysis of contest behaviour in ungulates, V. Geist [4] has given the concept of retaliation a more central position and has extended his [5] earlier argument. . . .

With the references to Gale and Eaves the author cites a difficulty with his favored argument, but it is one, he suggests, that can be resolved. Rather than pointing to unresolvable oppositions, he makes an attempt to reconcile all points of view on the topic.

Further, the three asides appearing in the "Spandrels" selections are designed not only to emphasize opposition between the authors and the supporters of the adaptationist program but also to encourage audience discontent. Gould and Lewontin claim that "of course" the adaptationists recognize "constraints upon the pervasive power of natural selection" and name these acknowledgments parenthetically. The implication is "how could not" the adaptationists acknowledge at least these constraints without appearing credulous. The authors also claim "for lack of space" to omit a "discussion of the vital issue: 'what is a trait'?" This cursory reference to an issue not discussed faults the very premise of adaptationist theory to explain organic functioning in terms of distinguishable traits. The authors again use an aside to emphasize the tendency of evolutionary biologists to honor Darwin's foundational views, quipping that "Darwin has attained sainthood (if not divinity) among evolutionary biologists." This aside directly challenges biologists to more closely examine the bases upon which premises are accepted. In contrast, the one aside that appears in the selected text from "Game Theory" simply notes the author's omission of a point not deemed relevant because no disagreement is associated with it: "Had there been space, the role of retaliation would have been treated . . . ; it was omitted only because it is observationally less controversial." The author offers this aside to assure that the audience remains content that he has covered all bases in presenting his argument. The asides in "Spandrels" create exactly the opposite effect: they raise issues about which the audience is likely to feel particularly touchy, thus ensuring an atmosphere of curious discontent. In short, Gould and Lewontin push the critique genre to limits more acceptable in humanistic argument than in the sciences by drawing clear boundaries between opposing points of view and exposing the individuals responsible for those perspectives.

Another feature that through closer analysis reveals a contrast between the social functions signalled in "Spandrels" and "Game Theory" is the use of idiosyncratic definitions. Ad hoc or idiosyncratic definitions highlight the text's ideational function of representing experience from the evaluative perspective of the author. In "Game Theory," the author redefines "interpreting" in the phrase "interpreting animal contests" to mean "giving an account of the selective forces responsible for the evolution of fighting behavior." This definition simply narrows the scope of what we commonly accept as interpretive activity. In contrast, here is how Gould and Lewontin redefine the adaptationist interpretation of evolution: "We call it the adaptationist programme, or the Panglossian paradigm. It is rooted in a notion popularized . . . towards the end of the nineteenth century: the near omnipotence of natural selection in forging organic design and fashioning the best among possible worlds." This wry "definition" is recognized as such because it employs the standard syntax for offering an definition; however, it substitutes for the primary term not a technical gloss, but rather an evaluative commentary: adaptationist theory is reconstrued as a program somewhat akin to a religious movement. This metaphor is extended by positing the "omnipotence of natural selection" in "forging" and "fashioning" the "best among possible worlds," an act attributed in other contexts to the Almighty. The irony of this association is exacerbated by the overlaid theme of Pangloss's folly. Gould and Lewontin's definition fails to apply standard criteria for defining contrasts and creates instead an imaginative alternative. Their technique of "defining" is closer to the humanists' method of inventing criteria as one goes than to the scientists' method of stating established criteria beforehand. In short, Gould and Lewontin not only make lexical and grammatical selections that signal the critique genre, but also they call attention to the social functions of this genre by using features ordinarily associated with straight scientific reporting for their own purposes, upsetting expectations and challenging accepted meanings.

The Participant-Exchange Structures of "Spandrels"

For my next look at "Spandrels," I selected a portion of the purpose paragraph of the article, that is, the paragraph that introduces the authors' argument against adaptationist theory as it will be developed in this text. My aim in analyzing the participant exchanges of the purpose paragraph is to show how the text sets up an expectation for reader response in all three of Halliday's semantic domains: the ideational, the interpersonal, and the textual.

My discussion refers to the analysis displayed in table 14.3. Column 1 of the table ("Text") shows a clause-by-clause breakdown of the text sample. Column 2 displays for each clause the textual information of the discourse (TD) and following that some questions that can be asked of this information (TQ); the questions elicit related information in the next clause. This "dialogic" technique of analyzing textual information is developed by Hoey in his 1983 text *On the Surface of Discourse*. Working with a theory of interclause relations developed by E. O. Winter, Hoey suggests that textual cohesion is perceived by readers when the information and logical relations expressed in clauses are anticipated by the information and logical relations expressed in previous clauses. One can clarify the ways in which clauses in monologue relate to one another (as do speakers' exchanges in conversation) by viewing "monologue as a specialised form of dialogue between the writer or speaker and the reader or listener" (Hoey, 27). To understand how one portion of a text responds to signals in another, one converts monologue to a controlled dialogue, "restricting the interlocutor's contributions to questions and requests for information" (28). In effect, one can establish whether the textual information is constructed cohesively through this questioning technique.

By following the series of questions and answers in the textual exchange column of table 14.3, we can conclude that the "Spandrels" sample communicates very effectively on the textual level, creating no gaps where the interlocutor's queries (TQ) are not satisfactorily answered in the text discourse following (TD). In an effort to prove a different case about the ideational and interpersonal functions of the discourse, I adapted Hoey's technique, assuming that the monologic text not only imitates dialogue in establishing textual relations but also does so in developing ideational relations between the message of the text and the message of other texts in the discourse field and interpersonal relations between the persona of the author and the persona of the reader. Hence, in columns 3 and 4 of table 14.3, I have created a display that captures

Barbara Couture

Table 14.3. Textual, ideational, and interpersonal exchange functions in purpose paragraph of "Spandrels"

Text	Textual exchange	Ideational exchange	Interpersonal exchange
We wish to question a deeply engrained habit of thinking among students of evolution.	TD: A "habit of thinking" is "questioned."	ID: A "deeply engrained habit of thinking" about "evolution" is to be "questioned."	PD: "We" have an authoritative position to take with you "students of evolution" who hold "deeply engrained" and "questionable" ideas.
	TQ: What is the "habit of thinking?" How is it "questioned?" Who has the "habit of thinking"? Where did it come from?	IQ: What ideas about "evolution" are "deeply engrained"? And what evidence is there that they are "habitually" endorsed? Who produced them? What evidence is there that they are "deeply engrained"?	PQ: I do/do not accept your authority and implication that I am a "student of evolution." Who are you to judge the "deeply engrained habits" of others? Further, why do you regard certain of us scholars of evolution as "students"?
We call it the adaptationist programme, or the Panglossian paradigm.	TD: The "habit" is the "adaptationist programme" or the "Pan. Parad." described above.	ID: The "adaptationist programme" is equated with the "Pan. Parad." and therefore both are improbable fictions.	PD: "We call" the "adapt. prog." the "Pan. Parad." to dub it as fanciful, so play along with us.
	TQ: What is the "adaptationist programme" and how did it become a "habit of thinking"?	IQ: How is the "adapt. prog.," as it is "called" here, described elsewhere? How is it analogous to the "Pan. Parad." as that is "called" here and described elsewhere? What criteria establish both as improbable fictions?	PQ: I do/do not "call" the "adapt. prog." by that name and I do/do not equate it with the "Pan. Parad." Consequently, I am pleased/displeased. I want to/will not play along.

(continued on the following page)

Table 14.3. Textual, ideational, and interpersonal exchange functions in purpose paragraph of "Spandrels" (*continued*)

Text	Textual exchange	Ideational exchange	Interpersonal exchange
It is rooted in a notion popularized by A. R. Wallace and A. Weismann . . . towards the end of the nineteenth century:	TD: The origin of the "adapt. prog." is an idea of "Wallace" and "Weismann."	ID: The "adapt. prog." is related to a "rooted" idea "popularized" by "Wallace" and "Weismann."	PD: We believe the idea is a mere "notion" that was "popularized" by "Wallace" and "Weismann" (whom we don't think much of) that has hung on beyond the "nineteenth century" to the present day.
	TQ: What is this idea? And how is it interpreted now? Who else promoted it in the "nineteenth century"?	IQ: What is the "rooted" idea and why is it to be regarded as "popularized"? As confirmed by what evidence in other texts, especially those of "Wallace" and "Weismann"?	PQ: You are right/ wrong about "Wallace" and "Weismann's" "notion"/theory. The "notion"/theory was/ was not indeed "popularized." I agree/ disagree that Weismann and Wallace were ne'er-do-wells.
. . . (but not, as we shall see, by Darwin) . . .	TD: "Darwin" did not promote it.	ID: The idea will not be found in the writings of "Darwin."	PD: Now that you are with us, together "we shall see" that "Darwin," not misled as were W. and W., had a better idea.
	TQ: What was the idea again? (What idea of "Darwin" will be discussed later?)	IQ: Again, what is the idea, where is it written, and how does it differ from one written elsewhere by "Darwin"?	PQ: I do/do not believe "we shall see." How will I "see"? I do/do not believe "Darwin" had another, perhaps better, idea. How will you convince me?
. . . the near omnipotence of natural selection in forging organic design and fashioning the best among possible worlds.	TD: "Natural selection" has created the "best among possible worlds."	ID: "Natural selection" is "near omnipotent" in making the "best among possible worlds."	PD: These adaptationists think "natural selection" is "near omnipotent" in "forging" and "fashioning the best among possible worlds," an idea we know to be fanciful.

(*continued on the following page*)

Barbara Couture

Table 14.3. Textual, ideational, and interpersonal exchange functions in purpose paragraph of "Spandrels" (*continued*)

Text	Textual exchange	Ideational exchange	Interpersonal exchange
	TQ: How does "natural selection" accomplish this? What is the "best among possible worlds"?	IQ: How is "natural selection" to be regarded as "near omnipotent" in the light of other written theory? What is defined in scientific literature as the "best of all possible worlds"?	PQ: You are right. This idea is not only hegemonic, but also fanciful! Or: You are wrong. This idea is democratic, fair, and sensible.
This programme regards natural selection as so powerful and the constraints upon it so few . . .	TD: "Natural selection" is "powerful" and has "few constraints."	ID: "Natural selection" is "powerful" and has "few constraints."	PD: These adaptationists "regard natural selection as so powerful" and with "so few constraints" that their idea is hardly credible.
	TQ: And what are the results?	IQ: What evidence is there of the "power" of "natural selection" and its "few constraints"? What are the results of this belief and do they concur with other written perspectives?	PQ: You are/are not exaggerating when you say that the adaptationists claim "natural selection" to be "so powerful" and "with so few constraints," but go on.
. . . that direct production of adaptation through its operation becomes the primary cause of nearly all organic form, function, and behaviour.	TD: "Adaptation" is the "primary cause of nearly all organic form, function, and behaviour."	ID: "Adaptation" is produced by "natural selection" and causes "organic form, function, and behaviour" to the exclusion of other causes.	PD: The adaptationists promote "natural selection" as "the primary cause of nearly all" organic development and this is not plausible.
	TQ: Any exceptions to this "primary cause"?	IQ: What other "cause" is documented? Does this statement of the adaptationists' position match others that are written?	PQ: The idea as you have stated it is/is not plausible. How are the adaptationists' related positions to be regarded?

(continued on the following page)

Table 14.3. Textual, ideational, and interpersonal exchange functions in purpose paragraph of "Spandrels" (*continued*)

Text	Textual exchange	Ideational exchange	Interpersonal exchange
Constraints upon the pervasive power of natural selection are recognized of course . . .	TD: "Constraints . . . are recognized."	ID: "Constraints . . . are recognized" in the adaptationist literature.	PD: The adaptationists "of course" see that "constraints . . . are recognized" (how could they not?), though they believe natural selection to have "pervasive power."
	TQ: What are the "recognized constraints"?	IQ: Are the "constraints" "recognized" in adaptationist reports as claimed?	PQ: I appreciate/do not appreciate the irony of this statement. The fact that adaptationists find "constraints" on the "pervasive power" of natural selection is/is not contradictory.
. . . (phyletic inertia primarily among them, . . .	TD: "Phyletic inertia" is one of the constraints.	ID: "Phyletic inertia" is established as a valid constraint in the adaptationist programme.	PD: We fault adaptationists for recognizing "phyletic inertia" (which as far as we're concerned is a given) as the only significant constraint, ignoring more important constraints.
	TQ: What other "constraints" are there?	IQ: Where is "phyletic inertia" validated as a constraint in other sources? Why is this constraint not regarded as sufficient?	PQ: I do/do not agree with your faulting the adaptationists for recognizing "phyletic inertia and few other constraints and am inclined/not inclined to agree with what else you have to say about "constraints" on natural selection.
. . . although immediate architectural constraints, as discussed in the last section, are rarely acknowledged).	TD: "Architectural constraints" exist and "will be discussed in the last section."	ID: "Immediate architectural constraints" are "rarely acknowledged" in adaptationist reports of evolution.	PD: "Architectural constraints" (we will tell you about them "later") are more "immediate" and yet the adaptationists "rarely acknowledge" them.

(*continued on the following page*)

Barbara Couture

Table 14.3. Textual, ideational, and interpersonal exchange functions in purpose paragraph of "Spandrels" (*continued*)

Text	Textual exchange	Ideational exchange	Interpersonal exchange
	TQ: What will be said now about "architectural constraints" and the consequences of not "acknowledging" them?	IQ: What are "architectural constraints" as defined in biologic science? How are "architectual constraints adknowledged" in adaptationist reports? Are they confirmed or unconfirmed as "immediate"?	PQ: I am sure/not sure that I wish to "acknowledge architectural constraints" or view them as "immediate" and will carry this view with me as I read the "last section." Meanwhile, what more are the adaptationists accused of?
But they are usually dismissed as unimportant or else, . . .	TD: They "are dismissed . . . or else"?	ID: They are dismissed as unimportant.	PD: Adaptationists "dismiss" the "important" consideration of architectural constraints and "what else" they do is worse.
	TQ: Or "what else"?	IQ: What evidence is there that these constraints are dismissed as opposed to rejected by sound criteria? Do other reports confirm these details as unimportant or do they agree with this text?	PQ: You are right; adaptationists make this mistake and what others do they make? Or: You are wrong; adaptationists do not dismiss important ideas, and why are you dismissing them?
. . . and more frustratingly, simply acknowledged and then not taken to heart and invoked.	TD: Architectural constraints are "acknowledged" and not "invoked."	ID: Architectural constraints must not only be "acknowledged" but also "invoked."	PD: We are "frustrated" by the adaptationists' failure to "invoke" architectural constraints as a check on the adaptationist imperative and feel it a personal failure that they have not "taken to heart" their obvious validity.
	TQ: And so what are the consequences?	IQ: Prove this idea has important consequences not cited elsewhere by criteria established in previous texts.	PQ: I have/have not "taken to heart" your criticism and as a result feel personally supported/dismissed.

these relations as a series of exchanges between the discourse and an imaginary interlocutor. Column 3 presents an imaginary dialogue that gives the "ideational" messages conveyed by the discourse (ID), followed by questions that might be asked of those messages by an interlocutor (IQ); and column 4 presents a parallel dialogue that represents the "interpersonal" relationship established by the discourse (PD) and questions that might be asked of that relationship (PQ).

The display reveals that relations between clauses that continuously develop textual meanings (or functions) somewhat discontinuously develop ideational and interpersonal meanings. In short, the message content of the text makes claims about messages of external texts without immediately substantiating these claims, and likewise the interpersonal tenor of the text assumes that readers accept the authors' superior authority without immediate evidence.

As one can see by following the imagined dialogue in columns 3 and 4, on the ideational and interpersonal levels the "Spandrels" purpose paragraph provokes far more questions than it answers. Further, the text features (in quotes) that elicit these questions include far more of those noted earlier as provoking critique. In the ideational domain, the "Spandrels" sample creates a tension between its claims and the claims of other texts on evolutionary biology as well as between its claims and the primary "text" that guides all scientific inquiry—the scientific method. In one statement after another, the sample puts forward arguments that are assumed to be supported by both kinds of contextual information. With its opening sentence, the purpose paragraph asserts that a questionable, "deeply engrained habit of thinking" is practiced by others. This statement's message immediately calls for verification: "Where is this questionable thinking documented? Who purports it? What evidence is there that it is 'deeply engrained'?" The text responds by naming the "habit of thinking" without providing evidence through citation that it exists, as is traditionally done in scientific reporting. Further, it equates the "habit" with an improbable fiction, the Panglossian paradigm.

The authors' attempt to develop a link between a claim of scientific theory and Voltaire's fictional tale creates dialogic chaos on the ideational plane. Not only does it beg the question of whether adaptationist theory is an improbable fiction, but it also equates the theory with a set of ideas to be evaluated for validity on the basis of criteria that lie outside the scientific paradigm. The message is: "Burn the books and forget what you believe; it was all fiction. And we have—get this—a new story to tell." By introducing the Pangloss metaphor, the authors suggest that humanist "logic" is at play here; at the same time Gould and Lewontin appeal to the audience either to play along or to stop while

Barbara Couture

they have the chance. Hence, the metaphor calls attention to the scientific community's belief in the infallibility of the scientific method by suggesting that at least one application of it has resulted in more fiction than fact.

On the interpersonal plane (column 4), the "Spandrels" text creates a tension between the level of authority that is assumed by the authors and the supposed authority of the audience. The tenor developed in the first sentence of the purpose paragraph is quite complex. At one level, the sentence boldly invites the reader to adopt the position of the author—both catholic and imperial—in avowing that "we wish to question. . . ." But on another level, the sentence implicitly suggests that the audience is less savvy than the authors would wish—mere "students of evolution." And on yet another level, the text invites its readers to regard themselves as far above ordinary ken, should they choose to agree with the authors' point of view and be unpersuaded by "deeply engrained" ideas.

A glance at the imaginary dialogue I have created to demonstrate the interpersonal relations that the purpose paragraph develops shows that the text can move the audience to stamp out adaptationist hegemony; after all, those demagogues support "the near omnipotence of natural selection in forging organic design and fashioning the best among possible worlds"! The commiserating reader is led to conclude: "You are right! This idea is not only hegemonic, but also fanciful!" (see table 14.3). Yet readers can be moved to take an opposite turn, provoked to ire by the very same features. In response to the opening claim, a reader might assert his or her own authority as is depicted in table 14.3: "I do not accept your authority and implication that I am a 'student of evolution.' Who are you to judge the 'deeply engrained habits' of others?" Coming to the next sentence of the paragraph, the reader must decide whether to be offended or amused: "We call it the adaptationist programme, or the Panglossian paradigm." Here the authors signal that this is their "call," that they intend to criticize and also to amuse, and that they are asking the readers to remove for a moment their professional hats and put on, say, their artists' berets. Even if readers disagree with the authors' assumptions, they risk branding themselves as being unwilling to "play"—a known criterion for all innovative thinking. Nonetheless, should readers rebel, yet continue to read, they will find themselves continually provoked to argue against the authors' claims. By the time readers reach the last clause shown in table 14.3, they will feel either like the authors' teammates or like odd men out, who must stick around to avoid losing by default.

In short, then, the textual, ideational, and interpersonal exchanges anticipated among clauses, between the discourse message and the messages of external discourses, and between the author and reader in the purpose paragraph of "Spandrels" all work together to prepare for the critique that follows. The cohesive interclause relations assure that at the textual level the critique will hang together and build a convincing argument. Like the internal frame of a building, the interclausal relations stabilize the structure against environmental assault. The complex intertextual relations developed on the ideational plan that relate "Spandrels" both to theories previously developed and to Voltaire's literary satire signal not only that the ideas presented here will counter previous arguments but also that they will upset the ground upon which those arguments have been built. And finally, the interpersonal tenor developed between the authors and the reader creates the tension of ambivalence: the reader must decide whether to side with the authors and debunk a traditional theory or reject the authors and be perceived as unwilling to play with new possibilities. In sum, the purpose paragraph is cleverly crafted to lay the groundwork for the provocative architecture that follows.

The Macrostructuring Devices of "Spandrels"

In its title and at several places in the text, "The Spandrels of San Marco" presents clear signals that it is to be considered as a critique and not a scientific report. At the same time, however, the text implicates scientific reporting as *the* genre that supports the tradition of the scientific method. A reminder of the authors' challenge to the adaptationists' application of the scientific method is forwarded in the macrostructuring devices of "Spandrels," which offer a subtle parody of the traditional macrostructuring of the research report.

Macrostructuring devices include titles, subtitles, lists, and the like. Expectations for structures such as these in scientific research reports are fairly standard. As Glassman and Pinelli explain, the research process in science and technology is systematic, the usual steps including the following: "state general problem," "conduct literature search," "state specific problem," "design methodology," "gather data," "analyze data," and "report results" (11). This process is idealized in the social milieu of science, as Latour and Woolgar have demonstrated in their sociological case study of laboratory science. Nevertheless, the scientific report is designed most often to support the myth of this standard methodology, with its sections generally ordered: 1. Intro-

Barbara Couture

duction, 2. Experimental Method, 3. Results, 4. Interpretation, 5. Discussion, and 6. Conclusion (see Bazerman, 175).

"Spandrels" points to the established organizational conventions of research reports as a way of highlighting its unconventional foray into critique. Following the standard Introduction (figure 14.2), the subtitles of "Spandrels" parallel the research report paradigm, yet develop a parody of the expected form. Instead of "2. Method" we have "2. The adaptationist programme." Instead of the expected "3. Results" we have "3. Telling stories," which by association with the expected subtitle casts suspicion on the adaptationists' questionable findings. Instead of "4. Interpretation" the authors invoke a higher authority—the unimpeachable Darwin in "4. The master's voice re-examined." Replacing the expected "5. Discussion" we get "5. A partial typology of alternatives to the adaptationist programme," an organized list of developed alternatives rather than the traditionally discursive discussion section. Finally, in place of "6. Conclusion," the reader finds "6. Another, and unfairly maligned, approach to evolution," which in essence begins a new argument rather than concluding one—a new argument that introduces, with Darwin's blessing, the concept of holistic interpretation of organic design.

Also interesting is the manner in which Gould and Lewontin use enumeration, a point also developed by Jeanne Fahnestock in chapter 9 of this book. Enumeration is conventionally used in scientific writing to highlight logical argument by listing points presented as evidence, indicating hierarchy of importance among ideas, or defining stages in a procedure. Figure 14.5 displays an instance of the use of enumeration in "Game Theory" alongside an instance taken from "Spandrels."

"Game Theory"

How is this apparent discrepancy between theory and observation to be reconciled? [on the interpretation of animal communication as a communication of 'intention']

No complete answer to this question can yet be given, because the data were often collected with no very clear functional question in mind. Some points are, however, worth making:

  (i) Much of the information transmitted concerns 'resource-holding potential' (r.h.p.) (Parker 1974b) and not intentions.
  (ii) Little information about long-term intentions is in fact transmitted.
  (iii) 'Lying' may be impossible, or may be punished.
  (iv) The advantage of a 'surrender' signal.
  (v) Communication is useful if resources are divisible.

"Spandrels"

The adaptationist programme can be traced through common styles of argument. We illustrate just a few; we trust they will be recognized by all:

(1) If one adaptive argument fails, try another.
(2) If one adaptive argument fails, assume that another must exist; a weaker version of the first argument.
(3) In the absence of a good adaptive argument in the first place, attribute failure to imperfect understanding of where an organism lives and what it does.
(4) Emphasize immediate utility and exclude other attributes of form.

Figure 14.5. Comparison of use of enumerated arguments in "Game Theory" and "Spandrels"

In "Game Theory" enumeration is employed to distinguish several reasons why discrepancies might appear to exist between game theory's interpretation of animal communication, as signalling "intent," and observed data on animal behavior. Each point covers contextual factors affecting the meaning of "intent" signals. These factors are expressed in the form of a principle (e.g., "Much of the information transmitted concerns 'resource-holding potential' [r.h.p.] [Parker 1974b] and not intentions"); and a reference supporting the principle follows immediately or shortly after. In "Spandrels," enumeration is used purportedly to list the "common styles of argument" among adaptationists. Yet after this introduction, the reader is presented not with a list of alternative styles, but with a single "non-procedure" for solving a problem. The procedure depicting the "style" of adaptationist argument lists one desperate probe after another, each repeating the same mistake made in the first move. The adaptationists are made to look a bit like the frustrated driver who can't get the car started and who nevertheless continues to crank the starter and flood the engine with gasoline, all the time failing to consider other causes of the problem. The authors play with the readers' expectations here in two ways: first, by presenting only one "style of argument" as opposed to the many they promised, they lead the reader to conclude that the adaptationists are monolithic in their approach; and second, by presenting an unsystematic circular process enumerated as a systematic procedure, they highlight the inability of the adaptationists to recognize their own tautologic reasoning.

Finally, let us look briefly at how the authors situate "Spandrels" among the other texts included in the Royal Society's discussion of "The Evolution of Adaptation by Natural Selection," organized by fellows Maynard Smith and Holliday. Figure 14.6 shows the titles of all ten contributions to the volume. A glance at the list even through Coke-bottle lenses would identify Gould and Lewontin's contribution as unusual. Its title is far longer than the others and is the only one with a colon, signaling its humanist bent (and perhaps a bit of duplicity as well). In place of straightforward terms like Maynard Smith's "Game Theory and the Evolution of Behaviour" we have the "Spandrels of San Marco" (secret code for "Architectural Constraints of Organic Design) and "the Panglossian paradigm" (code for "Adaptationist Theory"). The other titles are unremarkable by comparison. Though chapter 5 ("Arms Races between Species") also plays with a metaphor, the arms race image is consistent with traditional theories of "contest behavior" among animals.

---

1. Selection *in vitro*
2. Evolution of enzyme structure
3. The evolution of genetic diversity
4. Game theory and the evolution of behaviour
5. Arms races between and within species
6. The evolutionary genetics of sexual systems in flowering plants
7. Comparison and adaptation
8. The question of adaptive sex ratio in outcrossed vertebrates
9. The spandrels of San Marco and the Panglossian paradigm: a critique of the adaptationist programme
10. Introduction to general discussion

Figure 14.6. Titles of the ten contributions to "The Evolution of Adaptation by Natural Selection: A Discussion Organized by J. Maynard Smith, F. R. S., and R. Holliday, F. R. S."

In summary, the macrostructuring devices of "Spandrels" are designed to highlight its innovative and provocative character by suggesting meanings that contrast with the typical function of organizational devices in scientific reporting. By inviting the reader to compare its form with conventional forms, the text calls attention to its purpose—to critique—and to the function of conventions themselves. This latter strategy has the effect of implicating the adaptationist program as a misuse of the primary paradigm guiding all inquiry in science—the scientific method.

Conclusion

A structuralist analysis of "The Spandrels of San Marco and the Panglossian Paradigm" shows it to be artfully designed to disassemble the atomistic foundational premises of adaptationist theory and to clear the ground for a return to holistic interpretation of organic design, fully cognizant of the complementary roles of natural selection and architectural constraints. The features highlighted in my analysis provoke the reader to either critique the adaptationist position or risk being judged a dupe for "form without substance." They do so through evoking social meanings that stand in contrast to the "moral imperatives" of the scientific method. The "Spandrels" critique undermines the adaptationist program by equating its premise with the faulty philosophy of a renowned—though fictional—fool. Its scientific application to individual cases is thus made out to be seriously wrong-headed, a clear example of how the scientific method coupled with unimaginative thinking can go awry. Through intratextual, intertextual, and interpersonal exchanges among clauses within "Spandrels," between "Spandrels" and other texts, and between the authors and the reader, "The Spandrels of San Marco" builds the scene for critique. At the same time, it alters the foundation upon which ideas are developed in science—the scientific method—and openly invites the reader to respond.

All of the effects attributed to the structure of "Spandrels" are, of course, made possible only because its authors and audience share common ideas about how language works in the discipline of science. Structural analysis is simply one way to define the meanings that communities of authors and readers hold in common and thus to uncover the range of social functions that are conventionally invoked in scientific discourse.

NOTES

1. In *An Introduction to Functional Grammar,* Halliday articulates the "experiential" and "logical" functions of language as separate meaning systems. In the scheme that is outlined in this essay, the experiential and logical functions are combined under the descriptor "ideational" meaning, as defined in Halliday's original articulation of his theory. In my view, the original formulation is more compatible with common perceptions of the communication event as having three components: referential content, situational participants, and linguistic text.

WORKS CITED

Attridge, Derek. "Closing Statement: Linguistics and Poetics in Retrospect." In *The Linguistics of Writing: Arguments between Language and Literature,* ed.

Barbara Couture

Nigel Fabb, Derek Attridge, Alan Durant, and Colin MacCabe, 15–32. New York: Methuen, 1987.

Bazerman, Charles. *Shaping Written Knowledge: The Genre and Activity of the Experimental Article in Science.* Madison: University of Wisconsin Press, 1988.

Berry, Margaret. *Introduction to Systemic Linguistics: 2 Levels and Links.* New York: St. Martin's Press, 1976.

Couture, Barbara. "Attributing Value to Written Text: Some Implications for Linguistics and Interpretation." *Word* 40 (1989): 51–64.

Francis, Gill, and Anneliese Kramer-Dahl. "From Clinical Report to Clinical Story: Two Ways of Writing a Medical Case." In *Recent Systemic and Other Functional Views on Language,* ed. Eija Ventola, 339–68. Berlin: Mouton de Gruyter, 1991.

Glassman, Myron, and Thomas E. Pinelli. "Scientific Inquiry and Technical Communication: An Introduction to the Research Process." *Technical Communication* 32 (1985): 8–13.

Halliday, M. A. K. "Functions of Language." In *Language, Context, and Text: Aspects of Language in a Social-Semiotic Perspective,* ed. M. A. K. Halliday and Ruqaiya Hasan, 15–28. Victoria: Deakin University Press, 1985.

Halliday, M. A. K. *An Introduction to Functional Grammar.* London: Edward Arnold, 1985.

Halliday, M. A. K. "Language Structure and Language Function." In *New Horizons in Linguistics,* ed. John Lyons, 140–65. Baltimore: Penguin, 1970.

Halliday, M. A. K., and Ruqaiya Hasan. *Cohesion in English.* London: Longman, 1976.

Hawkes, Terence. *Structuralism and Semiotics.* Berkeley: University of California Press, 1977.

Hoey, Michael. *On the Surface of Discourse.* London: Allen and Unwin, 1983.

Jakobson, Roman. "Closing Statement: Linguistics and Poetics." In *Style in Language,* ed. Thomas A. Sebeok, 350–77. Cambridge: MIT Press, 1960.

Latour, Bruno, and Steve Woolgar. *Laboratory Life: The Social Construction of Scientific Facts.* Princeton: Princeton University Press, 1986.

Martin, James R., and Joan Rothery. "What a Functional Approach to the Writing Task Can Show Teachers about 'Good Writing.'" In *Functional Approaches to Writing: Research Perspectives,* ed. Barbara Couture, 241–65. Norwood, N.J.: Ablex, 1986.

Maynard Smith, John. "Game Theory and the Evolution of Behaviour." *Proceedings of the Royal Society of London B: Biological Sciences* 205 (1979): 475–88.

Merton, Robert K. "The Normative Structure of Science." In *The Sociology of Science,* ed. Norman W. Storer, 267–79. Chicago: University of Chicago Press, 1973.

Myers, Greg. "The Rhetoric of Irony in Academic Writing." *Written Communication* 7 (1990): 419–55.

Peters, Pamela. "Getting the Theme Across: A Study of Dominant Function in the Academic Writing of University Students." In *Functional Approaches to*

Writing: Research Perspectives, ed. Barbara Couture, 169–85. Norwood, N.J.: Ablex, 1986.

Rorty, Richard. "Science as Solidarity." In *The Rhetoric of the Human Sciences: Language and Argument in Scholarship and Public Affairs,* ed. John S. Nelson, Allan Megill, and Donald N. McCloskey, 38–52. Madison: University of Wisconsin Press, 1987.

Saussure, Ferdinand de. *Course in General Linguistics* [1972]. Trans. Roy Harris. London: Duckworth, 1983.

Snow, C. P. *The Two Cultures: And a Second Look.* London: New American Library, Mentor Book, 1963.

Stehr, Nico. "The Ethos of Science Revisited." *Sociological Inquiry* 48 (1978): 172–96.

Winter, E. O. "Replacement as a Function of Repetition: A Study of Some of Its Principal Features in the Clause Relations of Contemporary English." Ph.D. thesis, University of London, 1974.

15 FULFILLING THE SPANDRELS

OF WORLD AND MIND

STEPHEN JAY GOULD

1. The Historical Context of Adaptation and "Spandrels"

In the lyrical mode that he used rarely, but so effectively, Charles Darwin explained what an evolutionary theory must provide in order to persuade. He states, in the introduction to the *Origin of Species*, that the multifarious evidence of genealogical descent should establish the "bare bones" fact of evolution:

> It is quite conceivable that a naturalist, reflecting on the mutual affinities of organic beings, on their embryological relations, and other such facts, might come to the conclusion that each species had not been independently created, but had descended, like varieties, from other species. (3)

But we will achieve no emotional satisfaction from such a demonstration, for we long, above all, to understand the causes of adaptation—the most prominent and interesting of all biological phenomena:

> Nevertheless, such a conclusion, even if well founded, would be unsatisfactory, until it could be shown how the innumerable species inhabiting this world have been modified, so as to acquire that perfection of structure and coadaptation which most justly excites our admiration.

In so stating his priorities, Darwin expressed fealty with a distinctively English tradition in natural history—a preference long antedating evolutionary theory. In the standard argument of British "natural theology"—from John Ray's mid-seventeenth-century *Wisdom of God Manifested in the Works of Creation* to Paley's *Natural Theology* of 1802—God's goodness and character (not merely his existence) lie exposed in the exquisite designs of organisms for their modes of life. The natural theologians referred to this excellent match between form and function as "adaptation," using a vernacular word that the *Oxford English Dictionary* traces to the early seventeenth century. A standard (and accurate) epitome of Darwin's achievement states that he turned Paley on his head by continuing to accept the hegemony of adaptation, while utterly inverting its causal basis by substituting natural selection for God.

If adaptation had commanded universal assent in its various historical roles (first as a testimony to God's nature, later as a primary result of evolution), then the current debate might never have arisen—for adaptation would either be a self-evident truth of nature, or would be so universally mistaken for such unobtainable certainty that no biologist would have thought to raise questions. But debate on the meaning and primacy of adaptation is coextensive with the history of natural history itself—and "Spandrels" is only a salvo in the latest chapter.

Moreover, the debate has spawned distinctive "national styles," at least since the Western religious and political turmoils of the sixteenth and seventeenth centuries. If adaptationist (functionalist) thinking is traditionally British, then the effectively opposite formalist or structuralist mode of interpretation has been just as prevalent in France and Germany, with a distinguished pedigree running from Goethe and Geoffroy Saint-Hilaire (friend and confidant of Balzac and George Sand) to the scientists showcased in the last section of "Spandrels." (In this continental alternative, "laws of form" and rules for transformation of organic architecture generate the basic diversity of anatomy; adaptation is a confusing epiphenomenon, a secondary tinkering that often masks the lawlike regularity underneath—and certainly not, as in the English tradition, the underlying cause of order.)

Darwin's theory of natural selection is undeniably functionalist in its basic mechanism—and no prominent critic has ever denied this plain meaning. Natural selection is a theory devised to explain adaptation and committed to the hegemony of adaptation as a proper description of nature. The debate among biologists working in the Darwinian tradition—an identification that both Dick Lewontin and I accept with pride—focuses upon two issues: "hardline" versus nuanced versions of the basic formulation, and extent of applicability (for any version) in our complex and multifarious world of relative frequencies.

Prevailing opinion has shuttled back and forth throughout the nearly 150-year history of Darwinism. Hardliners have seen selection as a nearly exclusive mechanism and have therefore interpreted all but the most trivial or secondary structures as adaptations. (This is the position that we characterize as the "Panglossian paradigm"—not a statement about abstract perfection or optimality, for no one in this field of complex history could be such a foolish Pollyanna, but a claim for maximization of adaptation under prevailing constraints of phyletic history and the admitted impossibility of optimizing all traits simultaneously; truly, in other words, *le meilleur des mondes possibles*.) Pluralists have permitted a guiding sway to natural selection (and hence a major role for adaptation), but have honored constraints of development and history, and strictures of organic integrity, as coequal in importance, not as bothersome epiphenomena upon the primacy of adaptation.

Since Darwin is the profession's patron saint, both sides cite him in support. Passages friendly to both positions can be found in Darwin's writing, for his theory is basically functionalist, but he did allow an important scope for pluralism. Lewontin and I devoted a major section of "Spandrels" to documenting Darwin's severe and genuine distress when critics falsely identified his own position as "hardline."

Hardline adaptationism has experienced two periods of popularity. The first, associated primarily with the great German theorist August Weismann and with Darwin's codiscoverer, Alfred Russel Wallace, arose in the late nineteenth century (as the self-styled Neo-Darwinism that Romanes in 1900 so vigorously attacked as contrary to the pluralist spirit of the original Darwinism). Weismann spoke of the *Allmacht* ("all-might" or omnipotence) of natural selection, while Wallace (in 1890) wrote: "No special organ, no characteristic form or marking, no peculiarities of instinct or of habit, no relations between species or between groups of species, can exist but which must now be, or once have been, useful to the individuals or races which possess them." (Note Wallace's only category for an "exception"—one that actually strengthens the panselectionist position: features that arose as adaptations have ceased to so function, but have not yet been altered or eliminated!) Wallace denies spandrels and other primary nonadaptations a priori: "The assertion of 'inutility' in the case of any organ is not, and can never be, the statement of a fact, but merely an expression of our ignorance of its purpose or origin."

William Bateson, leading British formalist of the early twentieth century, attacked Wallace's panselectionism in much the same spirit that we would later invoke in "Spandrels." He too, while not denying a large scope for adaptation in nature, questions the "story-telling" mode of gathering supposed evidence (I did not know about Bateson's invocation of Voltaire when I wrote "Spandrels," but the convergence is scarcely surprising, as Dr. Pangloss is a standard synecdoche for this form of ridicule): Bateson's critique is all the more noteworthy, for he delivered it at the celebrated symposium, held at Darwin's Cambridge alma mater, to celebrate the centenary of his birth. Bateson satirized the panselectionists:

> All that had to be done to develop evolution theory was to discover the good in everything, a task which, in the complete absence of any control or test whereby to check the truth of the discovery, is not very onerous. The doctrine "que tout est au mieux" was therefore preached with fresh vigor, and examples of that illuminating principle were discovered with a facility that Pangloss himself might have envied, till at last even the spectators wearied of such dazzling performances. ("Heredity," 99–100)

Moreover, Bateson also developed the idea of spandrels as a counterweight to selection—and he also saw fit to illustrate the unfamiliar argument with a metaphor based on human manufacture. (I think that my spandrels work better than his toolmarks, because spandrels are so expansively available for later cooptation in the service of secondary utility, while toolmarks tend to be trivial scratches):

> I feel quite sure that we shall be rightly interpreting the facts of nature if we cease to expect to find purposefulness whenever we meet with definite structures or patterns. Such things are, as often as not, I suspect rather of the nature of toolmarks, mere incidents of manufacture, benefiting their possessor not more than the wire-marks in a sheet of paper, or the ribbing on the bottom of an oriental plate renders those objects more attractive in our eyes ("Heredity," 100–101)

The second episode of hardline success developed more slowly and to much greater popularity and import. Darwin convinced the thinking world that evolution had occurred, but no consensus about causes and mechanisms congealed in his time (the panselectionism of Wallace and Weismann represented a strong contender, but never a dominant, or even a majority, position—see Kellogg for a classic explication of the confusing plethora of alternatives). Bateson himself lamented the near-anarchy in a famous address delivered in 1922: "Less and less was heard about evolution in genetical circles, and now the topic is dropped. When students of other sciences ask us what is now currently believed about the origin of species we have no clear answer to give. Faith has given place to agnosticism" (*William Bateson*, 391).

This agnosticism lifted dramatically in the 1930s as the data and ideas of genetics and natural history finally meshed to form the theoretical consensus that has dominated evolutionary theory ever since—the Darwinian view that came to be known as the "Modern Synthesis" (from the title of Huxley's seminal book of 1942). Most early Mendelians, after rediscovery of the good abbot's work in 1900, had focussed on large mutations as a source of sudden origin for species—a dramatically anti-Darwinian view (de Vries, 1909). When the Mendelian basis for ubiquitous, small-scale, Darwinian variability became apparent by the 1920s, these two great literatures—Mendelism and Darwinism—could be sensibly fused, and the Modern Synthesis developed.

The early Synthesis (1930 through the late 1940s) was decidedly pluralistic on the subject of adaptation. Its aim was to integrate modern genetics with the Darwinian substrate of natural history, not to advocate one particular "take" on genetics over all others. Thus, Sewall Wright's random genetic drift (as a potential source for small-scale differences among populations) received equal respect with R. A. Fisher's panselectionism.

For reasons that are well documented (Provine; Gould, "Hardening") but that I do not fully understand, the Synthesis "hardened" during the 1950s to the most widespread consensus ever achieved on the centrality, indeed the virtual exclusivity, of adaptation (as the result of a pervasively powerful natural selection). (Occasional nonadaptations were not denied—natural history, as a science of relative frequencies, never works this way—but rather pushed to a corner of irrelevance.) The 1959 Darwin centennials (this time for the *Origin,* and the sesquicentenary of his birth) were, in sharp contrast to the anarchy of 1909, a set of remarkably congenial paeans of praise for Darwin and his inseparable, nearly all-powerful agent of natural selection (see Tax).

In his bible of the hard version, *Animal Species and Evolution,* Ernst Mayr writes on page 1: "In essence, it [the Synthetic Theory] is a two-factor theory, considering the diversity and harmonious adaptation of the organic world as the result of a steady production of variation and of the selective effects of the environment." Note the central claim of the hard version: the phenomenon to be explained is "the diversity and harmonious adaptation of the organic world"; the components of explanation are twofold, representing the two sides of the Synthesis—the internal supply of variation (from genetics) and selection by the external environment (from natural history, with an explicit nod to selection as *the* mechanism).

Mayr's summary statement on adaptation is equally revealing:

> Every species is the product of a long history of selection and is
> thus well adapted to the environment in which it lives. There is no

doubt that the phenotype as a whole, including its physical properties, is adaptive and is produced by a genotype that is the result of natural selection. This is not contradicted by the fact that an occasional component of the phenotype is adaptively irrelevant. (60)

We can have "no doubt" that all species are "well adapted" by a "long history of selection." Even the necessary and canonical hedge of the last sentence is carefully written to support the hardline version: exceptions are only "occasional," and they are described as passively as possible—as merely "adaptively irrelevant," rather than actively built by a mechanism that might challenge panselectionism.

This hardline version provoked many rumblings of dissent during the 1960s and 1970s, arising largely from domains both "below" (the developing field of molecular genetics) and "above" (paleontology) the traditional Darwinian focus on organisms in populations. Kimura's 1983 theory of neutralism proposed a predominant role (not just a small corner) for randomness as a source of evolutionary *change* (not only for the "raw material" molded by selection in classical Darwinism). Punctuated equilibria (Eldredge and Gould, 1972; Gould and Eldredge, 1977) and new views on the extent and catastrophic character of mass extinction (Alvarez et al., 1980) suggested that extrapolation of gradual and imperceptible change through time might not encompass all of evolution at grander scales.

In this context, sociobiology, by launching its "founding document" in the midst thereof (Wilson, 1975), must be seen as a counterreformation of sorts. Those who wish to consider this movement as "revolutionary" within the human sciences should consider what a peculiar rebellion they are characterizing. Sociobiology may eventually evolve into something else, but it began as an application of hardline adaptationism to the interpretation of behavior—in other words, as an orthodoxy enlarging its realm of operation. (In this sense, social scientists who view the movement as imperialistic are not being entirely parochial.) With its basic procedure of isolating "items" of behavior, positing selective values, and assuming genetic foundations, sociobiology reasserted the Panglossian form of adaptationism at a time when much of the excitement in evolutionary theory was breeding departure from this former orthodoxy. Thus, the several authors in this book who identified sociobiology as an impetus for "Spandrels" are entirely correct. But the order of concern must not be inverted. Lewontin and I did not write "Spandrels" as a *roman à clef* about a greater concern with sociobiology. Wilsonian sociobiology was an important part of the context—

and we did not try to hide this concern, what with our early reference to Wilson's support for the adaptive value of Aztec cannibalism, and our later choice of Barash, author of the leading popular text on sociobiology during these years, for our longest critique of a particular adaptationist story (and an explicitly sociobiological tale at that). But sociobiology is only a miniature or microcosm of the larger issue—one of the grandest themes in all biology, with a pedigree of argument far antedating evolutionary theory itself: the conflict between functional (adaptationist) and formalist approaches to the interpretation of morphology, physiology, and behavior. We had our eye on these higher stakes.

Against this background in both older and recent history, the interest and tension of the Royal Society meeting, held on Pearl Harbor Day in 1978, can be better appreciated. We met in the antique rooms of the oldest and most venerated scientific society of England—not only the land of Darwin but (more important) the nation with a three-hundred-year-old commitment to adaptationism. The meeting was convened by John Maynard Smith, the world's kindest man and also the most brilliant and celebrated of committed adaptationists. (He is still royally put out by "Spandrels," and recently devoted almost an entire article in the *New York Review of Books*, ostensibly on the disparate subject of dinosaurs, to a critique of my views on adaptation—see Maynard Smith, "Dinosaur Dilemmas.") The reinvigoration of adaptationism by sociobiology served as a primary impetus for the decision to hold such a meeting. I (as a surrogate for Dick Lewontin, who had received the original invitation) was the token opposition in a group of speakers who wished to explore the range and application of a basic truth. (In the best tradition of fairness, we were extended all courtesies and given the coveted last say in the resulting publication of 1979). Arthur Cain, the summer-up and final commentator, could scarcely be called a moderator (at least in the literal sense)—for he is universally recognized (and self-identified) as the strongest "pure" adaptationist among well-known and respected evolutionists. In 1964, he wrote a famous paper entitled "The Perfection of Animals"—and he wasn't kidding. I didn't feel quite like Daniel in the lions' den—for, after all, the future of the world and the course of human suffering did not depend upon the outcome of this meeting (and I was still feeling quite mellow even two months after my beloved New York Yankees had, following their most improbable pennant victory on Bucky Dent's homer, beaten the hated Dodgers in the World Series, for the second straight year, and after losing the first two games). Still, I was not exactly wading in a sea of approbation.

Nor was I wallowing in a slough of despond. I will cite one anecdote—primarily because I am proud of it, in the most shamefully petty sort of way—to illustrate the intellectual tension of the 1978 meeting for which Lewontin and I wrote "Spandrels." Arthur Cain and I are old friends and fellow workers on land snails, but he overstepped the accepted bounds in presenting his summary remarks at the conference—and I got angry. He devoted almost his entire time to "Spandrels" and argued, basically, that Lewontin and I had consciously betrayed the norms of science and intellectual decency by denying something that we knew to be true (adaptationism) because we so disliked the political implications of an argument (sociobiology) based upon it. (Arthur's published version is a very tame rewrite of his rather different oral remarks.)

I don't usually employ the debater's tactic of emotional appeal in oral argument but, as I said, I was angry. I was given an opportunity to reply. R. Holliday, coconvener of the meeting, was standing in front of the podium. I asked him to move aside. He seemed confused and a bit annoyed, but complied. This left the motto of the Royal Society—*Nullius in verba*—exposed on the emblem that adorns the podium.

Now I knew that most of the membership, if they had ever noted the motto at all, had a totally false view, based on a canonical mistranslation, of its import. Most scientists are convinced that the text must mean something like "Not by words [you have to do the experiment and get the data]"—for this reading fits so well with the standard riff about the nature of science. But *nullius* is genitive singular, not nominative—and such a reading will not work. The puzzle is solved by recognizing *Nullius in verba* as a standard shorthand—known to all "educated men" of the seventeenth century, when the Royal Society was founded—for a famous line from Horace:

Nullius addictus iurare in verba magistri
quo me cumque rapit tempestas, deferor hospes.
(I am not bound to swear allegiance to the words of any master;
Where the storm carries me, I go ashore and make myself at home)

(Incidentally, I never studied much Latin. I first saw this phrase on the masthead of a Bahamian newspaper when I was doing research on land snails, and I had to find out what it meant. I am no classical scholar, just a very curious man.)

When Holliday had stepped aside, I pointed to the motto, told the story of its true meaning and standard mistranslation, and then made the obvious point, directly to Arthur Cain, that my talk had merely tried to follow the venerable motto of the sponsoring society—by questioning a dogma and trying to expand the realm of alternatives—and had not been an exercise in political restriction. A brash Yankee move perhaps, but it was effective (and I did feel a bit like Bucky Dent surmounting the Green Monster two months earlier).

2. Some Personal Thoughts and Admissions

Call the beginning of this section "Confessions of a former adaptationist." I did my graduate work at Columbia University, strongest bastion of the hardline synthesis (the three leaders of the Modern Synthesis in America—Theodosius Dobzhansky, Ernst Mayr, and George Gaylord Simpson—all taught there or at the affiliated outpost of the American Museum of Natural History). I emerged as a philosophically committed adaptationist, dedicated to putting this orthodoxy into better empirical practice. I had focussed my training on statistics and multivariate biometry, hoping to supply the quantitative rigor that adaptationist tales in the "story-telling" mode had lacked. My first major series of papers treated the subject of allometry, or change of proportions during growth or evolutionary size increase. Allometry had been a traditional bastion of nonadaptationist thinking, for systematic changes in proportion could be attributed to incidental side consequences—select for increase in body size, and change of shape may come along for the developmental ride. I tried to invert this explanation, and win these cases for adaptationism, by arguing that selection could work as effectively upon rates of growth as upon features finally obtained.

These papers now embarrass me, especially when I face such juvenilia as this: "I acknowledge a nearly complete bias for seeking causes framed in terms of adaptation" ("Allometry," 588)—oh well, at least I labelled the preference as a bias! In another early paper ("Evolutionary Patterns," 385), I discussed "the fundamental problem of evolutionary paleontology—the explanation of form in terms of adaptation."

Show me a zealot for the banning of cigarettes in nearly any place that a person might puff, and I'll show you a former smoker (folks like me, who have never smoked, tend to be much more sympathetic toward people caught in the throes of a serious addiction). The "zeal of the convert" is a cardinal phenomenon of sociology and will certainly help to explain my own participation in "Spandrels" (I will spare you any further psychobabble about my inner self and deepest motivation, but I did need to make the bare connection). What, then, led me to change my mind?

Despite Cain's charges on Pearl Harbor Day, the roots of my change are complex, and long antedate sociobiology. I would identify three major sources, each from a different realm of evolutionary thought:

1. Starting in the late 1960s and culminating in the publication of *Ontogeny and Phylogeny* in 1977, I read extensively, and for the first time, in the continental, formalist literature on morphology and development. (I read more German than English in writing this book, for English functionalism implies a lack of interest in embryological pathways: if selection is so powerful that channels of growth can readily be broken, why worry about formalist constraints on development?) It took a while to sink in, but years of immersion in this literature finally spawned fascination and respect. I wrote in the preface: "I only began [the book] as a practice run to learn the style of lengthy exposition before embarking on my *magnum opus* about macroevolution. And I'm mighty glad I did because, in the meantime, my views on macroevolution have changed drastically and my original plan, had it been executed, would now be an embarrassment to me" (vii–viii).

2. At the invitation of the late Tom Schopf, and with paleontologist Dave Raup and ecologist Dan Simberloff, I participated in a self-conscious research effort to buck tradition in paleontology and to explore how far purely random models could be pushed to encompass the observed order of the fossil record (see Raup et al., "Stochastic Models"; Gould et al., "The Shape of Evolution"). (We were successful, at least in spawning controversy, and were generally known among our colleagues as "the gang of four.") I, at least, was surprised (Dave Raup was not—he had expected it) at the range of apparent order that could be generated by random models but that, unhesitatingly, had always been viewed as *prima facie* evidence, sufficient all by itself, for adaptation. This work made me reassess the methodological requirements for advancing a claim of adaptation.

3. The publication of Wilson's *Sociobiology* in 1975, though later than the other two influences, did help to cement my doubts. But I do think that Cain had the causal pathway backward. I considered the basic argument of sociobiology as flawed, and I was bothered by the political implications of its genetic determinism. One encounters flawed arguments by the score every day. A decision to grant special scrutiny and attention to such an argument may then be influenced by other factors—including dislike of its implications. (I need hardly mention that unpleasant consequences do not brand a claim as false. I cannot think of an argument I like less than "all men are mortal"; yet I do not dispute it.) When I located adaptationism as the central intellectual flaw of sociobiology, I gained more insight into the scope of misuse and came to regard the subject as sufficiently important for attention and ink.

(But the flaw exists quite independent of the political implications; Barash's bluebird argument is inadequate and unscientific, though *sans* impact on human·politics.) In short, I did not attack adaptationism because I disdained sociobiology; I disliked sociobiology because I regarded its central premise as fatally flawed (and regretted the social implications falsely drawn therefrom).

Thus, by 1978, when the invitation that resulted in "Spandrels" arrived, I had come to regard adaptationism as a conceptual impediment to evolutionary biology in several important domains; I also thought that I had a positive alternative to suggest. The invitation to the symposium came to Dick Lewontin. He hates to fly and I wanted a trip to England for other reasons. We taught (and still teach) the basic evolution course together at Harvard. We had similar views on adaptation and had worked together on the critique of sociobiology. We are good friends. I seemed an obvious surrogate and happily accepted.

One thing you learn very quickly about joint ventures in writing: never try either to compose together line by line, or to interdigitate finely—for you run up against the law of diminishing returns almost immediately. You must assign large and definite chunks to individuals and then respect each other's work. In this case, since Dick was busy and I had time, and since I was doing the oral presentation anyway, I wrote nearly the entire article. (Dick composed number 1 of the five-part typology of alternatives in section 5—pp. 156–57. I asked him to do this because I have no technical knowledge of population genetics.) We consulted very little before the writing and did even less revision thereafter. (I know that this sounds like a nasty and petty attempt to hog credit for something that turned out to be successful. Honest, folks, I don't mean it this way. Dick Lewontin is the most brilliant person I have ever had the privilege of knowing. We had talked adaptation for years. I can't even begin to sort out which of the ideas in this paper come from him, and which from me—but I know that his contribution is paramount in all but the last section on continental alternatives and the short history-of-science interlude on Darwin. For example, the key five-part typology in section 5 is entirely his, even if I wrote most of it. In short, the ideas are more his, the writing almost entirely mine. The metaphorical apparatus of the first section—the move from spandrels to Voltaire to Aztecs—is my device. In our little partnership, I am the pedant and he the innovator.)

Nearly every contributor to this book stresses the unconventional character of "Spandrels" as a scientific article. I agree that it is unconventional, but not for the reason most often stated. We are told that the format is unusual—that most scientific articles use a standardized sequence of introduction, presentation of data, and drawing of conclusions, while "Spandrels" is a partisan intellectual argument with commentary on data. (The contributors do recognize, of course, that a genre of metacommentary does exist in science—the "review article" that attempts a dispassionate and impersonal summary of a field. Obviously, "Spandrels" is not a review article either.)

I disagree; there is nothing unusual about the format of "Spandrels." It is, of course, neither a data paper nor a review article, but a third genre does exist in scientific writing, if not in the abundance of the other two—the "opinion piece," with its inevitable personality and point of view. Opportunities for this third genre are more limited; most standard journals will not accept such pieces, or will limit them to one or two per issue, in a separate section. But several forums exist for such articles—*Festschriften*, published symposia, special meetings, presidential addresses, etc. Moreover, complex fields like evolutionary biology, which must struggle with deep questions of an essentially philosophical nature, tolerate more writing in this mode; opinion pieces have a long history in the evolutionary literature. Since "Spandrels" was first presented at such a broad symposium and then appeared in the published proceedings of this two-day meeting, the chosen genre is entirely appropriate.

Something else about "Spandrels" is decidedly unusual, even provocative: its style—particularly its metaphors, literary and cultural allusions, and brashly personal language, including all those irreverent asides about beefing up the meat supply. This does fly in the face of the most cherished and widely obeyed convention that good science is impersonal and that the intrusion of self can only denote partiality and attendant flawed reasoning. I do what I do entirely on purpose and with definite aim. (At this point, having kicked around for some years, I generally get away with it by tacit special dispensation. People say: "That asshole again! Oh well, at least he can write.")

Why, then, do I persist in proceeding this way, especially since the style may alienate readers and thus be counterproductive? Three reasons occupy my conscious mind.

1. Vanity. I am lousy at math (basically numerate, but not innovative) and ham-handed at experiment (competent but inelegant, to put it mildly). I am so-so as an observer, and scarcely macho as a field worker. But I can think, and somehow I figured out an odd manner of writing, based on no formal training. I have spent years learning languages, and I cannot bear not to find out about anything interesting that I encounter—and I am quite catholic in what I consider interesting. In other words, I have none of the skills traditionally associated with success in

science, but I am competent (or at least as thrilled as a kid who can still see splendor in the grass) in the face of ideas and their expression. What can one do but make the most of a few lucky gifts? (They also serve who only stand and wait.) If I had a better voice, I would be singing Wotan; if I had a better body, I'd be playing center field for the Yankees. The Lord gave me an odd way with words and a passion for ideas and facts. To quote from "Spandrels" as I originally wrote it: "might as well use them" (or as the editor recast it in "properly" impersonal form: "use might as well be made of them").

2. Belief. Scratch a cynic and you find a gushing idealist. I believe in the dream of integrated knowledge, in the abolition of falsely restrictive intellectual boundaries. Above all, I believe in the commingling of the arts and sciences as the two greatest expressions of human creativity. I regret the petty and parochial boundaries that both domains have established—the impenetrable and sterile language of so much scholarship in the humanities, the dry, impersonal and barbarous passive voice of scientific prose. I want to break through. You (as readers of this book) can contribute by learning about science and by treasuring its conclusions as part of humanism. I can contribute by using the data and arguments of the humanities, not as window dressing for vain show, but as an intrinsic and central part of a scientific case (the intent, at least, of the spandrels metaphor).

I get such a kick when interpenetration works. I remember once reading one of T. H. Huxley's beautiful letters. He states that if such-and-such occurs, he will be ready happily to chant (they didn't split infinitives back then) his *Nunc dimittis*. I knew that these words are the Latin title for Simeon's song—the prayer of the old man who states that he can now die content, having seen the infant Jesus: "Lord, now lettest thou thy servant depart in peace" (Luke 2:29). Big deal, you might say. So what if I didn't know that, if I just read through the line in Huxley's letter? Well, it is a big deal, for you would then miss the full flavor of Huxley's prose and accomplishments, the full depth of his feeling and the elegance of his vision.

3. Efficacy. Interpenetration also has practical benefits. I once entitled an essay "Death before birth, or a mite's *Nunc dimittis*" (the odd, but sensible, evolutionary tale of a male mite that becomes sexually mature while still "developing" within his mother's body, fertilizes his sisters, and, his evolutionary work accomplished, literally dies before birth, at least as defined by exit from a parent's body). Simeon's prayer provided a good device for linking this bare story to an empathic context.

But I speak primarily of a different sort of utility in advocating humanistic reference, and attention to persuasive writing, in scientific prose—efficacy in advancing whatever argument you are trying to make. This efficacy arises primarily for an odd reason. I suppose that anyone skilled in the art of rhetoric is better off than an opponent not so endowed. But when you work in a field that acknowledges rhetoric as an explicit and universal aspect of argument, then you are always looking out for the components of persuasion, even when you can't use them effectively yourself. By contrast, any weapon will be vastly more effective if it can be wielded invisibly.

Scientists, for the most part, simply do not acknowledge that the form and language of an argument (as opposed to its logic and empirical content) could have anything to do with its effectiveness. Humanists probably do not know that scientists even define the word *rhetoric* in an exclusively pejorative way—as an attempt to bamboozle by words alone, when you don't have the goods in logic or data.[1] Thus, since good and honorable rhetoric works (scientists are generally smart enough to identify the dishonorable and effective mode), an addition of the element of surprise—in this case, a surprise never revealed—boosts utility enormously.

My own profession boasts the very best example. Charles Lyell's *Principles of Geology* (1830–33) is, without doubt, the most influential work ever written upon the subject in English. This book established the so-called uniformitarian world view—both a boon and a detriment to the profession (see Rudwick, *Great Devonian Controversy*; Gould, *Time's Arrow*). Most geologists regard *Principles* as a textbook and locate its persuasive power in Lyell's abundant documentation of the efficacy of slowly acting present processes. But *Principles* is a brief for a partisan argument, written by one of the great prose stylists of the Victorian age—a man who was, need we say more, a barrister by original profession. Lyell persuaded primarily by rhetoric (sometimes, but not usually, in the pejorative sense of that word); all modern commentators (Porter; Rudwick, *Great Devonian Controversy*; Rossi) have now recognized this primary characteristic of *Principles*; most geologists still do not. How much easier it must be to persuade when most people don't realize that you are trying (or, rather, mistake *how* you are trying).

Let me, in closing this section, try to integrate these three conscious reasons for a chosen style in discussing the most unconventional and probably most successful aspect of "Spandrels"—the opening metaphor of the spandrels of San Marco itself. The basic reason is highly personal and a product of my own ontogeny. I presented "Spandrels" as a talk in London in December 1978. Just three months earlier, I had visited Venice for the first time. I had been thinking about adaptation, for I knew that I had to prepare for the London talk. These things can only happen once or twice in an intellectual lifetime, but I had an epiphany of sorts, appropriately enough under the great dome of San Marco. I looked up at the spandrels, worked out the complex (and lovely) iconography of four evangelists above men personifying the four biblical rivers, and the whole argument hit me all at once—a strange feeling of almost manic exhilaration followed by the total calm of understanding. The two main supporting claims fell quickly into place as I looked at the other domes: all had well and sensibly decorated spandrels, and all domes (but one), though radially symmetrical and therefore not constrained intrinsically to any order of repetition, carried a quadripartite iconography, in clear conformity with the spandrels below. Spandrels are not "nooks and crannies left over," but potential determinants of the entire design.

I always carry a small pocket notebook with me (for recording epiphanies, of course). I don't think that I had looked at the 1978 volume since I wrote "Spandrels," but I dug it out while preparing this article. This is what I wrote on September 6, 1978:

> San Marco, Venezia. Much of mosaic decoration is based upon *bautechnische* [Seilacher's word for nonadaptive architectural constraint in organisms] necessity. Church is basilical with 5 domes. Each dome supported by arches, and this leaves necessarily triangular spaces (called squinches I think) [I didn't even know the right name for spandrels yet] 4 squinches for each dome. Therefore used (given they must be there) for all manner of ornament making sense by 4, and might be tempted to say adaptation lies in fitting the 4—and structure is adaptation to it. Would be backwards. Moreover, this 4 may then dictate radial division of dome above into 8, 12, 16, etc. For linking symmetry. . . . Great central dome has 4 evangelists in squinches. Each holds book with 1st words of his gospel and each surrounded by buildings to preserve triangle. Below each evangelist is a man holding pitcher over shoulder and pouring water out—the 4 biblical rivers. And below each is a tree.

This was just too delicious not to use in London three months later. Moreover, since all scientists illustrate talks with slides, I had a golden opportunity (quite literally) to begin my lecture with both surprise and beauty. Surely, all three of my criteria entered this decision: vanity, for I was damned proud of myself for developing this argument and not just being another tourist walking through and getting a neckache; belief, for the story is pretty and it does fit (intrinsically and not as window dressing) into the biological theme of adaptation; and utility, for nothing can be so clear as a well-chosen metaphor.

Yet, as St. Paul said about charity, the greatest of these is surely utility. I am convinced that the success of "Spandrels"—for this paper has become the standard citation for nonadaptationist alternatives (see section 4 following)—lies, above all, with this opening metaphor, perhaps the only truly original point in the paper. We faced a special and unusual sort of problem in gaining attention and understanding for alternatives to adaptation: most evolutionary biologists simply accepted adaptation as a nearly given universal of form. How can you challenge something if most people simply regard it as true and therefore haven't even conceptualized the possibility of another reading? You can't initiate this sort of reform from within.

I didn't think I had a chance of success if I tried to raise the argument head-on by labelling a set of biological structures as potential spandrels. Too many colleagues would have turned off right there—either by putting all their ingenuity to finding a proper adaptationist explanation for the examples or, much worse, by failing to understand the point because, after all, we know that well-designed parts of organisms are adaptations; what else could they be? But, by using a "neutral" architectural example, I could make an end run around these prejudices and compel attention. Scientists do try to be fair (and most are very smart); and we will consider the intricacies and subtleties of an argument on merit, provided we are not too hard-hit in the guts of our biases. The spandrels example worked beautifully. Anyone could see (literally—never doubt the value of pictures) that the mosaics of the spandrels were beautifully fit to the triangular space; while no one could doubt that the space came first (as a nonadaptive side consequence of a decision to mount a dome on arches) and the design followed.

But I now faced another problem: how to get back to biology. And I tried to solve it by the old trope (particularly favored by evolutionists for obvious professional reasons) of continuity in graded sequences. Thus, I went from architecture in Venice to architecture in nearby Cambridge, to the canonical literary allusion (Dr. Pangloss), to anthropology (Aztec cannibalism), and finally back home. The rhetoric (good sense) worked. Ten years later, my friend Dave Raup (coconspirator in the different criticism of adaptation that I discussed earlier) said to me, "We have all been spandrelized." When your example becomes both generic and a different part of speech, you have won. Call those San Marco spandrels "Kleenex," "Jell-O," and a most emphatically non-metaphorical "Band-Aid."

3. Random (and Therefore Nonadaptive) Jottings on Commentaries

In reading the contributions to this book, I must confess that I feel rather like the peasant who, dragged into court on suspicion of being the notorious Jean Valjean in disguise, is so overwhelmed by the majesty of architecture and the dignity of officialdom that he is finally inclined to agree that he must, after all, be Mr. Valjean if they say so with such assurance. I never suspected that so much was going on in the text of "Spandrels," yet I am forced to accept the majority of interpretations, not all flattering by any means, here offered.

Which is not to say that I agree with everything; yet my sizable list of demurrals includes nothing fundamental, and rests only (with God) in the details. For example, I am quite sure that my reference to Darwin's book as *"Origin"* was not, as one author argues, a conscious pun on Genesis. This is standard shorthand for my entire profession (the *"nullius in verba"* convention). We always call the book *"Origin"* (I will accept the charge of a profession-wide pun, but not the personal claim). St. Thomas wrote a *Summa;* Newton a *Principia;* Darwin an *Origin.* I am also confident that the sequence of heading titles in "Spandrels" is not, as another author suggests, a conscious parody of conventional sections in a research report—for I have never written a paper in this style, and I don't even know what the standard headings are.

I am particularly grateful for comments on aspects clearly present but quite unrecognized by me: I am an intuitive writer. I work bloody hard in doing the research. But then I wait for an outline to hit me (I should say *the* outline for, as a true Platonist, I believe that each piece has one and only one correct format; the outline resides somewhere in heaven, and you just have to wait until whoever runs things up there decides to beam it down). I then write the piece, without much revision and with little self-conscious attention to the formalisms of style—most of which I don't know. I don't say all this (or at least I don't think I do) either for vanity ("What a clever chap, to do this so unconsciously!") or for self-criticism ("What a frightful dunce!"), but for a quite specific reason centrally relevant to what I understand as the main purpose of this book—that is, to show the many fruitful ways in which a scientific text can be analyzed. For, surely, this theme is best applied to a naive document like my own—to one not written with any particular literary form in mind, or with any explicit knowledge of the range of forms. If I were writing as a deconstructionist, or a structuralist linguist, or a politeness theorist, then you would have a preferred channel for analysis. If I am writing as, consciously, nothing (in another domain, that is, for I fancy that I know exactly what I am doing in biological terms), then my text is neutral before your methods. And if they all work in the sense of yielding insight—as they did for me at least—then they must all be right, or all be useful, or all be suggestive, or whatever praise to fruitfulness you wish to offer. (Is this the way of reading and interpretation without limit, the "no preferred mode," of which deconstructionists speak? I'll be damned if I have ever been able to penetrate this movement, although twenty people have tried to explain it to me. If I ever comprehend Derrida, who knows—I might even be ready for *Finnegans Wake*).

I feel enlightened on three different levels. First, as a perennial student and virtual worshipper of detail, a person committed to the idea that big insights lie enclosed within small hints, I greatly appreciate the tutelage on what I had missed in my own efforts. (I will give, in each category, just an illustration or two, not a compendium; my apologies, therefore, to many others noted with pleasure but not listed here). Myers, for example, in citing my unconventional prose even in papers more technical than "Spandrels," noted that puns on researchers' names are "normally taboo," but that I had violated this precept (though perhaps more permissibly for a long-dead author) by writing: "This dubious claim remained impervious to future test because Dubois. . . ." I never saw it, but of course Myers is right. Fahnestock makes a major point of my spelling "programme" in the English manner throughout the text, claiming that I do so to dissociate a usage that I strongly criticize from my own American traditions. I was about to cite this example as a classic, silly overinterpretation, sealing my case with the incontrovertible evidence that I had spelled the word "program" in my text and that the English editor had made the change for obvious reasons. But I went to my original manuscript to check, just to make sure. I had spelled it "programme" throughout. Thanks for the insight, Dr. Fahnestock.

Second, I am just as grateful for the more comprehensive lessons on strategies pursued, and devices used, throughout "Spandrels." Some of the threads I sewed myself, and consciously. Yes, I knew that the German proclivity for speaking of formalist alternatives to adaptationism as theories of *Baupläne* (or building plans) gave an awfully nice balance (at the end) to a piece that had begun with an architectural metaphor—and I might not have quoted the German phrase at all otherwise. Yes, though more dimly this time, I was aware that I had employed the word *we* in several shifting senses. (As for the changing usage of *we, you,* and even *I* in this chapter—well, Dr. Selzer, boss man, this is just for you, as my way of saying thanks. Once is an illustration; twice, a generality. Have fun.)

But other, even quite central and unifying devices had passed me by. I am astonished (and here I must don the dunce cap) that I never consciously noted the integrative symmetry that Herndl, in his analysis by cultural studies, so correctly identifies in my sequence of citations with political import: Wilson, Barash, and Spencer, at beginning, middle, and end. Wilson, the major name behind sociobiology, in anecdote, for defense of a truly silly argument (almost a caricature) about Aztec cannibalism; Barash, author of the major textbook then available in sociobiology, *in extenso,* for developing an equally foolish, and conceptually similar, argument about bird behavior; Spencer, the major name associated with Social Darwinism, the closest Victorian analog to sociobiology, for political misuse of evolutionary theory, again (for closing symmetry) in anecdote, for an embarrassing and silly error about the adaptive value of fingerprints. The connecting theme of political misuse lies in a famous line with the most essential comma in English: "Whatever is, is right."

I still don't understand deconstruction, but I do grasp Journet's powerful point that intertextuality (a term I did not know) must "defer any interpretive certainty"—and that a literary mode of this sort is especially valuable in a paper that pleads for pluralism in attitudes towards organisms. Coincidence of form and content is intriguing. I have been known to do it on purpose—but not this time. Speaking of pervasive themes and Journet's insights, I also did not recognize, when I wrote "Spandrels," my propensity for choosing adaptationist stories that are not only wrong but nasty (about cannibalism, cuckoldry, and venereal disease), while using opposite language, even to the point of discussing Darwin's "resurrection," for favorable parts of my tale.

Third, and most important to the stated purpose of this book, it is fascinating to learn that, like a kind of giant *Rashomon,* a text can be read with perfect sense and coherence in so many uncorrelated, truly orthogonal, and sometimes even contradictory ways. "All are right; all are useful" sounds like such a wishy-washy conclusion, a whacko, blissed-out, pot-befuddled love feast, rather than a tough-minded analysis. But, as on the Oriental Express, sometimes the claim that "everyone shares equally" is a correct, principled, and genuine solution.

I may be bewildered by the array of interpretive strategies available, each with a name, no less—dramatism, structural linguistics, politeness theory, stasis theory, the "thinking-aloud" method, the "reader-response" perspective, social constructionism, intertextual and citation analysis, the cultural studies approach, deconstruction, feminism—but I see that each can fill a spandrel with a coherent design full of import. I am also struck by the degree of collegial cooperation obtainable in producing such a useful work, based on the common style for each critique: a beginning that sets forth the distinctive method, followed by an application to "Spandrels." Either Dr. Selzer is operating the most effective "favor bank" since Lyndon B. Johnson, or you have admirable traditions of cooperation that we cannot match. I could never cajole or compel a group of scientists to follow a common style on anything.

I learned a great deal from all these approaches, and had enormous fun seeing how a product of my own could be analyzed for itself, quite apart from my conscious intent. And it must be so if traditional scholarship, based on rational and interpretive principles, is to maintain its legitimate pride of place—for a text must be knowable (in multifarious ways) *in se* and by content, for what can we ever know for sure about intent?

I appreciated Fahnestock's analysis of classical modes of rhetoric in "Spandrels"; Gragson and Selzer on the "fictionalization" of audiences (I wouldn't have thought of using this word, but the point is well taken, for no piece can be effective unless it is directed to a particular group of readers); Wells, in part via Habermas, on the interplay of lyric and modernist critical writing in "Spandrels," for I struggle to link these two divergent modes in almost everything I write.

I learned a great deal from Charney's empirical study of how biologists read "Spandrels." I was not pleased with all I learned, particularly that the most professional and sophisticated subjects did not read the paper straight through (while most graduate students did), but browsed first in the disconnected sections (abstract, conclusions, figure captions) that quickly provide the gist and main points, and only then (if at all) looked at the entire text. This bothered me because "Spandrels" must be read as an integral and linear text for maximal effect; it is an essay, not a research report. In fact, I initially declined to submit an abstract for "Spandrels," because I didn't want readers to get an epitome on the cheap and then neglect the rest. (The editors wouldn't let me get away with this unconventionality.) But then I recognized—and therefore had to laugh—that I never read scientific papers straight through either. I always browse for the key bits first. So I have no right to complain, but must remember my own habits better when I try to persuade others. I also got some insight into the struggle that literacy faces, when I read Charney's quotation from one befuddled student who not only failed to grasp anything essential about "Spandrels," but also managed to use all four of the great empty phrases of contemporary speech—"like," "well," "you know," and "I mean"—in just a sentence or two of commentary! (I do, however, disagree with Charney's argument that "Spandrels" may be limited in effectiveness because colleagues do not read it as we intend. I doubt that the impact of any paper can be measured by amount retained at first reading. The analysis is right, but the scale is wrong. "Spandrels" succeeds or fails by hanging around as a "standard source" to be reconsulted and, above all, assigned to future generations of students. I think, and shall argue later, that its main success lies with the spandrel imagery itself—for this picture is the key to grasping an unconventional alternative viewpoint.)

I even accept the analysis of the paper that inspired my strongest general disagreement: Rosner and Rhoades are right in noting something of a paradox in our development—a defense of feminist principles by an almost antifeminist adversarial argument. I would reply that the defense of pluralism by adversarial devices is honorable (dishonor lies only in the frequent misuse of adversarial style), potentially effective, and practically necessary. How else could we proceed, given the opposition? And how else is cultural pluralism defended in America today?

So I thank you all for your insights. Now I fear only the paralysis of Cassius, for it may not only be dangerous to think too much about what you are doing, but also potentially immobilizing. However, I will take heart; Cassius was, after all, still able to act—not wisely, perhaps, but too well.

4. Aftermath and Epilogue

It is one of the great ironies and frustrations of intellectual life that effort and effect have such a poor correlation. I have spent years writing technical monographs on land snails, only to see the publications sink without a ripple or notice. I have published several hundred items, yet just over half the citations to my work in the *Science Citation Index* refer to six pieces (including "Spandrels"), only one of which, my book *Ontogeny and Phylogeny,* was the product of intense and prolonged effort in writing. (None of the others was a capricious or ill-thought-out throwaway; like "Spandrels," they emerged from concentrated thought, argument and reading, but the writing was quick and painless.)

"Spandrels" has been a rare success in a bibliography dominated by writings that, however personally satisfying, have had no recognizable impact. Professionally, it is the most widely cited of all my articles (only my book *Ontogeny and Phylogeny* exceeds it within my bibliography), with just over 550 total citations through 1990, the year that it finally edged out my second-favorite—the 1972 article on punctuated equilibrium that Niles Eldredge and I wrote, which had a seven-year head start on "Spandrels." It is, of course, impossible to translate quantity into quality, or number of citations into intellectual impact. (I don't think that the correlation is quite so low as that between effort and effect, but the two phenomena are incommensurate.) A certain percentage, and perhaps a high one, of references to "Spandrels" is of the knee-jerk variety—that is, "honorary" mentions from adaptationists who want to acknowledge an alternative for fairness' sake, but have no intention of considering the arguments explicitly. But even the knee-jerk citations indicate that "Spandrels" has become a standard source for a point of view.

There can be little doubt, I think, that the hardline adaptationist version of the Modern Synthesis has passed from orthodoxy. This would surely have happened even if "Spandrels" had never been written, for so much of the most exciting novelty in evolutionary theory during the last thirty years, from molecular genetics to mass extinction, cannot fit happily under the old, restrictive paradigm. But "Spandrels" did provide a focus and a terminology, and thereby helped to coordinate a disparate body of information and ideas.

Moreover, the structuralist or formalist alternative represented in "Spandrels" is beginning to get a hearing in popular sources as well. For example, I have, for years and publicly, criticized the "Ask the Globe" column of the weekly science section of the *Boston Globe* for answering every—and I do mean absolutely every—inquiry about "why organism *x* has feature *y*" with an adaptationist story. Imagine my surprise and delight when the August 5, 1991, issue carried a drawing of an architectural spandrel and the following commentary in response to the question: How did language begin?

> Language may well be a spandrel of the mind. . . . "Spandrel" turns out to be an architectural term for the space between two arches. . . . The first builders who supported domes with arches created spandrels by accident. In the same way . . . language . . . may have been a thinking and communicating "spandrel" accidentally created by the development of some cultural "arch."

But any scientist will argue—and I ally myself entirely with this central conviction—that the proper measure of any publication's success must lie in its future utility in advancing the fruitful work of a field. Any writing, no matter how brilliant or elegant, is sterile if it stands only as a monument to its moment. My greatest personal pleasure in "Spandrels" lies not in the document itself but in the role it played in furthering my own views on the subject. (I cannot speak for colleagues, but frequency of citation does indicate that "Spandrels" got other people thinking and reacting as well.) "Spandrels" pushed my own work on adaptation forward in at least two ways. First, the success of the "spandrel" terminology helped me to realize (along with my colleague Elisabeth Vrba) that a larger and more important terminological lacuna existed in the evolutionary literature—and concepts without names cannot be well formulated or even properly specified. No name existed for the general concept of a structure now useful in some particular role, but recruited or coopted from another context (either a nonadaptive spandrel or a structure originally developed for a different use), and not directly evolved for its current function. We called such coopted features "exaptations," restricting "adaptation" to features evolved directly for their current function (Gould and Vrba, "Exaptation—A Missing Term"). This terminology has been widely adopted (more than two hundred citations for "Exaptation—A Missing Term" through 1990), and has been very helpful in sharpening the cardinal distinction between historical origin and current utility. (When I read "Spandrels" today, my strongest criticism lies in the textual confusion still present because we did not then have a term like "exaptation.")

The concept of exaptation will probably prove most important in understanding the essential activities of the human mind. The brain is the most complex computing device ever evolved in nature. Most of what it does, and can do, must arise as architectural side consequences— as spandrels—of its structure, not as the direct adaptive reasons for its original increase in size. All of the vital activities of mind—reading, writing & 'rithmetic for starters—are (as spandrels) exaptations, not adaptations. Yet they are vital to our success and to the definition of our being (Gould, "Exaptation—A Crucial Tool").

Second, I have tried to integrate the theoretical pluralism of "Spandrels" with earlier work on the effect of chance in the history of life to gain a better appreciation for the dominant role of historical contingency in shaping our current living world and the pathways of its development through time (see my book *Wonderful Life*). What happened makes sense (it is not "random" in the standard misuse of "inexplicable"), but life's history could have cascaded down millions of other equally sensible alternative paths, none (or precious few) of which would have led to the evolution of self-conscious intelligence.

Finally, and perhaps most relevant for the enterprise of this book, when I reflect on "Spandrels" and ask myself why it became a standard reference for the formalist alternative and a pluralist approach to adaptation, I keep returning to its true iconoclasm—of presentation, more than of substance. "Spandrels" develops no new biology and presents no novel critique. But it does arrange a set of arguments in a different and insightful way (the objection to story-telling with the continental formalist alternative, for example), and it does provide a broader historical and disciplinary context for a particularly biological critique. (For this I must again largely credit my coauthor, Dick Lewontin, as so much of "Spandrels" is my acting as the amanuensis for his analysis. This paper may be full of personal arrogance and puffery, but I have my moments of critical self-reflection and recognition of limited effectiveness. It cannot be accidental that my two most widely cited papers—"Spandrels" and "Punctuated Equilibria"—are both, and such items are rare in my bibliography, examples of my attempt to write down and interpret the ideas of a coauthor). In addition, "Spandrels" succeeded because it developed a novel imagery—the picture of an architectural spandrel itself—that turned out to be striking, easy to understand, simple to convey and, above all, ripe for extension.[2] A critique cannot advance far when it is inchoate—and nothing focuses thought so well as a good picture.

In other words, I believe that the success of "Spandrels" arises not so much from its "pure" science, or even from the logic of its argument, but most of all from its rhetoric (in the honorable, not pejorative, sense) and its humanistic imagery. The very aspect of writing that rhetoricians treasure and analyze, but that we scientists ignore and disparage, has caught our colleagues unawares and won attention for "Spandrels."

I know no better argument for the most important point of all—that disciplinary boundaries are false and harmful, especially when we hide behind our barricades of jargon and self-reference and shoot arrows of scorn at occasional intruders. By turning their skills to the analysis of a scientific paper, albeit an unconventional item written in a literary style, the authors of this book have made a welcome contribution to this essential and fruitful breaching of barricades. I can give endless abstract and theoretical reasons for the value of such rupturing and commingling, but the greatest benefit is surely practical. We all get caught in unrecognized dogmas behind our own barricades. What could be more salutary as a source of new and challenging ideas than the insights of other professions? *Nullius in verba.*

NOTES

1. I'm trying to obey strict scientific style and avoid footnotes entirely. But then, you folks know how to deal with them, so permit me one or two. Differences in the meaning of a word—either between professions at one moment (as for *rhetoric*), or within a profession through time—can be very

Stephen Jay Gould

revealing. One example in the latter category is especially relevant to this paper and to "Spandrels." As the best illustration I can provide for the success of hardline adaptationism in the late 1950s and 1960s, when I was a graduate student at Columbia (1963-67), we used the word *adaptation* as a neutral description for any feature of an organism—quite independently of any thought or claim about its status. We might say, for example, in speaking of ribs on a brachiopod valve: "this adaptation was especially prominent in specimens collected from the Cincinnati Arch." We were not making a claim about the utility of the ribs, but merely describing them. It is a measure of the success of our arguments against panadaptationism that no student today would think of using the word in such a way. In fact, students are amused and puzzled when I describe this fossil biospeak.

2. In my most interesting personal extension, I got involved in a friendly argument with Pat Bateson of Cambridge University about my second architectural example: pendant roses and portcullises as ceiling bosses in the spandrels of King's College Chapel. (Bateson is a leading student of animal behavior; he has an adaptationist bent but is among the most critical of thinkers with such a preference.) He argued that the pendant ceiling bosses might be integral to the structure and not secondary (to adorn spandrels necessarily present as architectural byproducts). So we discovered the pathway to the space between the stone-vaulted ceiling and the external roof of King's College Chapel and took a walk on top of the vaults. We saw that the ceiling bosses are clearly secondary; the architects simply cut holes into the spandrels and dropped the bosses through into position. Bateson relented, but later sent me an analysis arguing that the weight of the bosses nonetheless had functional value in balancing the thrust of the heavy vault against the outer pillars. I replied that this may well be so—for secondary *utility* is a primary theme for the later use of spandrels—but that the function of bosses cannot be the reason for spandrels in a fan-vaulted ceiling, for spandrels must be present by architectural necessity in any case, and most other fan-vaulted ceilings (the retrochoir of Peterborough and the cloister of Gloucester, for example) have richly (but entirely superficially) ornamented spandrels, with no bosses or any other component that would add substantial weight.

WORKS CITED

Alvarez, Luis W., Walter Alvarez, F. Asaro, and H. Michel. "Extraterrestrial Cause for the Cretaceous-Tertiary Extinction." *Science* 208 (1980): 1095–1108.

Bateson, William. "Heredity and Variation in Modern Lights." In *Darwin and Modern Science,* ed. A. C. Seward, 85–101. Cambridge (England): Cambridge University Press, 1909.

Bateson, William. *William Bateson, F. R. S. Naturalist: His Essays and Addresses.* Cambridge (England): Cambridge University Press, 1928.

Cain, Arthur J. "The Perfection of Animals." *Viewpoints in Biology* 3 (1964): 36–63.

Darwin, Charles. *On the Origin of Species and the Descent of Man.* Facsimile of the first edition, with an introduction by Ernst Mayr. 1859. Cambridge: Harvard University Press, 1964.

De Vries, Hugo. *The Mutation Theory: Experiments and Observations on the Origin of Species in the Vegetable Kingdom.* Trans. J. B. Farmer and A. D. Darbishire. 2 vols. London: Kegan, Paul, French, and Trubner, 1909.

Eldredge, Niles, and Stephen Jay Gould. "Punctuated Equilibria: An Alternative to Phyletic Gradualism." In *Models in Paleobiology,* ed. T. J. M. Schopf, 82–115. San Francisco: Freeman, Cooper and Co., 1972.

Gould, Stephen Jay. "Allometry and Size in Ontogeny and Phylogeny." *Biological Review* 41 (1966): 587–640.

Gould, Stephen Jay. "Evolutionary Patterns in Pelycosaurian Reptiles: A Factor-Analytic Study." *Evolution* 21 (1967): 385–401.

Gould, Stephen Jay. "Exaptation: A Crucial Tool for an Evolutionary Psychology." *Journal of Social Issues* 47 (1991): 43–65.

Gould, Stephen Jay. "The Hardening of the Modern Synthesis." In *Dimensions of Darwinism,* ed. Marjorie Grene, 71–93. Cambridge (England): Cambridge University Press, 1983.

Gould, Stephen Jay. *Ontogeny and Phylogeny.* Cambridge: Harvard Belknap, 1977.

Gould, Stephen Jay. *Time's Arrow, Time's Cycle.* Cambridge: Harvard University Press, 1987.

Gould, Stephen Jay. *Wonderful Life: The Burgess Shale and the Nature of History.* New York: W. W. Norton, 1989.

Gould, Stephen Jay, and Niles Eldredge. "Punctuated Equilibria: The Tempo and Mode of Evolution Reconsidered." *Paleobiology* 3 (1977): 115–51.

Gould, Stephen Jay, D. M. Raup, J. J. Sepkoski, Jr., T. J. M. Schopf, and D. S. Simberloff. "The Shape of Evolution: A Comparison of Real and Random Clades." *Paleobiology* 3 (1977): 23–40.

Gould, Stephen Jay, and Elisabeth S. Vrba. "Exaptation—A Missing Term in the Science of Form." *Paleobiology* 8 (1982): 4–15.

Huxley, Julian S. *Evolution: The Modern Synthesis.* London: George Allen and Unwin, 1942.

Kellogg, Vernon Lyman. *Darwinism Today.* London: George Bell and Sons, 1907.

Kimura, Motoo. *The Neutral Theory of Molecular Evolution.* Cambridge (England): Cambridge University Press, 1983.

Lyell, Charles. *Principles of Geology: Being an Attempt to Explain the Former Changes of the Earth's Surface by Reference to Causes Now in Operation.* London: John Murray, 1830–33.

Maynard Smith, John. "Dinosaur Dilemmas." *New York Review of Books* 38 (25 April 1991): 5–7.

Mayr, Ernst. *Animal Species and Evolution.* Cambridge: Harvard University Press, 1963.

Stephen Jay Gould

Paley, William. *Natural Theology.* London, 1802.

Porter, Roy. "Charles Lyell and the Principles of the History of Geology." *British Journal for the History of Science* 9 (1976): 91–103.

Provine, William. *Sewall Wright and Evolutionary Biology.* Chicago: University of Chicago Press, 1986.

Raup, David M., S. J. Gould, T. J. M. Schopf, and D. S. Simberloff. "Stochastic Models of Phylogeny and the Evolution of Diversity." *Journal of Geology* 81 (1973): 525–42.

Romanes, George J. "The Darwinism of Darwin and of the Post-Darwinian Schools." In *Darwin, and after Darwin,* vol. 2. London: Longmans, Green and Co., 1900.

Rossi, Paolo. *The Dark Abyss of Time.* Chicago: University of Chicago Press, 1984.

Rudwick, Martin J. S. *The Great Devonian Controversy.* Chicago: University of Chicago Press, 1985.

Rudwick, Martin J. S. *The Meaning of Fossils.* London: Macdonald, 1972.

Tax, Sol, ed. *Evolution after Darwin.* 3 vols. Chicago: University of Chicago Press, 1960.

Wallace, Alfred R. *Darwinism.* London: MacMillan, 1890.

Wilson, Edward O. *Sociobiology: The New Synthesis.* Cambridge: Harvard University Press, 1975.

APPENDIX

THE SPANDRELS OF SAN MARCO

AND THE PANGLOSSIAN PARADIGM

Appendix: "Spandrels," p. 147

The spandrels of San Marco and the Panglossian paradigm: a critique of the adaptationist programme

By S. J. Gould and R. C. Lewontin

Museum of Comparative Zoology, Harvard University, Cambridge, Massachusetts 02138, U.S.A.

An adaptationist programme has dominated evolutionary thought in England and the United States during the past 40 years. It is based on faith in the power of natural selection as an optimizing agent. It proceeds by breaking an organism into unitary 'traits' and proposing an adaptive story for each considered separately. Trade-offs among competing selective demands exert the only brake upon perfection; non-optimality is thereby rendered as a result of adaptation as well. We criticize this approach and attempt to reassert a competing notion (long popular in continental Europe) that organisms must be analysed as integrated wholes, with *Baupläne* so constrained by phyletic heritage, pathways of development and general architecture that the constraints themselves become more interesting and more important in delimiting pathways of change than the selective force that may mediate change when it occurs. We fault the adaptationist programme for its failure to distinguish current utility from reasons for origin (male tyrannosaurs may have used their diminutive front legs to titillate female partners, but this will not explain *why* they got so small); for its unwillingness to consider alternatives to adaptive stories; for its reliance upon plausibility alone as a criterion for accepting speculative tales; and for its failure to consider adequately such competing themes as random fixation of alleles, production of non-adaptive structures by developmental correlation with selected features (allometry, pleiotropy, material compensation, mechanically forced correlation), the separability of adaptation and selection, multiple adaptive peaks, and current utility as an epiphenomenon of non-adaptive structures. We support Darwin's own pluralistic approach to identifying the agents of evolutionary change.

1. Introduction

The great central dome of St Mark's Cathedral in Venice presents in its mosaic design a detailed iconography expressing the mainstays of Christian faith. Three circles of figures radiate out from a central image of Christ: angels, disciples, and virtues. Each circle is divided into quadrants, even though the dome itself is radially symmetrical in structure. Each quadrant meets one of the four spandrels in the arches below the dome. Spandrels – the tapering triangular spaces formed by the intersection of two rounded arches at right angles (figure 1) – are necessary architectural by-products of mounting a dome on rounded arches. Each spandrel contains a design admirably fitted into its tapering space. An evangelist sits in the

"The Spandrels of San Marco and the Panglossian Paradigm" first appeared in the Proceedings of the Royal Society of London, Series B: Biological Sciences 205 (1979): 581–98. It is reprinted here courtesy of S. J. Gould, R. C. Lewontin, and the Royal Society of London.

upper part flanked by the heavenly cities. Below, a man representing one of the four Biblical rivers (Tigris, Euphrates, Indus and Nile) pours water from a pitcher into the narrowing space below his feet.

The design is so elaborate, harmonious and purposeful that we are tempted to view it as the starting point of any analysis, as the cause in some sense of the surrounding architecture. But this would invert the proper path of analysis. The

FIGURE 1. One of the four spandrels of St Mark's; seated evangelist above, personification of river below.

system begins with an architectural constraint: the necessary four spandrels and their tapering triangular form. They provide a space in which the mosaicists worked; they set the quadripartite symmetry of the dome above.

Such architectural constraints abound and we find them easy to understand because we do not impose our biological biases upon them. Every fan vaulted ceiling must have a series of open spaces along the mid-line of the vault, where the sides of the fans intersect between the pillars (figure 2). Since the spaces must exist, they are often used for ingenious ornamental effect. In King's College Chapel in Cambridge, for example, the spaces contain bosses alternately embellished with the Tudor rose and portcullis. In a sense, this design represents an 'adaptation', but the architectural constraint is clearly primary. The spaces arise as a necessary by-product of fan vaulting; their appropriate use is a secondary effect. Anyone who tried to argue that the structure exists because the alternation of rose and portcullis makes so much sense in a Tudor chapel would be inviting the same ridicule that Voltaire heaped on Dr Pangloss: 'Things cannot be other than they

FIGURE 2. The ceiling of King's College Chapel.

are... Everything is made for the best purpose. Our noses were made to carry spectacles, so we have spectacles. Legs were clearly intended for breeches, and we wear them.' Yet evolutionary biologists, in their tendency to focus exclusively on immediate adaptation to local conditions, do tend to ignore architectural constraints and perform just such an inversion of explanation.

As a closer example, recently featured in some important biological literature on adaptation, anthropologist Michael Harner has proposed (1977) that Aztec human sacrifice arose as a solution to chronic shortage of meat (limbs of victims were often consumed, but only by people of high status). E. O. Wilson (1978) has used this explanation as a primary illustration of an adaptive, genetic predisposition for carnivory in humans. Harner and Wilson ask us to view an elaborate

Appendix: "Spandrels," p. 150

social system and a complex set of explicit justifications involving myth, symbol, and tradition as mere epiphenomena generated by the Aztecs as an unconscious rationalization masking the 'real' reason for it all: need for protein. But Sahlins (1978) has argued that human sacrifice represented just one part of an elaborate cultural fabric that, in its entirety, not only represented the material expression of Aztec cosmology, but also performed such utilitarian functions as the maintenance of social ranks and systems of tribute among cities.

We strongly suspect that Aztec cannibalism was an 'adaptation' much like evangelists and rivers in spandrels, or ornamented bosses in ceiling spaces: a secondary epiphenomenon representing a fruitful use of available parts, not a cause of the entire system. To put it crudely: a system developed for other reasons generated an increasing number of fresh bodies; use might as well be made of them. Why invert the whole system in such a curious fashion and view an entire culture as the epiphenomenon of an unusual way to beef up the meat supply. Spandrels do not exist to house the evangelists. (Moreover, as Sahlins argues, it is not even clear that human sacrifice was an adaptation at all. Human cultural practices can be orthogenetic and drive towards extinction in ways that Darwinian processes, based on genetic selection, cannot. Since each new monarch had to outdo his predecessor in even more elaborate and copious sacrifice, the practice was beginning to stretch resources to the breaking point. It would not have been the first time that a human culture did itself in. And, finally, many experts doubt Harner's premise in the first place (Ortiz de Montellano 1978). They argue that other sources of protein were not in short supply, and that a practice awarding meat only to privileged people who had enough anyway, and who used bodies so inefficiently (only the limbs were consumed, and partially at that) represents a mighty poor way to run a butchery.)

We deliberately chose non-biological examples in a sequence running from remote to more familiar: architecture to anthropology. We did this because the primacy of architectural constraint and the epiphenomenal nature of adaptation are not obscured by our biological prejudices in these examples. But we trust that the message for biologists will not go unheeded: if these had been biological systems, would we not, by force of habit, have regarded the epiphenomenal adaptation as primary and tried to build the whole structural system from it?

2. THE ADAPTATIONIST PROGRAMME

We wish to question a deeply engrained habit of thinking among students of evolution. We call it the adaptationist programme, or the Panglossian paradigm. It is rooted in a notion popularized by A. R. Wallace and A. Weismann (but not, as we shall see, by Darwin) towards the end of the nineteenth century: the near omnipotence of natural selection in forging organic design and fashioning the best among possible worlds. This programme regards natural selection as so powerful and the constraints upon it so few that direct production of adaptation through the catechism about genetic drift: it can only be important in populations so small that they are likely to become extinct before playing any sustained evolutionary role (but see Lande 1976).

The admission of alternatives in principle does not imply their serious consideration in daily practice. We all say that not everything is adaptive; yet, faced with an organism, we tend to break it into parts and tell adaptive stories as if trade-offs among competing, well designed parts were the only constraint upon perfection for each trait. It is an old habit. As Romanes complained about A. R. Wallace in 1900: 'Mr. Wallace does not expressly maintain the abstract impossibility of laws and causes other than those of utility and natural selection... Nevertheless, as he nowhere recognizes any other law or cause..., he practically concludes that, on inductive or empirical grounds, there *is* no such other law or cause to be entertained.'

The adaptationist programme can be traced through common styles of argument. We illustrate just a few; we trust they will be recognized by all:

(1) If one adaptive argument fails, try another. Zig-zag commissures of clams and brachiopods, once widely regarded as devices for strengthening the shell, become sieves for restricting particles above a given size (Rudwick 1964). A suite of external structures (horns, antlers, tusks) once viewed as weapons against predators, become symbols of intraspecific competition among males (Davitashvili 1961). The eskimo face, once depicted as 'cold engineered' (Coon *et al.* 1950), becomes an adaptation to generate and withstand large masticatory forces (Shea 1977). We do not attack these newer interpretations; they may all be right. We do wonder, though, whether the failure of one adaptive explanation should always simply inspire a search for another of the same general form, rather than a consideration of alternatives to the proposition that each part is 'for' some specific purpose.

(2) If one adaptive argument fails, assume that another must exist; a weaker version of the first argument. Costa & Bisol (1978), for example, hoped to find a correlation between genetic polymorphism and stability of environment in the deep sea, but they failed. They conclude (1978, pp. 132, 133): 'The degree of genetic polymorphism found would seem to indicate absence of correlation with the particular environmental factors which characterize the sampled area. The results suggest that the adaptive strategies of organisms belonging to different phyla are different.'

(3) In the absence of a good adaptive argument in the first place, attribute failure to imperfect understanding of where an organism lives and what it does. This is again an old argument. Consider Wallace on why all details of colour and form in land snails must be adaptive, even if different animals seem to inhabit the same environment (1899, p. 148): 'The exact proportions of the various species of plants, the numbers of each kind of insect or of bird, the peculiarities of more or less exposure to sunshine or to wind at certain critical epochs, and other slight differences which to us are absolutely immaterial and unrecognizable, may be of

its operation becomes the primary cause of nearly all organic form, function, and behaviour. Constraints upon the pervasive power of natural selection are recognized of course (phyletic inertia primarily among them, although immediate architectural constraints, as discussed in the last section, are rarely acknowledged). But they are usually dimissed as unimportant or else, and more frustratingly, simply acknowledged and then not taken to heart and invoked.

Studies under the adaptationist programme generally proceed in two steps:

(1) An organism is atomized into 'traits' and these traits are explained as structures optimally designed by natural selection for their functions. For lack of space, we must omit an extended discussion of the vital issue: 'what is a trait?' Some evolutionists may regard this as a trivial, or merely a semantic problem. It is not. Organisms are integrated entities, not collections of discrete objects. Evolutionists have often been led astray by inappropriate atomization, as D'Arcy Thompson (1942) loved to point out. Our favourite example involves the human chin (Gould 1977, pp. 381–382; Lewontin 1978). If we regard the chin as a 'thing', rather than as a product of interaction between two growth fields (alveolar and mandibular), then we are led to an interpretation of its origin (recapitulatory) exactly opposite to the one now generally favoured (neotenic).

(2) After the failure of part-by-part optimization, interaction is acknowledged via the dictum that an organism cannot optimize each part without imposing expenses on others. The notion of 'trade-off' is introduced, and organisms are interpreted as best compromises among competing demands. Thus, interaction among parts is retained completely within the adaptationist programme. Any suboptimality of a part is explained as its contribution to the best possible design for the whole. The notion that suboptimality might represent anything other than the immediate work of natural selection is usually not entertained. As Dr Pangloss said in explaining to Candide why he suffered from venereal disease: 'It is indispensable in this best of worlds. For if Columbus, when visiting the West Indies, had not caught this disease, which poisons the source of generation, which frequently even hinders generation, and is clearly opposed to the great end of Nature, we should have neither chocolate nor cochineal.' The adaptationist programme is truly Panglossian. Our world may not be good in an abstract sense, but it is the very best we could have. Each trait plays its part and must be as it is.

At this point, some evolutionists will protest that we are caricaturing their view of adaptation. After all, do they not admit genetic drift, allometry, and a variety of reasons for non-adaptive evolution? They do, to be sure, but we make a different point. In natural history, all possible things happen sometimes; you generally do not support your favoured phenomenon by declaring rivals impossible in theory. Rather, you acknowledge the rival, but circumscribe its domain of action so narrowly that it cannot have any importance in the affairs of nature. Then, you often congratulate yourself for being such an undogmatic and ecumenical chap. We maintain that alternatives to selection for best overall design have generally been relegated to unimportance by this mode of argument. Have we not all heard the highest significance to these humble creatures, and be quite sufficient to require some slight adjustments of size, form, or colour, which natural selection will bring about.'

(4) Emphasize immediate utility and exclude other attributes of form. Fully half the explanatory information accompanying the full-scale Fibreglass *Tyrannosaurus* at Boston's Museum of Science reads: 'Front legs a puzzle: how *Tyrannosaurus* used its tiny front legs is a scientific puzzle; they were too short even to reach the mouth. They may have been used to help the animal rise from a lying position.' (We purposely choose an example based on public impact of science to show how widely habits of the adaptationist programme extend. We are not using glass beasts as straw men; similar arguments and relative emphases, framed in different words, appear regularly in the professional literature.) We don't doubt that *Tyrannosaurus* used its diminutive front legs for something. If they had arisen *de novo*, we would encourage the search for some immediate adaptive reason. But they are, after all, the reduced product of conventionally functional homologues in ancestors (longer limbs of allosaurs, for example). As such, we do not need an explicitly adaptive explanation for the reduction itself. It is likely to be a developmental correlate of allometric fields for relative increase in head and hindlimb size. This non-adaptive hypothesis can be tested by conventional allometric methods (Gould (1974) in general; Lande (1978) on limb reduction) and seems to us both more interesting and fruitful than untestable speculations based on secondary utility in the best of possible worlds. One must not confuse the fact that a structure is used in some way (consider again the spandrels, ceiling spaces and Aztec bodies) with the primary evolutionary reason for its existence and conformation.

3. TELLING STORIES

'All this is a manifestation of the rightness of things, since if there is a volcano at Lisbon it could not be anywhere else. For it is impossible for things not to be where they are, because everything is for the best' (Dr Pangloss on the great Lisbon earthquake of 1755 in which up to 50000 people lost their lives).

We would not object so strenuously to the adaptationist programme if its invocation, in any particular case, could lead in principle to its rejection for want of evidence. We might still view it as restrictive and object to its status as an argument of first choice. But if it could be dismissed after failing some explicit test, then alternatives would get their chance. Unfortunately, a common procedure among evolutionists does not allow such definable rejection for two reasons. First, the rejection of one adaptive story usually leads to its replacement by another, rather than to a suspicion that a different kind of explanation might be required. Since the range of adaptive stories is as wide as our minds are fertile, new stories can always be postulated. And if a story is not immediately available, one can always plead temporary ignorance and trust that it will be forthcoming, as did Costa & Bisol (1978), cited above. Secondly, the criteria for acceptance of a story are so loose that many pass without proper confirmation. Often, evolutionists use *consistency* with natural selection as the sole criterion and consider their work done when they concoct a plausible story. But plausible stories can always be told. The key to historical research lies in devising criteria to identify proper explanations among the substantial set of plausible pathways to any modern result.

We have, for example (Gould 1978) criticized Barash's (1976) work on aggression in mountain bluebirds for this reason. Barash mounted a stuffed male near the nests of two pairs of bluebirds while the male was out foraging. He did this at the same nests on three occasions at 10 day intervals: the first before eggs were laid, the last two afterwards. He then counted aggressive approaches of the returning male towards both the model and the female. At time one, aggression was high towards the model and lower towards females but substantial in both nests. Aggression towards the model declined steadily for times two and three and plummeted to near zero towards females. Barash reasoned that this made evolutionary sense since males would be more sensitive to intruders before eggs were laid than afterwards (when they can have some confidence that their genes are inside). Having devised this plausible story, he considered his work as completed (1976, pp. 1099, 1100):

> 'The results are consistent with the expectations of evolutionary theory. Thus aggression toward an intruding male (the model) would clearly be especially advantageous early in the breeding season, when territories and nests are normally defended...The initial aggressive response to the mated female is also adaptive in that, given a situation suggesting a high probability of adultery (i.e. the presence of the model near the female) and assuming that replacement females are available, obtaining a new mate would enhance the fitness of males... The decline in male–female aggressiveness during incubation and fledgling stages could be attributed to the impossibility of being cuckolded after the eggs have been laid...The results are consistent with an evolutionary interpretation.'

They are indeed consistent, but what about an obvious alternative, dismissed without test by Barash? Male returns at times two and three, approaches the model, tests it a bit, recognizes it as the same phoney he saw before, and doesn't bother his female. Why not at least perform the obvious test for this alternative to a conventional adaptive story: expose a male to the model for the *first* time after the eggs are laid.

Since we criticized Barash's work, Morton *et al.* (1978) repeated it, with some variations (including the introduction of a female model), in the closely related eastern bluebird *Sialia sialis*. 'We hoped to confirm', they wrote, that Barash's conclusions represent 'a widespread evolutionary reality, at least within the genus *Sialia*. Unfortunately, we were unable to do so.' They found no 'anticuckoldry' behaviour at all: males never approached their females aggressively after testing the model at any nesting stage. Instead, females often approached the male model and, in any case, attacked female models more than males attacked male models.

'This violent response resulted in the near destruction of the female model after presentations and its complete demise on the third, as a female flew off with the model's head early in the experiment to lose it for us in the brush' (1978, p. 969). Yet, instead of calling Barash's selected story into question, they merely devise one of their own to render both results in the adaptationist mode. Perhaps, they conjecture, replacement females are scarce in their species and abundant in Barash's. Since Barash's males can replace a potentially 'unfaithful' female, they can afford to be choosy and possessive. Eastern bluebird males are stuck with uncommon mates and had best be respectful. They conclude: 'If we did not support Barash's suggestion that male bluebirds show anticuckoldry adaptations, we suggest that both studies still had "results that are consistent with the expectations of evolutionary theory" (Barash 1976, p. 1099), as we presume any careful study would.' But what good is a theory that cannot fail in careful study (since by 'evolutionary theory', they clearly mean the action of natural selection applied to particular cases, rather than the fact of transmutation itself).

4. THE MASTER'S VOICE RE-EXAMINED

Since Darwin has attained sainthood (if not divinity) among evolutionary biologists, and since all sides invoke God's allegiance, Darwin has often been depicted as a radical selectionist at heart who invoked other mechanisms only in retreat, and only as a result of his age's own lamented ignorance about the mechanisms of heredity. This view is false. Although Darwin regarded selection as the most important of evolutionary mechanisms (as do we), no argument from opponents angered him more than the common attempt to caricature and trivialize his theory by stating that it relied exclusively upon natural selection. In the last edition of the *Origin*, he wrote (1872, p. 395):

'As my conclusions have lately been much misrepresented, and it has been stated that I attribute the modification of species exclusively to natural selection, I may be permitted to remark that in the first edition of this work, and subsequently, I placed in a most conspicuous position – namely at the close of the Introduction – the following words: "I am convinced that natural selection has been the main, but not the exclusive means of modification." This has been of no avail. Great is the power of steady misinterpretation.'

Romanes, whose once famous essay (1900) on Darwin's pluralism versus the panselectionism of Wallace and Weismann deserves a resurrection, noted of this passage (1900, p. 5): 'In the whole range of Darwin's writings there cannot be found a passage so strongly worded as this: it presents the only note of bitterness in all the thousands of pages which he has published.' Apparently, Romanes did not know the letter Darwin wrote to *Nature* in 1880, in which he castigated Sir Wyville Thomson for caricaturing his theory as panselectionist (1880, p. 32):

Appendix: "Spandrels," p. 156

'I am sorry to find that Sir Wyville Thomson does not understand the principle of natural selection...If he had done so, he could not have written the following sentence in the Introduction to the Voyage of the Challenger: "The character of the abyssal fauna refuses to give the least support to the theory which refers the evolution of species to extreme variation guided only by natural selection." This is a standard of criticism not uncommonly reached by theologians and metaphysicians when they write on scientific subjects, but is something new as coming from a naturalist...Can Sir Wyville Thomson name any one who has said that the evolution of species depends only on natural selection? As far as concerns myself, I believe that no one has brought forward so many observations on the effects of the use and disuse of parts, as I have done in my "Variation of Animals and Plants under Domestication"; and these observations were made for this special object. I have likewise there adduced a considerable body of facts, showing the direct action of external conditions on organisms.'

We do not now regard all of Darwin's subsidiary mechanisms as significant or even valid, though many, including direct modification and correlation of growth, are very important. But we should cherish his consistent attitude of pluralism in attempting to explain Nature's complexity.

5. A PARTIAL TYPOLOGY OF ALTERNATIVES TO THE ADAPTATIONIST PROGRAMME

In Darwin's pluralistic spirit, we present an incomplete hierarchy of alternatives to immediate adaptation for the explanation of form, function, and behaviour.

(1) No adaptation and no selection at all. At present, population geneticists are sharply divided on the question of how much genetic polymorphism within populations and how much of the genetic differences between species is, in fact, the result of natural selection as opposed to purely random factors. Populations are finite in size and the isolated populations that form the first step in the speciation process are often founded by a very small number of individuals. As a result of this restriction in population size, frequencies of alleles change by *genetic drift*, a kind of random genetic sampling error. The stochastic process of change in gene frequency by random genetic drift, including the very strong sampling process that goes on when a new isolated population is formed from a few immigrants, has several important consequences. First, populations and species will become genetically differentiated, and even fixed for different alleles at a locus in the complete absence of any selective force at all.

Secondly, alleles can become fixed in a population *in spite of natural selection*. Even if an allele is favoured by natural selection, some proportion of population, depending upon the product of population size N and selection intensity s, will become homozygous for the less fit allele because of genetic drift. If Ns is large this random fixation for unfavourable alleles is a rare phenomenon, but if selection coefficients are on the order of the reciprocal of population size ($Ns = 1$) or smaller, fixation for deleterious alleles is common. If many genes are involved in influencing a metric character like shape, metabolism or behaviour, then the intensity of selection on each locus will be small and Ns per locus may be small. As a result, many of the loci may be fixed for non-optimal alleles.

Thirdly, new mutations have a small chance of being incorporated into a population, even when selectively favoured. Genetic drift causes the immediate loss of most new mutations after their introduction. With a selection intensity s, a new favourable mutation has a probability of only $2s$ of ever being incorporated. Thus, one cannot claim that, eventually, a new mutation of just the right sort for some adaptive argument will occur and spread. 'Eventually' becomes a very long time if only one in 1000 or one in 10000 of the 'right' mutations that do occur ever get incorporated in a population.

(2) No adaptation and no selection on the part at issue; form of the part is a correlated consequence of selection directed elsewhere. Under this important category, Darwin ranked his 'mysterious' laws of the 'correlation of growth'. Today, we speak of pleiotropy, allometry, 'material compensation' (Rensch 1959, pp. 179–187) and mechanically forced correlations in D'Arcy Thompson's sense (1942; Gould 1971). Here we come face to face with organisms as integrated wholes, fundamentally not decomposable into independent and separately optimized parts.

Although allometric patterns are as subject to selection as static morphology itself (Gould 1966), some regularities in relative growth are probably not under immediate adaptive control. For example, we do not doubt that the famous 0.66 interspecific allometry of brain size in all major vertebrate groups represents a selected 'design criterion,' though its significance remains elusive (Jerison 1973). It is too repeatable across too wide a taxonomic range to represent much else than a series of creatures similarly well designed for their different sizes. But another common allometry, the 0.2 to 0.4 intraspecific scaling among homeothermic adults differing in body size, or among races within a species, probably does not require a selectionist story though many, including one of us, have tried to provide one (Gould 1974). R. Lande (personal communication) has used the experiments of Falconer (1973) to show that selection upon *body size alone* yields a brain–body slope across generations of 0.35 in mice.

More compelling examples abound in the literature on selection for altering the timing of maturation (Gould 1977). At least three times in the evolution of arthropods (mites, flies and beetles), the same complex adaptation has evolved, apparently for rapid turnover of generations in strongly r-selected feeders on superabundant but ephemeral fungal resources: females reproduce as larvae and grow the next generation within their bodies. Offspring eat their mother from inside and emerge from her hollow shell, only to be devoured a few days later by their own progeny. It would be foolish to seek adaptive significance in paedomorphic morphology *per se*; it is primarily a by-product of selection for rapid cycling of generations. In more interesting cases, selection for small size (as in animals of the interstitial fauna) or rapid maturation (dwarf males of many crustaceans) has occurred by progenesis (Gould 1977, pp. 324–336), and descendant adults contain a mixture of ancestral juvenile and adult features. Many biologists have been tempted to find primary adaptive meaning for the mixture, but it probably arises as a by-product of truncated maturation, leaving some features 'behind' in the larval state, while allowing others, more strongly correlated with sexual maturation, to retain the adult configuration of ancestors.

(3) The decoupling of selection and adaptation.

(i) Selection without adaptation. Lewontin (1979) has presented the following hypothetical example: 'A mutation which doubles the fecundity of individuals will sweep through a population rapidly. If there has been no change in efficiency of resource utilization, the individuals will leave no more offspring than before, but simply lay twice as many eggs, the excess dying because of resource limitation. In what sense are the individuals or the population as a whole better adapted than before? Indeed, if a predator on immature stages is led to switch to the species now that immatures are more plentiful, the population size may actually decrease as a consequence, yet natural selection at all times will favour individuals with higher fecundity.'

(ii) Adaptation without selection. Many sedentary marine organisms, sponges and corals in particular, are well adapted to the flow régimes in which they live. A wide spectrum of 'good design' may be purely phenotypic in origin, largely induced by the current itself. (We may be sure of this in numerous cases, when genetically identical individuals of a colony assume different shapes in different microhabitats.) Larger patterns of geographic variation are often adaptive and purely phenotypic as well. Sweeney & Vannote (1978), for example, showed that many hemimetabolous aquatic insects reach smaller adult size with reduced fecundity when they grow at temperatures above and below their optima. Coherent, climatically correlated patterns in geographic distribution for these insects – so often taken as *a priori* signs of genetic adaptation – may simply reflect this phenotypic plasticity.

'Adaptation' – the good fit of organisms to their environment – can occur at three hierarchical levels with different causes. It is unfortunate that our language has focused on the common result and called all three phenomena 'adaptation': the differences in process have been obscured and evolutionists have often been misled to extend the Darwinian mode to the other two levels as well. First, we have what physiologists call 'adaptation': the phenotypic plasticity that permits organisms to mould their form to prevailing circumstances during ontogeny. Human 'adaptations' to high altitude fall into this category (while others, like resistance of sickling heterozygotes to malaria, are genetic and Darwinian). Physiological adaptations are not heritable, though the capacity to develop them presumably is. Secondly, we have a 'heritable' form of non-Darwinian adaptation in humans (and, in rudimentary ways, in a few other advanced social species):

cultural adaptation (with heritability imposed by learning). Much confused thinking in human sociobiology arises from a failure to distinguish this mode from Darwinian adaptation based on genetic variation. Finally, we have adaptation arising from the conventional Darwinian mechanism of selection upon genetic variation. The mere existence of a good fit between organism and environment is insufficient evidence for inferring the action of natural selection.

(4) Adaptation and selection but no selective basis for differences among adaptations. Species of related organisms, or subpopulations within a species, often develop different adaptations as solutions to the same problem. When 'multiple adaptive peaks' are occupied, we usually have no basis for asserting that one solution is better than another. The solution followed in any spot is a result of history; the first steps went in one direction, though others would have led to adequate prosperity as well. Every naturalist has his favourite illustration. In the West Indian land snail *Cerion*, for example, populations living on rocky and windy coasts almost always develop white, thick and relatively squat shells for conventional adaptive reasons. We can identify at least two different developmental pathways to whiteness from the mottling of early whorls in all *Cerion*, two paths to thickened shells and three styles of allometry leading to squat shells. All 12 combinations can be identified in Bahamian populations, but would it be fruitful to ask why – in the sense of optimal design rather than historical contingency – *Cerion* from eastern Long Island evolved one solution, and *Cerion* from Acklins Island another?

(5) Adaptation and selection, but the adaptation is a secondary utilization of parts present for reasons of architecture, development or history. We have already discussed this neglected subject in the first section on spandrels, spaces and cannibalism. If blushing turns out to be an adaptation affected by sexual selection in humans, it will not help us to understand why blood is red. The immediate utility of an organic structure often says nothing at all about the reason for its being.

6. Another, and unfairly maligned, approach to evolution

In continental Europe, evolutionists have never been much attracted to the Anglo-American penchant for atomizing organisms into parts and trying to explain each as a direct adaptation. Their general alternative exists in both a strong and a weak form. In the strong form, as advocated by such major theorists as Schindewolf (1950), Remane (1971), and Grassé (1977), natural selection under the adaptationist programme can explain superficial modifications of the *Bauplan* that fit structure to environment: why moles are blind, giraffes have long necks, and ducks webbed feet, for example. But the important steps of evolution, the construction of the *Bauplan* itself and the transition between *Baupläne*, must involve some other unknown, and perhaps 'internal', mechanism. We believe that English biologists have been right in rejecting this strong form as close to an appeal to mysticism.

Appendix: "Spandrels," p. 160

But the argument has a weaker – and paradoxically powerful – form that has not been appreciated, but deserves to be. It also acknowledges conventional selection for superficial modifications of the *Bauplan*. It also denies that the adaptationist programme (atomization plus optimizing selection on parts) can do much to explain *Baupläne* and the transitions between them. But it does not therefore resort to a fundamentally unknown process. It holds instead that the basic body plans of organisms are so integrated and so replete with constraints upon adaptation (categories 2 and 5 of our typology) that conventional styles of selective arguments can explain little of interest about them. It does not deny that change, when it occurs, may be mediated by natural selection, but it holds that constraints restrict possible paths and modes of change so strongly that the constraints themselves become much the most interesting aspect of evolution.

Rupert Riedl, the Austrian zoologist who has tried to develop this thesis for English audiences (1977 and 1975, now being translated into English by R. Jefferies), writes:

'The living world happens to be crowded by universal patterns of organization which, most obviously, find no direct explanation through environmental conditions or adaptive radiation, but exist primarily through universal requirements which can only be expected under the systems conditions of complex organization itself. . . This is not self-evident, for the whole of the huge and profound thought collected in the field of morphology, from Goethe to Remane, has virtually been cut off from modern biology. It is not taught in most American universities. Even the teachers who could teach it have disappeared.'

Constraints upon evolutionary change may be ordered into at least two categories. All evolutionists are familiar with *phyletic* constraints, as embodied in Gregory's classic distinction (1936) between habitus and heritage. We acknowledge a kind of phyletic inertia in recognizing, for example, that humans are not optimally designed for upright posture because so much of our *Bauplan* evolved for quadrupedal life. We also invoke phyletic constraint in explaining why no molluscs fly in air and no insects are as large as elephants.

Developmental constraints, a subcategory of phyletic restrictions, may hold the most powerful rein of all over possible evolutionary pathways. In complex organisms, early stages of ontogeny are remarkably refractory to evolutionary change, presumably because the differentiation of organ systems and their integration into a functioning body is such a delicate process, so easily derailed by early errors with accumulating effects. Von Baer's fundamental embryological laws (1828) represent little more than a recognition that early stages are both highly conservative and strongly restrictive of later development. Haeckel's biogenetic law, the primary subject of late nineteenth century evolutionary biology, rested upon a misreading of the same data (Gould 1977). If development occurs in integrated packages, and cannot be pulled apart piece by piece in evolution, then the adaptationist programme cannot explain the alteration of developmental programmes underlying nearly all changes of *Bauplan*.

The German palaeontologist A. Seilacher, whose work deserves far more attention than it has received, has emphasized what he calls '*bautechnischer*', or *architectural*, constraints (Seilacher 1970). These arise not from former adaptations retained in a new ecological setting (phyletic constraints as usually understood), but as architectural restrictions that never were adaptations, but rather the necessary consequences of materials and designs selected to build basic *Baupläne*. We devoted

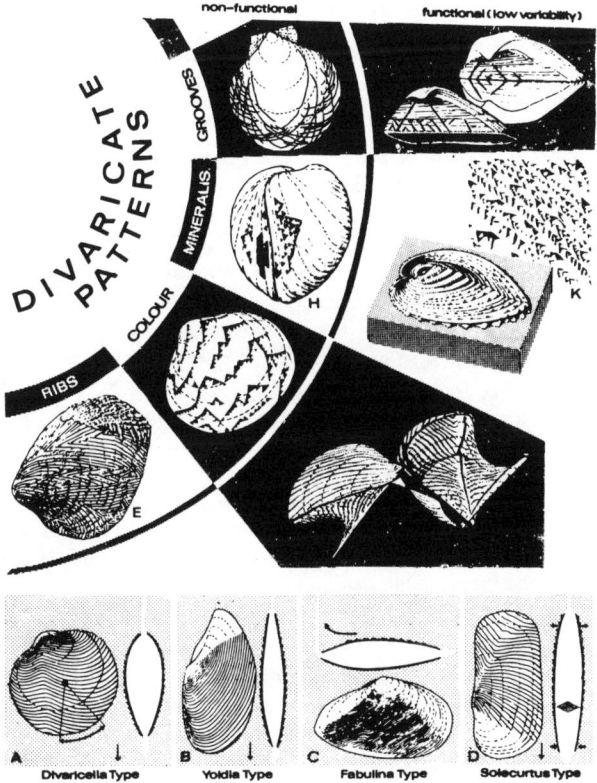

FIGURE 3. The range of divaricate patterns in molluscs. E, F, H, and L are non-functional in Seilacher's judgement. A–D are functional ribs (but these are far less common than non-functional ribs of the form E). G is the mimetic *Arca zebra*. K is *Corculum*. See text for details.

the first section of this paper to non-biological examples in this category. Spandrels must exist once a blueprint specifies that a dome shall rest on rounded arches. Architectural constraints can exert a far-ranging influence upon organisms as well. The subject is full of potential insight because it has rarely been acknowledged at all.

In a fascinating example, Seilacher (1972) has shown that the divaricate form of architecture (figure 3) occurs again and again in all groups of molluscs, and in brachiopods as well. This basic form expresses itself in a wide variety of structures: raised ornamental lines (not growth lines because they do not conform to the mantle margin at any time), patterns of coloration, internal structures in the mineralization of calcite, and incised grooves. He does not know what generates this pattern and feels that traditional and nearly exclusive focus on the adaptive value of each manifestation has diverted attention from questions of its genesis in growth and also prevented its recognition as a general phenomenon. It must arise from some characteristic pattern of inhomogeneity in the growing mantle, probably from the generation of interference patterns around regularly spaced centres; simple computer simulations can generate the form in this manner (Waddington & Cowe 1969). The general pattern may not be a direct adaptation at all.

Seilacher then argues that most manifestations of the pattern are probably non-adaptive. His reasons vary, but seem generally sound to us. Some are based on field observations: colour patterns that remain invisible because clams possessing them either live buried in sediments or remain covered with a periostracum so thick that the colours cannot be seen. Others rely on more general principles: presence only in odd and pathological individuals, rarity as a developmental anomaly, excessive variability compared with much reduced variability when the same general structure assumes a form judged functional on engineering grounds.

In a distinct minority of cases, the divaricate pattern becomes functional in each of the four categories (figure 3). Divaricate ribs may act as scoops and anchors in burrowing (Stanley 1970), but they are not properly arranged for such function in most clams. The colour chevrons are mimetic in one species (*Pteria zebra*) that lives on hydrozoan branches; here the variability is strongly reduced. The mineralization chevrons are probably adaptive in only one remarkable creature, the peculiar bivalve *Corculum cardissa* (in other species, they either appear in odd specimens or only as post-mortem products of shell erosion). This clam is uniquely flattened in an anterio-posterior direction. It lies on the substrate, posterior up. Distributed over its rear end are divaricate triangles of mineralization. They are translucent, while the rest of the shell is opaque. Under these windows dwell endosymbiotic algae!

All previous literature on divaricate structure has focused on its adaptive significance (and failed to find any in most cases). But Seilacher is probably right in representing this case as the spandrels, ceiling holes and sacrificed bodies of our first section. The divaricate pattern is a fundamental architectural constraint. Occasionally, since it is there, it is used to beneficial effect. But we cannot understand the pattern or its evolutionary meaning by viewing these infrequent and secondary adaptations as a reason for the pattern itself.

Galton (1909, p. 257) contrasted the adaptationist programme with a focus on constraints and modes of development by citing a telling anecdote about Herbert Spencer's fingerprints:

'Much has been written, but the last word has not been said, on the rationale of these curious papillary ridges; why in one man and in one finger they form whorls and in another loops. I may mention a characteristic anecdote of Herbert

Spencer in connection with this. He asked me to show him my Laboratory and to take his prints, which I did. Then I spoke of the failure to discover the origin of these patterns, and how the fingers of unborn children had been dissected to ascertain their earliest stages, and so forth. Spencer remarked that this was beginning in the wrong way; that I ought to consider the purpose the ridges had to fulfil, and to work backwards. Here, he said, it was obvious that the delicate mouths of the sudorific glands required the protection given to them by the ridges on either side of them, and therefrom he elaborated a consistent and ingenious hypothesis at great length. I replied that his arguments were beautiful and deserved to be true, but it happened that the mouths of the ducts did not run in the valleys between the crests, but along the crests of the ridges themselves.

We feel that the potential rewards of abandoning exclusive focus on the adaptationist programme are very great indeed. We do not offer a council of despair, as adaptationists have charged; for non-adaptive does not mean non-intelligible. We welcome the richness that a pluralistic approach, so akin to Darwin's spirit, can provide. Under the adaptationist programme, the great historic themes of developmental morphology and *Bauplan* were largely abandoned; for if selection can break any correlation and optimize parts separately, then an organism's integration counts for little. Too often, the adaptationist programme gave us an evolutionary biology of parts and genes, but not of organisms. It assumed that all transitions could occur step by step and underrated the importance of integrated developmental blocks and pervasive constraints of history and architecture. A pluralistic view could put organisms, with all their recalcitrant, yet intelligible, complexity, back into evolutionary theory.

REFERENCES (Gould & Lewontin)

Baer, K. E. von 1828 *Entwicklungsgeschichte der Tiere.* Königsberg: Bornträger.
Barash, D. P. 1976 Male response to apparent female adultery in the mountain bluebird: an evolutionary interpretation. *Am. Nat.* **110**, 1097-1101.
Coon, C. S., Garn, S. M. & Birdsell, J. B. 1950 *Races.* Springfield, Ohio: C. Thomas.
Costa, R. & Bisol, P. M. 1978 Genetic variability in deep-sea organisms. *Biol. Bull.* **155**, 125-133.
Darwin, C. 1872 *The origin of species.* London: John Murray.
Darwin, C. 1880 Sir Wyville Thomson and natural selection. *Nature, Lond.* **23**, 32.
Davitashvili, L. S. 1961 *Teoriya polovogo otbora* [Theory of sexual selection]. Moscow: Akademii Nauk.
Falconer, D. S. 1973 Replicated selection for body weight in mice. *Genet. Res.* **22**, 291-321.
Galton, F. 1909 *Memories of my life.* London: Methuen.
Gould, S. J. 1966 Allometry and size in ontogeny and phylogeny. *Biol. Rev.* **41**, 587-640.
Gould, S. J. 1971 D'Arcy Thompson and the science of form. *New Literary Hist.* **2**(2), 229-258.
Gould, S. J. 1974 Allometry in primates, with emphasis on scaling and the evolution of the brain. In *Approaches to primate paleobiology. Contrib. Primatol.* **5**, 244-292.
Gould, S. J. 1977 *Ontogeny and phylogeny.* Cambridge, Mass.: Belknap Press.
Gould, S. J. 1978 Sociobiology: the art of storytelling. *New Scient.* **80**, 530-533.

356

Appendix: "Spandrels," p. 164

Grassé, P.-P. 1977 *Evolution of living organisms*. New York: Academic Press.
Gregory, W. K. 1936 Habitus factors in the skeleton of fossil and recent mammals. *Proc. Am. phil. Soc.* **76**, 429–444.
Harner, M. 1977 The ecological basis for Aztec sacrifice. *Am. Ethnologist* **4**, 117–135.
Jerison, H. J. 1973 *Evolution of the brain and intelligence*. New York: Academic Press.
Lande, R. 1976 Natural selection and random genetic drift in phenotypic evolution. *Evolution* **30**, 314–334.
Lande, R. 1978 Evolutionary mechanisms of limb loss in tetrapods. *Evolution* **32**, 73–92.
Lewontin, R. C. 1978 Adaptation. *Scient. Am.* **239** (3), 156–169.
Lewontin, R. C. 1979 Sociobiology as an adaptationist program. *Behav. Sci.* (In the press.)
Morton, E. S., Geitgey, M. S. & McGrath, S. 1978 On bluebird 'responses to apparent female adultery'. *Am. Nat.* **112**, 968–971.
Ortiz de Montellano, B. R. 1978 Aztec cannibalism: an ecological necessity? *Science N.Y.* **200**, 611–617.
Remane, A. 1971 *Die Grundlagen des natürlichen Systems der vergleichenden Anatomie und der Phylogenetik*. Königstein-Taunus: Koeltz.
Rensch, B. 1959 *Evolution above the species level*. New York: Columbia University Press.
Riedl, R. 1975 *Die Ordnung des Lebendigen*. Hamburg: Paul Parey.
Riedl, R. 1977 A systems-analytical approach to macro-evolutionary phenomena. *Q. Rev. Biol.* **52**, 351–370.
Romanes, G. J. 1900 The Darwinism of Darwin and of the post-Darwinian schools. In *Darwin, and after Darwin*, vol. 2, new edn. London: Longmans, Green & Co.
Rudwick, M. J. S. 1964 The function of zig-zag deflections in the commissures of fossil brachiopods. *Palaeontology* **7**, 135–171.
Sahlins, M. 1978 Culture as protein and profit. *New York review of books*, 23 Nov., pp. 45–53.
Schindewolf, O. H. 1950 *Grundfragen der Paläontologie*. Stuttgart: Schweizerbart.
Seilacher, A. 1970 Arbeitskonzept zur Konstruktionsmorphologie. *Lethaia* **3**, 393–396.
Seilacher, A. 1972 Divaricate patterns in pelecypod shells. *Lethaia* **5**, 325–343.
Shea, B. T. 1977 Eskimo craniofacial morphology, cold stress and the maxillary sinus. *Am. J. phys. Anthrop.* **47**, 289–300.
Stanley, S. M. 1970 Relation of shell form to life habits in the Bivalvia (Mollusca). *Mem. geol. Soc. Am.* no. 125, 296 pp.
Sweeney, B. W. & Vannote, R. L. 1978 Size variation and the distribution of hemimetabolous aquatic insects: two thermal equilibrium hypotheses. *Science, N.Y.* **200**, 444–446.
Thompson, D. W. 1942 *Growth and form*. New York: Macmillan.
Waddington, C. H. & Cowe, J. R. 1969 Computer simulation of a molluscan pigmentation pattern. *J. theor. Biol.* **25**, 219–225.
Wallace, A. R. 1899 *Darwinism*. London: Macmillan.
Wilson, E. O. 1978 *On human nature*. Cambridge, Mass.: Harvard University Press.

BIBLIOGRAPHY

INDEX

BIBLIOGRAPHY

Althusser, Louis. *Lenin and Philosophy and Other Essays*. Trans. Ben Brewster. New York: Monthly Review Press, 1971.

Altman, Jeanne. *Baboon Mothers and Infants*. Cambridge: Harvard University Press, 1980.

Alvarez, Luis W., Walter Alvarez, F. Asaro, and H. Michel. "Extraterrestrial Cause for the Cretaceous-Tertiary Extinction." *Science* 208 (1980): 1095–1108.

Andrews, James R. *The Practice of Rhetorical Criticism*. New York: Macmillan, 1988.

Aristotle. *The Rhetoric and Poetics of Aristotle*. Trans. W. R. Roberts and I. Bywater. New York: Modern Library, 1984.

Aronowitz, Stanley. *Science as Power: Discourse and Ideology in Modern Society*. Minneapolis: University of Minnesota Press, 1988.

Attridge, Derek. "Closing Statement: Linguistics and Poetics in Retrospect." In *The Linguistics of Writing: Arguments between Language and Literature*, ed. Nigel Fabb, Derek Attridge, Alan Durant, and Colin MacCabe, 15–32. New York: Methuen, 1987.

Bakhtin, M. M. *The Dialogic Imagination*. Austin: University of Texas Press, 1981.

Barash, David. "Predictive Sociobiology: Damselfishes and Sparrows." In *Sociobiology: Beyond Nature/Nurture?* ed. G. W. Barlow and J. Silverberg, 209–26. Boulder: Westview, 1980.

Barber, C. L. "Some Measurable Characteristics of Modern Scientific Prose." In *Contributions to English Syntax and Philology*. Stockholm: Almquist and Wiksell, 1962. Reprinted, with commentary, in *Episodes in ESP*, ed. John Swales, 1–16. Oxford: Pergamon, 1985.

Barthes, Roland. *Image, Music, Text*. New York: Hill and Wang, 1977.

Bateson, William. *William Bateson, F. R. S. Naturalist: His Essays and Addresses*. Cambridge: Cambridge University Press, 1928.

Bateson, William. "Heredity and Variation in Modern Lights." In *Darwin and Modern Science*, ed. A. C. Seward, 85–101. Cambridge: Cambridge University Press, 1909.

Bazerman, Charles. "How Natural Philosophers Can Cooperate." In *Textual Dynamics of the Professions*, ed. Charles Bazerman and James Paradis, 13–44. Madison: University of Wisconsin Press, 1991.

Bazerman, Charles. "Scientific Writing as a Social Act: A Review of the Literature." In *New Essays in Technical and Scientific Communication: Research, Theory, Practice*, ed. Paul V. Anderson, R. John Brockman, and Carolyn R. Miller, 156–84. Farmingdale, N.Y.: Baywood, 1983.

Bazerman, Charles. *Shaping Written Knowledge: The Genre and Activity of the*

Bibliography

Experimental Article in Science. Madison: University of Wisconsin Press, 1988.

Belenky, Mary Field, Blythe McVicher Clinchy, Nancy Rule Goldberger, and Jill Mattuck Tarule. *Women's Way of Knowing: The Development of Self, Voice, and Mind*. New York: Basic Books, 1986.

Benhabib, Seyla. *Critique, Norm, and Utopia: A Study of the Foundations of Critical Theory*. New York: Columbia University Press, 1986.

Bereiter, Carl, and Marlene Bird. "Use of Thinking Aloud in Identification and Teaching of Reading Comprehension Strategies." *Cognition and Instruction* 2 (1985): 131–56.

Berry, Margaret. *Introduction to Systemic Linguistics: 2 Levels and Links*. New York: St. Martin's Press, 1976.

Billig, Michael. *Arguing and Thinking: A Rhetorical Approach to Social Psychology*. Cambridge: Cambridge University Press, 1987.

Biology and Gender Study Group. "The Importance of Feminist Critique for Contemporary Cell Biology." In *Feminism and Science*, ed. Nancy Tuana, 172–87. Bloomington: Indiana University Press, 1989.

Bitzer, Lloyd, and Edwin Black. *The Prospect of Rhetoric*. Englewood Cliffs, N.J.: Prentice-Hall, 1971.

Black, Edwin. *Rhetorical Criticism: A Study in Method*. New York: Macmillan, 1965.

Black, Edwin. "The Second Persona." *Quarterly Journal of Speech* 56 (1970): 109–19.

Black, Max. *Models and Metaphors: Studies in Language and Philosophy*. Ithaca: Cornell University Press, 1962.

Bleich, David. *Subjective Critisism*. Baltimore: Johns Hopkins University Press, 1978.

Bleier, Ruth. "Introduction." In *Feminist Approaches to Science*, ed. Ruth Bleier, 1–17. New York: Pergamon Press, 1986.

Bleier, Ruth. "Lab Coat: Robe of Innocence or Klansman's Sheet?" In *Feminist Studies/Critical Studies*, ed. Teresa de Laurentis, 55–66. New York: Mac-Millan, 1988.

Bleier, Ruth. *Science and Gender: A Critique of Biology and Its Theories on Women*. New York: Pergamon Press, 1984.

Bokeno, R. Michael. "The Rhetorical Understanding of Science: An Explication and Critical Commentary." *Southern Speech Communication Journal* 52 (1987): 285–311.

Bolinger, Dwight. *Aspects of Language*. 2nd ed. New York: Harcourt and Brace, 1973.

Booth, Wayne. *The Rhetoric of Fiction*. Chicago: University of Chicago Press, 1961.

Brantlinger, Patrick. *Crusoe's Footprint: Cultural Studies in Britain and America*. London: Routledge, 1990.

Brodkey, Linda. "The Discourses of Difference and Consensus." In *Academic Writing as a Social Practice*, 108–66. Philadelphia: Temple University Press, 1984.

Brooke, Robert. "Control in Writing: Flower, Derrida, and the Images of the Writer." *College English* 51 (1989): 405–17.

Brooks, Cleanth, and Robert Penn Warren. *Understanding Poetry.* New York: Holt, Rinehart, and Winston, 1938.

Brown, Penelope, and Stephen Levinson. "Politeness." In *Questions and Politeness,* ed. Esther Goody. Cambridge: Cambridge University Press, 1978. Reprinted, with updated bibliography, as *Politeness.* Cambridge: Cambridge University Press, 1987.

Brown, Roger, and Albert Gilman. "Pronouns of Power and Solidarity." In *Style in Language,* ed. T. A. Sebeok, 253–76. Cambridge: MIT Press, 1960.

Brueggemann, Brenda. "The Collapsing Structure of Thomas Kuhn's (R)Evolutionary Text." Paper presented at the 1990 Twentieth Century Literature Conference, Louisville, Ky.

Burke, Kenneth. *A Grammar of Motives.* 3rd ed. Berkeley: University of California Press, 1969.

Cain, Arthur. J. "The Perfection of Animals." *Viewpoints in Biology* 3 (1964): 36–63.

Campbell, John A. "The Polemical Mr. Darwin." *Quarterly Journal of Speech* 61 (1975): 375–90.

Campbell, John A. "Scientific Discovery and Rhetorical Invention: The Path to Darwin's *Origin.*" In *The Rhetorical Turn: Invention and Persuasion in the Conduct of Inquiry,* ed. Herbert W. Simon, 58–90. Chicago: University of Chicago Press, 1990.

Campbell, John A. "Scientific Revolution and the Grammar of Culture: The Case of Darwin's *Origin.*" *Quarterly Journal of Speech* 72 (1989): 351–76.

Cannon, W. Faye. "Charles Lyell and the Principles of the History of Geology." *British Journal for the History of Science* 9 (1976): 91–103.

Carter, Ronald, and Walter Nash. *Seeing Through Language: A Guide to Styles of English Writing.* Oxford: Basil Blackwell, 1990.

Cherry, Roger. "Politeness in Written Persuasion." *Journal of Pragmatics* 12 (1988): 63–81.

Cicero. *Brutus, Orator.* Trans. H. M. Hubbell. Cambridge: Harvard University Press, 1952.

[Cicero]. *Rhetorica ad Herennium.* Trans. H. Caplan. Cambridge: Harvard University Press, 1954.

Collins, Harry. *Changing Order: Replication and Induction in Scientific Practice.* Beverly Hills: Sage, 1985.

Corbett, Edward P. J., ed. *Rhetorical Analysis of Literary Works.* New York: Oxford University Press, 1969.

Council of Biology Editors Style Manual Committee. *Council of Biology Editors Style Manual.* 5th ed. Bethesda, Md.: Council of Biology Editors, Inc., 1983.

Couture, Barbara. "Attributing Value to Written Text: Some Implications for Linguistics and Interpretation." *Word* 40 (1989): 51–64.

Cozzens, Susan. "Comparing the Sciences: Citation Context Analysis of Papers from Neuropharmacology and the Sociology of Science." *Social Studies of Science* 14 (1985): 127–53.

Crookes, Graham. "Towards a Validated Analysis of Scientific Text." *Applied Linguistics* 7 (1986): 57–70.

Crowley, Sharon. *A Teacher's Introduction to Deconstruction.* Urbana, Ill.: National Council of Teachers of English, 1989.

Culler, Jonathon. *On Deconstruction: Theory and Criticism after Poststructuralism.* Ithaca: Cornell University Press, 1982.

Darwin, Charles. *On the Origin of Species and the Descent of Man.* Facsimile of the first edition, with an introduction by Ernst Mayr. 1859. Cambridge: Harvard University Press, 1964.

Dawkins, Richard. *The Blind Watchmaker.* New York: Norton, 1986.

Dawkins, Richard. *The Selfish Gene.* New York: Oxford University Press, 1976.

Dear, Peter, ed. *The Literary Structure of Scientific Argument.* Philadelphia: University of Pennsylvania Press, 1991.

Delillo, Don. *White Noise.* New York: Viking, 1986.

Derrida, Jacques. "Limited Inc ABC." *Glyph* 2 (1977): 162–254.

Derrida, Jacques. *Margins of Philosophy.* Trans. Alan Bass. Chicago: University of Chicago Press, 1982.

Derrida, Jacques. *Of Grammatology.* Trans. Gayatri C. Spivak. Baltimore: Johns Hopkins University Press, 1976.

Derrida, Jacques. "Signature Event Context." *Glyph* 1 (1977): 172–97.

Derrida, Jacques. *Writing and Difference.* Trans. Alan Bass. Chicago: University of Chicago Press, 1978.

de Queiroz, Kevin. "Systematics and the Darwinian Revolution." *Philosophy of Science* 55 (1988): 238–59.

Devitt, Amy. "Intertextuality in Tax Accounting." In *Textual Dynamics of the Professions,* ed. Charles Bazerman and James Paradis, 336–57. Madison: University of Wisconsin Press, 1991.

DeVries, Hugo. *The Mutation Theory: Experiments and Observations on the Origin of Species in the Vegetable Kingdom.* 2 vols. Trans. J. B. Farmer and A. D. Darbishire. London: Kegan, Paul, French, and Trubner, 1909.

Dews, Peter. *Logics of Disintegration: Post-Structuralist Thought and the Claims of Critical Theory.* London: Verso, 1987.

Dillon, George L. *Constructing Texts: Elements of a Theory of Composition and Style.* Bloomington: Indiana University Press, 1981.

Dudley-Evans, Tony, and Willie Henderson. "The Organisation of Article Introductions: Evidence of Change in Academic Writing." In *The Language of Economics: The Analysis of Economics Discourse,* ed. Tony Dudley-Evans and Willie Henderson, 67–78. London: Modern English Publications and the British Council, 1990.

Duncan, Hugh Dalziel. *Communication and Social Order.* New York: Bedminster, 1962. Reprint. New York: Oxford University Press, 1968.

Eagleton, Terry. *Literary Theory: An Introduction.* Minneapolis: University of Minnesota Press, 1983.

Ede, Lisa, and Andrea Lunsford. "Audience Addressed/Audience Invoked: The Role of Audience in Composition Theory and Pedagogy." *College Composition and Communication* 35 (1984): 155–71.

Edge, David. "Quantitative Measures of Communication in Science: A Critical Review." *History of Science* 17 (1979): 102–34.

Eldredge, Niles, and Stephen Jay Gould. "Punctuated Equilibria: An Alternative to Phyletic Gradualism." In *Models in Paleobiology,* ed. T. M. J. Schopf, 82–115. San Francisco: Freeman, Cooper and Co., 1972.

Eldredge, Niles, and Stephen Jay Gould. "Punctuated Equilibrium Prevails." *Nature* 322 (1988): 211–12.

Ericsson, Anders, and Herbert A. Simon. *Protocol Analysis: Verbal Reports as Data.* Cambridge: MIT Press, 1984.

Fahnestock, Jeanne. "Accommodating Science: The Rhetorical Life of Scientific Facts." *Written Communication* 3 (1986): 275–96.

Fahnestock, Jeanne. "Arguing in Different Forums: The Bering Crossover Controversy." *Science, Technology and Human Values* 14 (1989): 26–42.

Fahnestock, Jeanne, and Marie Secor. "The Stases in Scientific and Literary Argument." *Written Communication* 5 (1988): 427–43.

Fausto-Sterling, Anne. "Woman and Science." *Women's Studies International Quarterly* 4 (1981): 41–50.

Fee, Elizabeth. "Critiques of Modern Science: The Relationship of Feminism to Other Radical Epistemologies." In *Feminist Approaches to Science,* ed. Ruth Bleier, 42–56. New York: Pergamon Press, 1986.

Fee, Elizabeth. "Is Feminism a Threat to Scientific Objectivity?" *International Journal of Women's Studies* 4 (1981): 378–92.

Fish, Stanley. *Is There a Text in This Class? The Authority of Interpretative Communities.* Cambridge: Harvard University Press, 1980.

Fish, Stanley. *Self-Consuming Artifacts.* Berkeley: University of California Press, 1972.

Fish, Stanley. *Surprised by Sin: The Reader in "Paradise Lost."* New York: St. Martin's Press, 1967.

Flax, Jane. "Postmodernism and Gender Relations in Feminist Theory." *Signs* 12 (1987): 621–43.

Flower, Linda, and John R. Hayes. "The Pregnant Pause: An Inquiry into the Nature of Planning." *Research in the Teaching of English* 15 (1981): 229–43.

Foss, Sonja, ed. *Rhetorical Criticism: Exploration and Practice.* Prospect Heights, Ill.: Waveland, 1989.

Foucault, Michel. "What Is an Author?" In *The Foucault Reader,* ed. Raul Rabinow and trans. Josue V. Harari, 101–20. New York: Pantheon, 1979, 1984.

Fowler, Roger. *Linguistic Criticism.* Oxford: Oxford University Press, 1986.

Francis, Gill, and Anneliese Kramer-Dahl. "From Clinical Report to Clinical Story: Two Ways of Writing a Medical Case." In *Recent Systemic and Other Functional Views on Language,* ed. Eija Ventola, 339–68. Berlin: Mouton de Gruyter, 1991.

Frey, Olivia. "Beyond Literary Darwinism: Women's Voices and Critical Discourse." *College English* 52 (1990): 507–26.

Garfield, Eugene. "The Articles Most Cited in the SCI from 1961 to 1982. 7. Another 100 *Citation Classics:* The Watson-Crick Double Helix Has Its Turn." In *Essays of an Information Scientist* 8, 187–96. Philadelphia: Institute for Scientific Information, 1985.

Bibliography

Garfield, Eugene. "The 'Obliteration Phenomenon' in Science—and the Advantage of Being Obliterated!" In *Essays of an Information Scientist* 2, 396–98. Philadelphia: Institute for Scientific Information, 1977.

Geertz, Clifford. *Works and Lives: The Anthropologist as Author.* Palo Alto: Stanford University Press, 1988.

Gibson, Walker. "Authors, Speakers, Readers, and Mock Readers." *College English,* 11 (1950): 265–69.

Giglioli, P., ed. *Language and Social Context.* Harmondsworth, England: Penguin, 1972.

Gilbert, G. Nigel. "Referencing as Persuasion." *Social Studies of Science* 7 (1977): 113–22.

Gilbert, G. Nigel. "The Transformation of Research Findings into Scientific Knowledge." *Social Studies of Science* 9 (1979): 281–306.

Gilbert, G. Nigel, and Michael Mulkay. *Opening Pandora's Box: A Sociological Analysis of Scientists' Discourse.* Cambridge: Cambridge University Press, 1984.

Gilligan, Carol. *In a Different Voice: Psychological Theory and Women's Development.* Cambridge: Harvard University Press, 1982.

Ginzberg, Ruth. "Uncovering Gynocentric Science." In *Feminism and Science,* ed. Nancy Tuana, 69–84. Bloomington: Indiana University Press, 1989.

Glassman, Myron, and Thomas Pinelli. "Scientific Inquiry and Technical Communication: An Introduction to the Research Process." *Technical Communication* 32 (1985): 8–13.

Gortzen, Rene. *Jurgen Habermas: Eine Bibliographie Seiner Schriften und der Sekundarliteratur, 1952–81.* Frankfurt: Suhrkamp, 1982.

Gopnik, Myrna. *Linguistic Structures in Scientific Texts.* The Hague: Mouton, 1972.

Goudge, Thomas Anderson. *The Ascent of Life: A Philosophical Study of the Theory of Evolution.* Toronto: University of Toronto Press, 1961.

Gould, Stephen Jay. "Allometry and Size in Ontogeny and Phylogeny." *Biological Review* 41 (1966): 587–640.

Gould, Stephen Jay. "Allometry in Primates, with Emphasis on Scaling and the Evolution of the Brain." In *Approaches to Primate Paleobiology,* ed. F. S. Szaley, 5244–92. Basel: Karger, 1975.

Gould, Stephen Jay. "D'Arcy Thompson and the Science of Form." *New Literary History* 2 (1971): 229–58.

Gould, Stephen Jay. "Darwinism and the Expansion of Evolutionary Theory." *Science* 216 (1982): 380–87.

Gould, Stephen Jay. *Ever Since Darwin: Reflections in Natural History.* New York: W. W. Norton, 1977.

Gould, Stephen Jay. "Evolution and the Triumph of Homology, or Why History Matters." *American Scientist* 74 (1986): 50–69.

Gould, Stephen Jay. "Evolutionary Patterns in Pelycosaurian Reptiles: A Factor-Analytic Study." *Evolution* 21 (1967): 385–401.

Gould, Stephen Jay. "Exaptation: A Crucial Tool for an Evolutionary Psychology." *Journal of Social Issues* 47 (1991): 43–65.

Gould, Stephen Jay. *The Flamingo's Smile.* New York: W. W. Norton, 1985.

Gould, Stephen Jay. "The Hardening of the Modern Synthesis." In *Dimensions of Darwinism,* ed. Marjorie Grene, 71–93. Cambridge: Cambridge University Press, 1983.

Gould, Stephen Jay. *Hen's Teeth and Horse's Toes.* New York: W. W. Norton, 1983.

Gould, Stephen Jay. "Is a New and General Theory of Evolution Emerging?" *Paleobiology* 6 (1980): 119–30.

Gould, Stephen Jay. *The Mismeasure of Man.* New York: W. W. Norton, 1981.

Gould, Stephen Jay. *Ontogeny and Phylogeny.* Cambridge: Harvard Belknap Press 1977.

Gould, Stephen Jay. *The Panda's Thumb.* New York: W. W. Norton, 1980.

Gould, Stephen Jay. "Sociobiology and the Theory of Natural Selection." In *Sociobiology: Beyond Nature/Nurture?* ed. G. W. Barlow and J. Silverberg, 257–69. Boulder: Westview, 1980.

Gould, Stephen Jay. "Sociobiology: The Art of Storytelling." *New Scientist* 80 (1978): 530–33.

Gould, Stephen Jay. *Time's Arrow, Time's Cycle.* Cambridge: Harvard University Press, 1987.

Gould, Stephen Jay. *Wonderful Life: The Burgess Shale and the Nature of History.* New York: W. W. Norton, 1989.

Gould, Stephen Jay, and Niles Eldredge. "Punctuated Equilibria: The Tempo and the Mode of Evolution Reconsidered." *Paleobiology* 3 (1977): 115–51.

Gould, Stephen Jay, and Niles Eldredge. "Punctuated Equilibrium at the Third Stage." *Systematic Zoology* 35 (1986): 143–48.

Gould, Stephen Jay, and Richard C. Lewontin. "The Spandrels of San Marco and the Panglossian Paradigm: A Critique of the Adaptationist Programme." *Proceedings of the Royal Society of London, B: Biological Sciences* 205 (1979): 581–98.

Gould, Stephen Jay, David M. Raup, J. John Sepkoski, Jr., T. J. M. Schopf, and D. S. Simberloff. "The Shape of Evolution: A Comparison of Real and Random Clades." *Paleobiology* 3 (1977): 23–40.

Gould, Stephen, and Elisabeth S. Vrba. "Exaptation—A Missing Term in the Science of Form." *Paleobiology* 8 (1982): 4–15.

Gragson, Gay, and Jack Selzer. "Fictionalizing the Readers of Scholarly Articles in Biology." *Written Communication* 7 (1990): 25–58.

Gross, Alan. "The Origin of Species: Evolutionary Taxonomy as an Example of the Rhetoric of Science." In *The Rhetorical Turn: Invention and Persuasion in the Conduct of Inquiry,* 91–115. Chicago: University of Chicago Press, 1990.

Gross, Alan. *The Rhetoric of Science.* Cambridge: Harvard University Press, 1989.

Gusfield, Joseph. "The Literary Rhetoric of Science." *American Sociological Review* 3 (1984): 70–83.

Habermas, Jürgen. *Knowledge and Human Interests.* Boston: Beacon, 1971.

Habermas, Jürgen. *Legitimation Crisis.* Boston: Beacon, 1975.

Habermas, Jürgen. "Modernity: An Incomplete Project." In *The Anti-Aesthetic,* ed. Hal Foster, 3–15. Seattle: Bay Press, 1983.

Bibliography

Habermas, Jürgen. The *Philosophical Discourse of Modernity.* Cambridge: MIT Press, 1987.

Habermas, Jürgen. *Theory and Practice.* Boston: Beacon, 1973.

Habermas, Jürgen. *The Theory of Communicative Action.* 2 vols. Boston: Beacon, 1983.

Habermas, Jürgen. *Toward a Rational Society: Student Protest, Science, and Politics.* Boston: Beacon, 1970.

Hall, Stuart. "Cultural Studies: Two Paradigms." *Media, Culture and Society* 2 (1980): 57–72.

Hall, Stuart. "The Emergence of Cultural Studies and the Crisis of the Humanities." *October* 53 (1990): 11–23.

Halliday, M. A. K. *An Introduction to Functional Grammar.* London: Edward Arnold, 1985.

Halliday, M. A. K. "Functions of Language." In *Language, Context, and Text: Aspects of Language in a Social-Semiotic Perspective,* ed. M. A. K. Halliday and Ruqaiya Hason. Victoria: Deakin University Press, 1985.

Halliday, M. A. K. "Language Structure and Language Function." In *New Horizons in Linguistics,* ed. John Lyons, 140–65. Baltimore: Penguin, 1970.

Halliday, M. A. K. "On the Language of Physical Science." In *Registers of Written English,* ed. M. Ghadessy, 162–78. London: Frances Pinter, 1988.

Halliday, M. A. K., and Ruqaiya Hasan. *Cohesion in English.* London: Longman, 1976.

Halloran, S. Michael. "Aristotle's Concept of Ethos, or If Not His, Someone Else's." *Rhetoric Review* 1 (1982): 58–63.

Halloran, S. Michael. "The Birth of Molecular Biology: An Essay in the Rhetorical Criticism of Scientific Discourse." *Rhetoric Review* 3 (1984): 70–83.

Halloran, S. Michael, and Annette N. Bradford. "Figures of Speech in the Rhetoric of Science and Technology." In *Classical Rhetoric and Modern Discourse,* ed. Robert Connors, Lisa Ede, and Andrea Lunsford, 179–92. Carbondale: Southern Illinois University Press, 1984.

Haraway, Donna. *Primate Visions: Gender, Race, and Nature in the World of Modern Science.* New York: Routledge, 1989.

Haraway, Donna. "Primatology Is Politics by Other Means." In *Feminist Approaches to Science,* ed. Ruth Bleier, 77–118. New York: Pergamon Press, 1986.

Haraway, Donna. "Situated Knowledge: The Science Question in Feminism and the Privilege of Partial Perspective." *Feminist Studies* 14 (1988): 575–99.

Harding, Sandra. "The Instability of the Analytical Categories of Feminist Theory." *Signs* 11 (1986): 645–64.

Harding, Sandra. "Introduction: Is There a Feminist Method?" In *Feminism and Methodology: Social Science Issues,* ed. Sandra Harding, 1–14. Bloomington: Indiana University Press, 1987.

Harré, Rom. *Varieties of Realism: A Rationale for the Natural Sciences.* New York: Basil Blackwell, 1986.

Harris, R. Allen. "Rhetoric of Science." *College English* 53 (1991): 282–307.

Harris, Wendell. "Toward an Ecological Criticism: Contextual Versus Unconditioned Literary Theory." *College English* 48 (1986): 116–31.

Hart, Roderick. *Modern Rhetorical Criticism*. Glenview, Ill.: Scott Foresman, 1990.

Hawkes, Terence. *Structuralism and Semiotics*. Berkeley: University of California Press, 1977.

Hayes, John R., and Linda Flower. "Uncovering Cognitive Processes in Writing: An Introduction to Protocol Analysis." In *Research on Writing: Principles and Methods*, ed. P. Mosenthal, L. Tamor, and S. Walmsley, 207–20. New York: Longman, 1983.

Hebridge, Dick. *Subcultures: The Meaning of Style*. London: Methuen, 1979.

Held, David. *Introduction to Critical Theory*. Berkeley: University of California Press, 1980.

Herrington, Anne. "Composing One's Self in a Discipline: Students' and Teachers' Negotiations." In *Constructing Rhetorical Education*, ed. Marie Secor and Davida Charney, 91–115. Carbondale: Southern Illinois University Press, 1992.

Hesse, Mary. "The Explanatory Function of Metaphor." In *Revolutions and Reconstructions in the Philosophy of Science*, 111–24. Bloomington: Indiana University Press, 1980.

Hicks, Diana, and Jonathon Potter. "Sociology of Scientific Knowledge: A Reflexive Citation Analysis *or* Scientific Disciplines and Disciplining Science." *Social Studies of Science* 21 (1991): 459–501.

Hoey, Michael. *On the Surface of Discourse*. London: Allen and Unwin, 1983.

Hoffman, Robert. "Metaphor in Science." In *Cognitive and Figurative Language*, ed. R. P. Honeck and R. R. Hoffman, 393–423. Hillsdale, N.J.: Erlbaum, 1980.

Holland, Norman. *The Dynamics of Literary Response*. New York: Oxford University Press, 1968.

Holland, Norman. *Five Readers Reading*. New Haven: Yale University Press, 1975.

Hooker, C. A. *A Realistic Theory of Science*. Albany: State University of New York Press, 1987.

Hubbard, Ruth. "The Emperor Doesn't Wear Any Clothes: The Impact of Feminism on Biology." In *Men's Studies Modified: The Impact of Feminism on the Academic Disciplines*, ed. Dale Spender, 213–35. New York: Pergamon Press, 1981.

Hubbard, Ruth. "Science, Facts, and Feminism." In *Feminism and Science*, ed. Nancy Tuana, 109–31. Bloomington: Indiana University Press, 1989.

Huddleston, R. D. *The Sentence in Written English*. Cambridge: Cambridge University Press, 1971.

Hull, David L. "Darwinism as a Historical Entity: A Historiographic Proposal." In *The Darwinian Heritage*, ed. David Kohn, 773–812. Princeton: Princeton University Press, 1985.

Hull, David L. *The Metaphysics of Evolution*. Albany: State University of New York Press, 1989.

Hull, David L. *Philosophy of Biological Science*. Englewood Cliffs, N.J.: Prentice-Hall, 1974.

Bibliography

Huxley, Julian S. *Evolution: The Modern Synthesis.* London: George Allen and Unwin, 1942.

Iser, Wolfgang. *The Act of Reading: A Theory of Aesthetic Response.* Baltimore: Johns Hopkins University Press, 1974.

Jakobson, Roman. "Closing Statement: Linguistics and Poetics." In *Style in Language,* ed. Thomas A. Sebeok, 350–71. Cambridge: MIT Press, 1960.

Jameson, Frederic. *Postmodernism: Or, the Cultural Logic of Late Capitalism.* Durham: Duke University Press, 1991.

Jauss, Hans R. *Towards an Aesthetic of Reception.* Trans. Timothy Bahti. Minneapolis: University of Minnesota Press, 1982.

Jay, Martin. *Marxism and Totality: The Adventures of a Concept from Lukacs to Habermas.* Berkeley: University of California Press, 1984.

Jeffords, Susan. *The Remasculization of America: Gender and the Vietnam War.* Bloomington: Indiana University Press, 1989.

Johnson, Barbara. *The Critical Difference: Essays in the Contemporary Rhetoric of Reading.* Baltimore: Johns Hopkins University Press, 1980.

Johnson, Nan. "Ethos and the Aims of Rhetoric." In *Essays on Classical Rhetoric and Modern Discourse,* ed. Robert J. Connors, Lisa Ede, and Andrea Lunsford, 179–92. Carbondale: Southern Illinois University Press, 1984.

Johnson, Nan. "Reader-Response and the *Pathos* Principle." *Rhetoric Review* 6 (1988): 152–66.

Johnson, Richard. "What is Cultural Studies Anyway?" *Social Text* 16 (1986–87): 38–80.

Journet, Debra. "Ecological Theories as Cultural Narratives: F. E. Clements's and H. A. Gleason's 'Stories' of Community Succession." *Written Communication* 8 (1991): 446–71.

Journet, Debra. "Forms of Discourse and the Sciences of the Mind: Luria, Sacks, and the Role of Narrative in Neurological Case Histories." *Written Communication* 7 (1990): 171–99.

Kaye, Howard L. *The Meaning of Modern Biology: From Social Darwinism to Sociobiology.* New Haven: Yale University Press, 1986.

Keller, Evelyn Fox. *A Feeling for the Organism: The Life and Work of Barbara McClintock.* New York: W. H. Freeman & Co., 1983.

Keller, Evelyn Fox. "The Gender/Science System; or, Is Sex to Gender as Nature Is to Science?" In *Feminism and Science,* ed. Nancy Tuana, 33–44. Bloomington: Indiana University Press, 1989.

Keller, Evelyn Fox. "Making Gender Visible in the Pursuit of Nature's Secrets." In *Feminist Studies/Critical Studies,* ed. Tereasa de Courentis, 66–77. New York: Macmillan, 1988.

Keller, Evelyn Fox. *Reflections on Gender and Science.* New Haven: Yale University Press, 1985.

Keller, Evelyn Fox. "Women Scientists and Feminist Critiques of Science." *Daedalus* 116 (1987): 77–91.

Kellogg, Vernon Lyman. *Darwinism Today.* London: George Bell and Sons, 1907.

Kimura, Motoo. *The Neutral Theory of Molecular Evolution.* Cambridge: Cambridge University Press, 1983.

Kinneavy, James. *A Theory of Discourse.* Englewood Cliffs, N.J.: Prentice-Hall, 1971. Reprint. New York: Norton, 1980.

Kitcher, Philip. *Vaulting Ambition: Sociobiology and the Quest for Human Nature.* Cambridge: MIT Press, 1985.

Kolodny, Annette. "Dancing through the Minefield: Some Observations on the Theory, Practice and Politics of a Feminist Literary Criticism." In *Feminist Literary Theory: A Reader,* ed. Mary Eagleton, 184–88. New York: Basil Blackwell, 1986.

Kristeva, Julia. *Desire in Language.* Ed. Leon S. Roudiez. Trans. T. Gora, A. Jardine, and L. Roudiez. New York: Columbia University Press, 1980.

Kristeva, Julia. "Stabat Mater." In *The Kristeva Reader,* ed. Toril Moi, 160–86. New York: Columbia University Press, 1986.

Kuhn, Thomas J. *The Structure of Scientific Revolutions.* 2nd ed. Chicago: University of Chicago Press, 1970.

Lakoff, Robin. "The Logic of Politeness, or, Minding Your P's and Q's." *Papers from the Ninth Regional Meeting of the Chicago Linguistics Society* 9 (1974): 292–305.

Landau, Misia. "Human Evolution as Narrative." *American Scientist* 72 (1984): 262–67.

Latour, Bruno. "Give Me a Laboratory." In *Sciences Observed,* ed. Karin D. Knorr-Cetina and Michael Mulkay, 141–70. Beverly Hills: Sage Publications, 1983.

Latour, Bruno. *Science in Action: How to Follow Scientists and Engineers through Society.* Cambridge: Harvard University Press, 1987.

Latour, Bruno, and Steve Woolgar. *Laboratory Life: The Construction of Scientific Facts.* Princeton: Princeton University Press, 1986.

Law, John, and R. J. Williams. "Putting Facts Together: A Study of Scientific Persuasion." *Social Studies of Science* 12 (1982): 535–58.

Leech, Geoffrey. *Principles of Pragmatics.* Harlow, England: Longman, 1983.

Leech, Geoffrey, and Michael Short. *Style in Fiction.* Harlow, England: Longman, 1983.

Leff, Michael, ed. "Rhetorical Criticism: The State of the Art." Special report, *Western Journal of Speech Communication* 44 (1980): 264–349.

LeGuin, Ursula. "The Eye Altering." In *The Compass Rose,* 156–70. New York: Harper and Row, 1982.

Lenoir, Timothy. "The Darwin Industry." *Journal of the History of Biology* 20 (1987): 115–30.

Levins, Richard, and Richard C. Lewontin. *The Dialectical Biologist.* Cambridge: Harvard University Press, 1985.

Lewontin, Richard C. "Adaptation." *Scientific American* 239 (1978): 213–30.

Lewontin, Richard C. "The Bases of Conflict in Biological Explanation." *Journal of the History of Biology* 2 (1969): 35–45.

Lewontin, Richard C. "The Corpse in the Elevator." *New York Review of Books* 29 (20 January 1983): 34–37.

Lewontin, Richard C. "Darwin's Revolution." *New York Review of Books* 30 (16 June 1983): 21–27.

Lewontin, Richard C. "Fallen Angels" (review of *Wonderful Life*). *New York Review of Books* 37 (14 June 1990): 3–7.

Lewontin, Richard C. "Is Nature Probable or Capricious?" *BioScience* 16 (1966): 25–27.

Lewontin, Richard C. "Sociobiology: Another Biological Determinism." In *Home and Health*, ed. Elizabeth Fee, 243–59. Farmingdale, New York: Baywood, 1982.

Longino, Helen. "Can there Be a Feminist Science?" In *Feminism and Science*, ed. Nancy Tuana, 45–56. Bloomington: Indiana University Press, 1989.

Longino, Helen. *Science as Social Knowledge*. Princeton: Princeton University Press, 1990.

Longino, Helen, and Ruth Doell. "Body, Bias, and Behavior: A Comparative Analysis of Reasoning in Two Areas of Biological Science." In *Sex and Scientific Inquiry*, ed. Sandra Harding and Jean O'Barr, 165–186. Chicago: University of Chicago Press, 1975, 1987.

Lukacs, Georg. *History and Class Consciousness*. Trans. Rodney Livingstone. Cambridge: MIT Press, 1968.

Lyell, Charles. *Principles of Geology: Being an Attempt to Explain the Former Changes of the Earth's Surface by Reference to Causes Now in Operation*. London: John Murray, 1830–33.

Lyne, John, and Henry F. Howe. "'Punctuated Equilibria': Rhetorical Dynamics of a Scientific Controversy." *Quarterly Journal of Speech* 72 (1986): 132–47.

Lyons, John. *Language and Linguistics: An Introduction*. Cambridge: Cambridge University Press, 1981.

Lyotard, Jean-François. *The Postmodern Condition: A Report on Knowledge*. Trans. Geoff Bennington and Brian Massumi. Minneapolis: University of Minnesota Press, 1984.

Mailloux, Steven. *Rhetorical Power*. Ithaca: Cornell University Press, 1989.

Mailloux, Steven. "The Turns of Reader-Response Criticism." In *Conversations: Contemporary Critical Theory and the Teaching of Literature*, ed. Charles Moran and Elizabeth Penfield, 38–54. Urbana, Ill.: National Council of Teachers of English, 1990.

Martin, James R., and Joan Rothery. "What a Functional Approach to the Writing Task Can Show Teachers about 'Good Writing.'" In *Functional Approaches to Writing: Research Perspectives*, ed. Barbara Couture, 241–65. Norwood, N.J.: Ablex, 1986.

Marx, Karl. *The Eighteenth Brumaire of Louis Bonaparte*. Hamburg, 1869. Reprint. New York: International, 1972.

Marx, Karl. *Grundrisse: Foundations of the Critique of Political Economy*. Trans. Martin Niehaus. New York: Random House, 1973.

Mayer, Richard. "The Sequencing of Instruction and the Concept of Assimilation-To-Schema." *Instructional Science* 6 (1977): 369–88.

Maynard Smith, John. "Darwinism Stays Unpunctuated." *Nature* 330 (1987): 516.

Maynard Smith, John. "Dinosaur Dilemmas." *New York Review of Books* 38 (25 April 1991): 5–7.

Maynard Smith, John. "Game Theory and the Evolution of Behavior." *Proceedings of the Royal Society of London, B: Biological Sciences* 205 (1979): 41–54.

Maynard Smith, John. "Optimization Theory in Evolution." *Annual Review of Ecology and Systematics* 9 (1978): 31–56.

Maynard Smith, John. "Punctuation in Perspective." *Nature* 332 (1988): 311–12.

Maynard Smith, John. *The Theory of Evolution.* Harmondsworth, England: Penguin Books, 1975.

Maynard Smith, John, and R. Holliday, organizers. "The Evolution of Adaptation by Natural Selection: A Discussion." *Proceedings of the Royal Society of London, B: Biological Sciences* 205 (1979): 435–604.

Mayr, Ernst. *Animal Species and Evolution.* Cambridge: Harvard University Press, 1963.

Mayr, Ernst. "Cause and Effect in Biology." *Science* 134 (10 November 1961): 1501–6.

Mayr, Ernst. "Darwin's Five Theories of Evolution." In *The Darwinian,* ed. David Kohn, 755–72. Princeton: Princeton University Press, 1985.

Mayr, Ernst. *The Growth of Biological Thought: Diversity, Evolution, and Inheritance.* Cambridge: Harvard Belknap, 1982.

Mayr, Ernst. "How to Carry Out the Adaptationist Programme?" *American Naturalist* 121 (1983): 324–34.

Mayr, Ernst. *Toward a New Philosophy of Biology: Observations of an Evolutionist.* Cambridge: Harvard Belknap, 1988.

McCarthy, Thomas. *The Critical Theory of Jurgen Habermas.* Cambridge: MIT Press, 1978.

McElroy, Joseph. *Plus.* New York: Knopf, 1976.

McElroy, Joseph. *Women and Men.* New York: Knopf, 1987.

Meadows, Arthur Jack. "The Citation Characteristics of Astronomical Research Literature." *Journal of Documentation* 23 (1967): 28–33.

Merton, Robert K. *The Sociology of Science,* ed. Norman Storer. Chicago: University of Chicago Press, 1973.

Messeri, Peter. "Obliteration by Incorporation." Paper delivered at the meeting of the American Sociological Association, San Francisco, September 1978.

Messing, Karen. "The Scientific Mystique: Can a White Lab Coat Guarantee Purity in the Search for Knowledge about the Nature of Women?" In *Woman's Nature: Rationalizations of Inequality,* ed. Marian Lowe and Ruth Hubbard, 75–88. New York: Pergamon Press, 1983.

Miller, Carolyn. "A Humanistic Rationale for Technical Writing." *College English* 40 (1979): 610–17.

Miller, Carolyn, and Jack Selzer. "Special Topics of Argument in Engineering Reports." In *Writing in Nonacademic Settings,* ed. Lee Odell and Dixie Goswami, 309–42. New York: Guilford, 1985.

Muhlhausler, Peter, and Rom Harré. *Pronouns and People: The Linguistic Construction of Social and Personal Identity.* Oxford: Blackwell, 1990.

Mulkay, Michael. *Science and the Sociology of Knowledge.* London: George Allen and Unwin, 1979.

Mullins, Nicholas, Lowell Hargens, P. Hecht, and E. Kick. "The Group Structure of Cocitation Clusters: A Comparative Study." *American Sociological Review* 42 (1977): 552–62.

Myers, Greg. "Politeness and Certainty: The Language of Collaboration in an A-I Group." *Social Studies of Science* 21 (1991): 37–73.

Myers, Greg. "The Pragmatics of Politeness in Scientific Writing." *Applied Linguistics* 10 (1989): 1–35.

Myers, Greg. "The Rhetoric of Irony in Academic Writing." *Written Communication* 7 (1990): 419–55.

Myers, Greg. "The Social Construction of Two Biologists' Proposals." *Written Communication* 2 (1985): 219–45.

Myers, Greg. "Stories and Styles in Two Molecular Biology Articles." In *Textual Dynamics of the Professions,* ed. Charles Bazerman and James Paradis, 45–75. Madison: University of Wisconsin Press, 1991.

Myers, Greg. "Text as Knowledge Claims: The Social Construction of Two Biology Articles." *Social Studies of Science* 15 (1985): 593–630.

Myers, Greg. *Writing Biology: Texts in the Social Construction of Scientific Knowledge.* Madison: University of Wisconsin Press, 1990.

Namenwirth, Marion. "Science through a Feminist Prism." In *Feminist Approaches to Science,* ed. Ruth Bleier, 18–41. New York: Pergamon Press, 1986.

National Research Council. *Mapping and Sequencing the Human Genome.* Washington, D.C.: National Academy Press, 1988.

Neel, Jasper. *Plato, Derrida, and Writing.* Carbondale: Southern Illinois University Press, 1988.

Norris, Christopher. *Deconstruction: Theory and Practice.* London: Methuen, 1982.

Norris, Christopher. *Derrida.* Cambridge: Harvard University Press, 1987.

O'Hara, Robert. "Homage to Clio, or, toward an Historical Philosophy for Evolutionary Biology." *Systematic Zoology* 37 (1988): 142–55.

O'Hara, Robert. "Telling the Tree." *Biology and Philosophy* 7 (1992): 135–60.

Olson, Gary, Susan Duffy, and Robert Mack. "Cognitive Aspects of Genre." *Poetics* 10 (1981): 283–315.

Olson, Gary, Susan Duffy, and Robert Mack. "Thinking-Out-Loud as a Method for Studying Real-Time Comprehension Processes." In *New Methods in Reading Comprehension Research,* ed. David Kieras and Marcel Just, 253–86. Hillsdale, N.J.: Lawrence Erlbaum Associates, 1984.

Ong, Walter. "The Writer's Audience Is Always a Fiction." *Publications of the Modern Language Association* 90 (1975): 9–21.

Overington, Michael. "The Scientific Community as Audience: Toward a Rhetorical Analysis of Science." *Philosophy and Rhetoric* 10 (1977): 143–64.

Paley, William. *Natural Theology.* London: 1802.

Perelman, Chaim. *The Realm of Rhetoric.* Notre Dame, Ind.: University of Notre Dame Press, 1982.

Perelman, Chaim, and Lucie Olbrechts-Tyteca. *The New Rhetoric.* Trans. J. Wilkinson and P. Weaver. Notre Dame, Ind.: Notre Dame University Press, 1969.

Peters, Pamela. "Getting the Theme Across: A Study of Dominant Function in the Academic Writing of University Students." In *Functional Approaches to Writing: Research Perspectives*, ed. Barbara Couture, 169–85. Norwood, N.J.: Ablex, 1986.

Pinelli, Thomas, Virginia Cordle, and Raymond Vondran. "The Function of Report Components in the Screening and the Reading of Technical Reports." *Journal of Technical Writing and Communication* 14 (1984): 87–94.

Popper, Karl. *The Logic of Scientific Discovery.* London: Routledge and Kegan Paul, 1959.

Porter, Roy. "Charles Lyell and the Principles of the History of Geology." *British Journal for the History of Science* 9 (1976): 91–103.

Prelli, Lawrence. *A Rhetoric of Science: Inventing Scientific Discourse.* Columbia: University of South Carolina Press, 1989.

Price, Derek J. de Solla. *Little Science, Big Science . . . and Beyond.* New York: Columbia University Press, 1986.

Price, Derek J. de Solla. "Networks of Scientific Papers: The Pattern of Bibliographic References Indicates the Nature of the Scientific Research Front." *Science* 149 (1965): 510–15.

Prince, Gerald. "Introduction to the Study of the Narratee." *Poetique* 14 (1973): 177–96. Reprinted in *Reader-Response Criticism*, ed. Jane Tompkins, 7–25. Baltimore: Johns Hopkins University Press, 1990.

Provine, William. *Sewall Wright and Evolutionary Biology.* Chicago: University of Chicago Press, 1986.

Punter, David, ed. *Introduction to Contemporary Cultural Studies.* London: Longman, 1986.

Radway, Janice. *Reading the Romance: Women, Patriarchy and Popular Literature.* Chapel Hill: University of North Carolina Press, 1984.

Raup, David M., Stephen Jay Gould, T. J. M. Schopf, and D. S. Simberloff. "Stochastic Models of Phylogeny and the Evolution of Diversity." *Journal of Geology* 81 (1973): 525–42.

Rich, Adrienne. "Frame." In *The Fact of a Doorframe: Poems Selected and New,* 303–5. New York: Norton, 1981.

Roger, Jacques. "Darwinism Today." In *The Darwinian Heritage*, ed. David Kohn, 813–23. Princeton: Princeton University Press, 1985.

Romanes, George J. "The Darwinism of Darwin and of the Post-Darwinian Schools." In *Darwin, and after Darwin*, vol. 2. London: Longmans, Green and Co., 1900.

Rorty, Richard. "Philosophy as a Kind of Writing: An Essay on Derrida." In *Consequences of Pragmatism*, 90–109. Minneapolis: University of Minnesota Press, 1982.

Rorty, Richard. "Science as Solidarity." In *The Rhetoric of the Human Sciences: Language and Argument in Scholarship and Public Affairs*, ed. John S. Nelson, Allan Megill, and Donald N. McClosky, 38–52. Madison: University of Wisconsin Press, 1987.

Rose, Hilary. "Beyond Masculinist Realities: A Feminist Epistemology for the Sciences." In *Feminist Approaches to Science*, ed. Ruth Bleier, 57–75. New York: Pergamon Press, 1986.

Rosenblatt, Louise. *Literature as Exploration*. New York: Appleton-Century-Crofts, 1938.

Rossi, Paolo. *The Dark Abyss of Time*. Chicago: University of Chicago Press, 1984.

Rouse, Joseph. "The Narrative Reconstruction of Science." *Inquiry* 33 (1990): 179–96.

Rudwick, Martin J. S. *The Great Devonian Controversy*. Chicago: University of Chicago Press, 1985.

Rudwick, Martin J. S. *The Meaning of Fossils*. London: Macdonald, 1972.

Sahlins, Marshall. *The Use and Abuse of Biology: An Anthropological Critique of Sociobiology*. Ann Arbor: University of Michigan Press, 1976.

Saussure, Ferdinand de. *Course in General Linguistics*, ed. Charles Bally and Albert Sechehaye, trans. Wade Baskin. New York: Philosophical Library, 1959.

Schilb, John. "Deconstructing Didion: Poststructuralist Rhetorical Theory in the Composition Class." In *Literary Nonfiction: Theory, Criticism, Pedagogy*, ed. Chris Anderson, 262–86. Carbondale: Southern Illinois University Press, 1989.

Scholes, Robert. *Protocols of Reading*. New Haven: Yale University Press, 1989.

Scholes, Robert. *Textual Power: Literary Theory and the Teaching of English*. New Haven: Yale University Press, 1985.

Schwartz, Barry. *The Battle for Human Nature*. New York: Norton, 1986.

Schweber, Silvan S. "Darwin and the Political Economist: Divergence of Character." *Journal of the History of Biology* 13 (1980): 195–290.

Schweber, Silvan S. "The Origin of the *Origin* Revisited." *Journal of the History of Biology* 10 (1977): 229–316.

Schwegler, Robert, and Linda Shamoon. "Meaning Attribution in Ambiguous Texts in Sociology." In *Textual Dynamics of the Professions*, ed. Charles Bazerman and James Paradis, 216–33. Madison: University of Wisconsin Press, 1991.

Searle, John. "Reiterating the Differences." *Glyph* 1 (1977): 198–208.

Segestrade, Ullica. "Colleagues in Conflict." *Biology and Philosophy* 1 (1986): 53–87.

Selzer, Jack. "Intertextuality and the Writing Process: An Overview." In *Writing in the Workplace: New Research Perspectives*, ed. Rachel Spilka. Carbondale: Southern Illinois University Press. In press.

Selzer, Jack. "More Meanings of *Audience*." In *A Rhetoric of Doing*, ed. Roger Cherry, Neil Nakadate, and Steven Witte. Carbondale: Southern Illinois University Press. In press.

Shapiro, Michael J. "The Rhetoric of Social Science: The Political Responsibilities of the Scholar." In *Rhetoric of the Human Sciences: Language and Argument in Scholarship and Public Affairs*, ed. John Nelson, Allan Megill, and Donald McCloskey, 363–80. Madison: University of Wisconsin Press, 1987.

Simons, Herbert W., ed. *Rhetoric in the Human Sciences*. Newbury Park, Calif.: Sage, 1989.

Simons, Herbert W., ed. *The Rhetorical Turn: Invention and Persuasion in the Conduct of Inquiry*. Chicago: University of Chicago Press, 1990.

Simons, Herbert W., and Trevor Melia, eds. *The Legacy of Kenneth Burke*. Madison: University of Wisconsin Press, 1989.

Simpson, George Gaylord. *This View of Life: The World of an Evolutionist*. New York: Harcourt, 1964.

Simpson, Paul. "Politeness Phenomena in Ionesco's *The Lesson*." In *Language, Discourse, and Literature: An Introductory Reader in Discourse Stylistics*, ed. Ronald Carter and Paul Simpson, 171-94. London: George Allen and Unwin, 1989.

Slobodkin, L. B. "Foxes and Hedgehogs: A Look at Four Books by Celebrated Scientists." *American Scientist* 76 (September–October 1988): 503-4.

Small, Henry G. "Cited Documents as Concept Symbols." *Social Studies of Science* 8 (1978): 327-40.

Smith, Paul. *Discerning the Subject*. Minneapolis: University of Minnesota Press, 1988.

Snow, C. P. *The Two Cultures: And a Second Look*. London: New American Library, Mentor Book, 1963.

Sober, Elliott. *Reconstructing the Past: Parsimony, Evolution, and Inference*. Cambridge: MIT Press, 1988.

Sociobiology Study Group of Sciences for the People. "Sociobiology—Another Biological Determinism." *BioScience* 26 (1976): 182-86.

Stehr, Nico. "The Ethos of Science Revisited." *Sociological Inquiry* 48 (1978): 172-96.

Suleiman, Susan, and Inge Crosman. *The Reader in the Text*. Princeton: Princeton University Press, 1990.

Swales, John. "Citation Analysis and Discourse Analysis." *Applied Linguistics* 7 (1986): 39-56.

Swales, John. *Genre Analysis: Analysing Academic and Research Texts*. Cambridge: Cambridge University Press, 1990.

Tax, Sol, ed. *Evolution after Darwin*. 3 vols. Chicago: University of Chicago Press, 1960.

Thompson, John, and David Held, eds. *Habermas: Critical Debates*. Cambridge: MIT Press, 1982.

Todd, Janet. *Feminist Literary History*. New York: Routledge, 1988.

Tompkins, Jane. "Fighting Words: Unlearning to Write the Critical Essay." *Georgia Review* 42 (1988): 585-90.

Tompkins, Jane, ed. *Reader-Response Criticism: From Formalism to Post-Structuralism*. Baltimore: Johns Hopkins University Press, 1990.

Toulmin, Stephen. *Human Understanding: The Collective Use and Evolution of Concepts*. Princeton: Princeton University Press, 1972.

Toulmin, Stephen. *The Return to Cosmology*. Berkeley: University of California Press, 1982.

Wallace, Alfred R. *Darwinism*. London: MacMillan, 1890.

Wander, Philip. "The Rhetoric of Science." *Western Journal of Speech Communication* 40 (1976): 226-35.

Watson, J. D., and F. H. C. Crick. "Molecular Structure of Nucleic Acids: Structure for Dioxyribose Nucleic Acid." *Nature* 171 (1953): 737-38.

Wellek, Rene, and Austin Warren. *Theory of Literature*. New York: Harcourt Brace, 1942.

White, Hayden. *The Content of the Form: Narrative Discourse and Historical Representation*. Baltimore: Johns Hopkins University Press, 1987.

White, Hayden. "The Fictions of Factual Representation." In *Tropics of Discourse: Essays in Cultural Criticism*, 121–34. Baltimore: Johns Hopkins University Press, 1978.

Widdowson, Henry. *Explorations in Applied Linguistics*. Oxford: Oxford University Press, 1979.

Willis, Paul. *Learning to Labor: How Working Class Kids Get Working Class Jobs*. Farnborough, England: Saxon House, 1977.

Wilson, Edward O. *On Human Nature*. Cambridge: Harvard University Press, 1978.

Wilson, Edward O. *Sociobiology: The New Synthesis*. Cambridge: Harvard University Press, 1975.

Wilson, W. Daniel. "Readers in Texts." *Publications of the Modern Language Association* 96 (1981): 848–63.

Winsor, Dorothy. "The Construction of Knowledge in Organizations: Asking the Right Questions about the Challenger." *Journal of Business and Technical Communications* 4 (1990): 7–20.

Winter, E. O. "Replacement as a Function of Repetition: A Study of Some of Its Principal Features in the Clause Relations of Contemporary English." Ph.D. Thesis, University of London, 1974.

Woolf, Virginia. *A Room of One's Own*. New York: Harcourt Brace Jovanovitch, 1929.

Woolgar, Steve. *Science: The Very Idea*. London: Tavistock, 1988.

Yates, Frances A. *The Art of Memory*. Chicago: University of Chicago Press, 1966.

Ziman, John. *Public Knowledge*. Cambridge: Cambridge University Press, 1968.

INDEX

AAAS (American Academy for the Advancement of Science), 50, 56
Absence: and deconstructionism, 233, 239
Abstractions, 83, 86
Active voice, 92, 117, 193, 194, 285
Adaptation: as Royal Society topic, 14–15; and sexism, 101n4; continuing debate on, 137–38; origins of word, 311–12; Gould's changing views on, 318–20. See also "Adaptationist programme"; Allometry Heterochrony
"Adaptationist programme": Gould and Lewontin's criticism of, 10, 15–16, 21, 22, 24–25, 33–38, 42, 56, 78, 90–96, 111, 318–20; beliefs of, 15; Gould and Lewontin's alternatives to, 16, 25, 73–74, 78, 94–97, 240; fairness of Gould and Lewontin's characterization of, 28–33, 218, 222–23, 226–27; as theme of Royal Society Symposium, 30; and Social Darwinism, 149–50. See also "Evolution of Adaptation by Natural Selection, The" (Symposium Topic); Narrative; Reductionism
Ad hominem arguments, 205; Cain's, 32–33, 36, 258, 263, 317. See also Pangloss, Dr.
Adverbials and adverbs, 259–60, 269
Affective fallacy, 4, 11, 181
Agents: in "Spandrels," 146, 148–50
Aggression, 35. See also Argumentation: Bluebird example; Connection vs. conflict; Fighting behavior
Aging, 31
Allometry, 29–30, 318
"Allometry in Primates" (Gould), 266–69
Allusions, 205, 246, 321. See also Citations
Althusser, Louis, 9, 65
Altmann, Jeanne, 87, 89, 98
Alvarez, Luis, 110
American Academy for the Advancement of Science (AAAS), 50, 56
American Museum of Natural History, 318

American Scientist, 108
AN (faculty member), 212–13, 214, 218–21, 223, 225
Analogies, 28; architectural, in "Spandrels," 16, 22, 25–26, 117, 118–19, 135, 152, 165, 193, 211, 276. Gould's explanation of, 320, 322–24, 327, 330, 333; other, in "Spandrels," 23, 42, 161, 213, 225, 263–38; between adaptation and political campaigns, 153–55. See also Architecture
Analytical matrix: in "Spandrels," 9, 45, 47, 54. See also Interpretation
Andrews, James R., 4
Animal Species and Evolution (Mayr), 314–15
Anthropology, 26, 56, 91. See also Cannibalism
Anthropomorphism: Gould's avoidance of, 43, 264; Barash's use of, 50, 51; difficulties of avoiding, 146–47
Approaches to Primate Paleobiology, 266, 269
Architecture, 35, 56, 73, 91, 118. See also Analogies: architectural
Argumentation: scientific discourse as, 5, 99, 128, 203–6, 272; scientists' refusal to acknowledge, 323; from authority, 109, 129–33, 148, 175; from definition, 109, 218; Gould and Lowentin's, 117–18, 158–79, 211, 256–75, 285–86, 289–94; Maynard Smith's, 189–91; from value, 218. See also Connection vs. conflict; Hierarchical arguments; Ideology; Inventional strategies; Parody; Part/whole arguments; Ridicule; Sarcasm; "Spandrels of San Marco, The": strategies of; "Us" vs "them" arguments; Wit
Aristotle, 150–51, 160, 161, 165, 171, 174. See also Rhetoric: classical
"Arms Races between and Within Species" (Dawkins and Krebs), 32, 264–65, 306
Aronowitz, Stanley, 75, 77

377

RHETORIC OF THE HUMAN SCIENCES

Lying Down Together: Law, Metaphor,
and Theology
Milner S. Ball

Shaping Written Knowledge: The Genre and
Activity of the Experimental Article in Science
Charles Bazerman

Textual Dynamics of the Professions: Historical
and Contemporary Studies of Writing in
Professional Communities
Charles Bazerman and James Paradis, editors

Politics and Ambiguity
William E. Connolly

Philosophy, Rhetoric, and the End of Knowledge:
The Coming of Science and Technology Studies
Steve Fuller

Machiavelli and the History of Prudence
Eugene Garver

Language and Historical Representation: Getting
the Story Crooked
Hans Kellner

The Rhetoric of Economics
Donald N. McCloskey

Therapeutic Discourse and Socratic Dialogue:
A Cultural Critique
Tullio Maranhao

The Rhetoric of the Human Sciences: Language
and Argument in Scholarship and Public Affairs
*John S. Nelson, Allan Megill, and
Donald N. McCloskey, editors*

What's Left? The Ecole Normale Supérieure and
the Right
Diane Rubenstein

Understanding Scientific Prose
Jack Selzer, editor

The Politics of Representation: Writing Practices in
Biography, Photography, and Policy Analysis
Michael J. Shapiro

The Legacy of Kenneth Burke
Herbert Simons and Trevor Melia, editors

The Unspeakable: Discourse, Dialogue, and
Rhetoric in the Postmodern World
Stephen A. Tyler

Heracles' Bow: Essays on the Rhetoric and the
Poetics of the Law
James Boyd White